T0230143

Lecture Notes in Computer Science 530

Edited by G. Goos and J. Hartmanis
Advisory Board: W. Brauer D. Gries J. Stoer

D. H. Pitt P.-L. Curien S. Abramsky
A. M. Pitts A. Poigné D. E. Rydeheard (Eds.)

Category Theory and Computer Science

Paris, France, September 3-6, 1991
Proceedings

Springer-Verlag
Berlin Heidelberg New York
London Paris Tokyo
Hong Kong Barcelona
Budapest

Series Editors

Gerhard Goos
GMD Forschungsstelle
Universität Karlsruhe
Vincenz-Priessnitz-Straße 1
W-7500 Karlsruhe, FRG

Juris Hartmanis
Department of Computer Science
Cornell University
Upson Hall
Ithaca, NY 14853, USA

Volume Editors

David H. Pitt
Department of Mathematics, University of Surrey
Guildford, Surrey GU2 5XH, UK

Pierre-Louis Curien
LIENS, 45, Rue d'Ulm, 75230 Paris Cedex 05, France

Samson Abramsky
Department of Computing
Imperial College of Science, Technology and Medicine
University of London, 180 Queen's Gate, London SW7 2BZ, UK

Andrew M. Pitts
Computer Laboratory, University of Cambridge
New Museum Site, Pembroke Street, Cambridge, CB2 3QG, UK

Axel Poigné
GMD, F2 G2, Schloß Birlinghoven
Postfach 1240, W-5205 St. Augustin 1, FRG

David E. Rydeheard
Department of Computer Science, The University
Manchester M13 9PL, UK

CR Subject Classification (1991): F.3, F.4.1, D.2.1, D.3.1, D.3.3

ISBN 3-540-54495-X Springer-Verlag Berlin Heidelberg New York
ISBN 0-387-54495-X Springer-Verlag New York Berlin Heidelberg

This work is subject to copyright. All rights are reserved, whether the whole or part of the material is concerned, specifically the rights of translation, reprinting, re-use of illustrations, recitation, broadcasting, reproduction on microfilms or in other ways, and storage in data banks. Duplication of this publication or parts thereof is only permitted under the provisions of the German Copyright Law of September 9, 1965, in its current version, and a copyright fee must always be paid. Violations fall under the prosecution act of the German Copyright Law.

© Springer-Verlag Berlin Heidelberg 1991
Printed in Germany

Typesetting: Camera ready by author
Printing and binding: Druckhaus Beltz, Hemsbach/Bergstr.
2145/3140-543210 - Printed on acid-free paper

Preface

The papers of this volume were presented at the fourth biennial Summer Conference on Category Theory and Computer Science, held in Paris, September 3-6, 1991.

The proceedings of the previous three conferences (Guildford 1985, Edinburgh 1987 and Manchester 1989) also appear in the Lecture Notes in Computer Science, as Volumes 240, 283 and 389.

Category theory, as well as other more specific algebraic, topological or geometric structures, continues to be an important tool in foundational studies in computer science.

Category theory, after having played a major role in the development of mathematics (e.g. in algebraic geometry), has been widely applied by logicians to get concise interpretations of many logical concepts. On the other hand, links between logic and computer science have been developed now for over twenty years, notably via the so called Curry–Howard isomorphism, which identifies programs with proofs, types with propositions. Together, the triangle category theory–logic–programming presents a rich world of interconnections.

Selected topics covered in this volume are the following:

– Type theory: Stratification of types and propositions, in relation to the distinction compile time - run time, can be discussed with precision in a categorical setting.

– Domain theory: Synthetic domain theory develops domain theory internally in the constructive universe of the effective topos: thereby many properties come for free. Stone Duality applied to computer science gives a duality between properties and denotations of programs. This approach is now applied to stable functions, an intermediate between parallel and sequential semantics notions.

– Linear Logic: This reconstruction of logic based on propositions as resources leads to alternatives to traditional syntaxes. Links with MacLane–Kelly coherence in monoidal closed categories have been found, deepening the understanding of both coherence and linear logic.

The organisers would like to thank the Ministère de la Recherche et de la Technologie, the Centre de la Recherche Scientifique and the Ministère de l'Education Nationale of France for their support. We would like to thank Chantal Butery for the local organisation of the meeting, and Debbie Murrells for invaluable editorial assistance in the preparation of this volume.

July 1991

D. H. Pitt
P.-L. Curien
S. Abramsky
A. Pitts
A. Poigné
D. E. Rydeheard

Organising and Programme Committee

S. Abramsky, P.-L. Curien, P. Dybjer, G. Longo, G.E.Mints, J.C. Mitchell, D.H. Pitt, A.M. Pitts, A. Poigné, D.E. Rydeheard, D. Sannella, E. Wagner.

Referees

S. Abramsky, P. Aczel, R. Amadio, S. Bechhofer, C. Brown, P.-L. Curien, R. Di Cosmo, P. Dybjer, T. Ehrhard, D. Gurr, M. Hedberg, B. Hilken, R. Hindley, M. Hyland, B. Jacobs, C. B. Jay, E. Kazmierczak, Y. Lafont, G. Longo, Z. Luo, N. Marti-Oliet, N. Mendler, G. E. Mints, J. C. Mitchell, E. Moggi, B. Monahan, F. J. Oles, B. Pierce, D. H. Pitt, A. M. Pitts, A. Poigné, J. Power, M. Rittri, E. Robinson, D. Rydeheard, D. Sannella, A. Scedrov, H. Simmons, P. Taylor, S. Vickers, F. J. de Vries, E. Wagner, J. van der Woude, G. Wraith.

Contents

Stone Duality for Stable Functions

Thomas Ehrhard*   Pasquale Malacaria†

Introduction

The problem of finding an algebraic structure for stable open subsets of a suitable domain has been recently raised by several authors. Specifically, it could be useful to have, in the stable case, a notion similar to the one of "frame", in order to develop something similar to pointless topology, that is an algebraic insight of spaces.

In domain theory, an interesting application of frame theory is the logic of domains developped by Abramsky (see [Ab]) in the topological case.

The idea of constructing a logic for "stable properties" has been originated by Zhang (see [Zha]), who got very interesting results about "stable opens" of dI domains. However, his notions remain concrete, and it is not clear whether they give rise to a duality. In that sense, they lack the "localic" properties which justify the canonicity of Abramsky's approach.

We introduce the notion of S-structure as the structure of the algebra of stable open sets intended to correspond to the concept of frame in the stable case. These S-structures have properties which are very similar to the ones of frames, from the point of view of duality. So we may hope to achieve a logic of domains as natural as Abramsky's one, but expressing properties of programs which are not captured by the continuous approach.

In this paper, we give the fundamentals of S-structures theory. We prove first general duality results which do not involve any domain theoretical assumption about spaces. Actually we introduce the S-spaces which play wrt S-structures the same role as topological spaces wrt frames. These duality results can be specialized to the case of domains, and then we obtain a result similar to known Scott-topological duality in domain theory. The corresponding notion of domain widely subsumes the usual dI domains to which stabilty theory is usually restricted. Indeed these domains are the most general ones where stability makes sense. By duality method it is rather simple to treat function spaces and we retrieve cartesian closedness of the category of L-domains and stable functions which has been recently proved by P. Taylor.

*Université Paris VII, L.I.T.P., Couloir 55-56, 1er étage, 2 Place Jussieu, 75251 PARIS CEDEX 05, FRANCE.
email: ehrhard@litp.ibp.fr

†Université Paris VII, Equipe de Logique Mathématique, Couloir 45-55, 5ème étage, 2 Place Jussieu, 75251 PARIS CEDEX 05, FRANCE.
email: pasquale@FRMAP711.BITNET
and Università di Torino, Dipartimento di Informatica, Corso Svizzera 185, 10149 TORINO, ITALY

1 Basic definitions and results

This section is intended to provide an abstract definition of the algebraic structure of stable open sets. Let us first give some intuition for the forthcoming definitions. Let us consider a dI domain D. We call stable open subset of D the inverse image of $\top$ under a stable map from D to the two points domain $\mathbf{O} = \{\bot < \top\}$. This is a natural definition since a Scott open subset is the inverse image of $\top$ under a continous function. The study of stable open sets has been initiated by Zang [Zha] and in this sense we follow his approach. As Zhang remarked stable opens are closed under finite intersection and disjoint union. Moreover they are closed under union of directed families, but not under union of arbitrary sets. In order to keep distributivity these kinds of join are the only ones we will consider.

Definition 1 *In a meet-semilattice with* 0, *we say that a subset* D *is disjoint-directed (dd for short) if for any* $u, v \in D$, *if* $u \wedge v \neq 0$ *then there exists a* $w \in D$ *such that* $u \leq w$ *and* $v \leq w$.

Lemma 1 *If* D *is dd not containing* 0, *then the binary relation* $\sim_D$ *defined by:* $u \sim_D v$ *iff there exists* $w \in D$ *such that* $u, v \leq w$ *is an equivalence relation the classes of which are directed.*

Proof: We have just to check transitivity of $\sim_D$; if $u \sim_D v$ and $v \sim_D u'$ then we have $w, w' \in D$ which verify $u, v \leq w$ and $v, u' \leq w'$; it follows that $0 \neq v \leq w \wedge w'$ so there exists w'' such that $u, u' \leq w''$. ∎

Definition 2 *A* S-space X *is a pair* $(X, \Omega_S(X))$ *where* X *is a set (the set of points) and* $\Omega_S(X)$ *is a subset of* $\mathcal{P}(X)$ *containing* $\emptyset$, X *and which is closed under finite intersections and dd unions (*$\Omega_S(X)$ *is the set of S-open subsets of* X, *a* S-topology *on* X*).*

If X, Y *are S-spaces, a function* $f : X \to Y$ *is a* S-map *if it preserves S-opens under inverse image.*

The category of S-spaces and S-maps is noted **Ssp**.

Definition 3 *A meet-semilattice* $(\mathcal{L}, \wedge, 0)$ *is an S-structure if*

i) any dd subset D *of* $\mathcal{L}$ *has a lub noted* $\bigvee D$

ii) finite glbs distribute over dd lubs

iii) there is a top element noted 1.

If $\mathcal{L}$ *and* $\mathcal{M}$ *are two S-structures, a function* $f : \mathcal{L} \to \mathcal{M}$ *is an S-morphism if it preserves the structure.*

The category of S-structures and S-morphisms is noted **Sstr**.

Definition 4 *An element* $u \neq 0$ *of an S-structure* $\mathcal{L}$ *is*

- connected *if for any* $v, w \in \mathcal{L}$ *such that* $v \wedge w = 0$, *if* $v \vee w \geq u$ *then* $v \geq u$ *or* $w \geq u$

- compact *if for any directed* $D \subseteq \mathcal{L}$, *if* $\bigvee D \geq u$ *then there exists a* $v \in D$ *such that* $v \geq u$

- dd-prime *if for any dd* $D \subseteq \mathcal{L}$, *if* $\bigvee D \geq u$ *then there exists a* $v \in D$ *such that* $v \geq u$.

Typographical convention: we use letters $u, v, w \ldots$ to denote arbitrary elements of $\mathcal{L}$, letters $a, b, c, \ldots$ to denote connected elements and $\alpha, \beta, \gamma \ldots$ to denote dd-prime elements. From now on, we shall not specify otherwise the nature of elements of $\mathcal{L}$.

Lemma 2 *The lub of any directed set of connected elements of an S-structure is connected.*

Proof: Let C be a directed set of connected elements, $w = \bigvee C$, and u, v such that $u \wedge v = 0$ and suppose $w \leq u \vee v$. If $a \in C$ then $a \leq u \vee v$, so $a \leq u$ or $a \leq v$; now if for $a, a' \in C$, $a \leq u$ and $a' \leq v$ then for $b \in C$ such that $b \geq a, a'$ if $b \leq u$ then $a' \leq u \wedge v$ and if $b \leq v$ then $a \leq u \wedge v$, both cases contradicting the hypothesis. We conclude that any $a \in C$ is less than u or any $a \in C$ is less than v, so $w \leq u$ or $w \leq v$. ∎

Definition 5 *A subset of a meet-semilattice with* 0 *is* disjoint *if the glb of any two different elements of this set is* 0.

Lemma 3 *Let* $D \subseteq \mathcal{L}$ *be a dd set (not containing* 0*). Then the set* $\{\bigvee \Delta \mid \Delta \in D/\!\sim_D\}$ *is disjoint.*

Proof: Let $\Delta, \Delta' \in D/\!\sim_D$. If $(\bigvee \Delta) \wedge (\bigvee \Delta') \neq 0$, there exist by distributivity a $u \in \Delta$ and a $u' \in \Delta'$ such that $u \wedge u' \neq 0$. Since D is dd, we have $u \sim_D u'$ and hence $\Delta = \Delta'$. ∎

Lemma 4 *An element* u *of an S-structure* $\mathcal{L}$ *is connected iff for any disjoint subset* E *of* $\mathcal{L}$, *if* $\bigvee E \geq u$ *there exists a* $v \in E$ *such that* $v \geq u$.

Proof: Let $\dot{\bigvee}$ denote the sup of a disjoint set, and suppose $u \leq \dot{\bigvee} E$; then, by distributivity, $u = \dot{\bigvee} A$ where A is $\{e \wedge u \mid e \in E\}$, hence there exists $e' \in E$ such that $0 \neq e' \wedge u \leq u$; by defining $E_0 = E - \{e'\}$ we get $u \leq \dot{\bigvee} E_0 \,\dot{\vee}\, e'$; the fact that u is connected implies $u \leq \dot{\bigvee} E_0$ or $u \leq e'$; but if $u \leq \dot{\bigvee} E_0$ then $0 \neq u \wedge e' \leq \dot{\bigvee} \{e \wedge e' \mid e \in E_0\} = 0$, which is a contradiction, hence $u \leq e'$. ∎

Lemma 5 *An element* u *of* $\mathcal{L}$ *is dd-prime iff it is compact and connected.*

Proof: The "if" part is trivial.
In order to show the "only if" part we remark that from lemma 1 a dd-set D can be factorized in a family of disjoint directed sets $(\Delta_i)_{i \in I}$, such that $\bigcup_{i \in I} \Delta_i = D$; moreover by defining $\delta_i = \bigvee \Delta_i$ we have $\bigvee D = \dot{\bigvee}_{i \in I} \delta_i$; so if $u \leq \bigvee D = \dot{\bigvee}_{i \in I} \delta_i$ then $\exists i \in I (u \leq \delta_i)$ (by using lemma 4 because u is connected) and finally that $\exists v \in \Delta_i (u \leq v)$ because u is compact. ∎

Definition 6 *An S-structure $\mathcal{L}$ is* algebraic *iff for any $u \in \mathcal{L}$ the set $\pi(u)$ of prime lower bounds of u is dd and $u = \bigvee \pi(u)$.*

Lemma 6 *An element u of an algebraic S-structure is connected iff $\pi(u)$ is directed.*

Proof: By using lemma 1 and by definition of algebricity we have $u = \dot{\bigvee}_{i\in I} (\bigvee \Delta_i)$ where each Δ_i is a directed set, $i \neq j$ implies $\Delta_i \cap \Delta_j = \emptyset$ and $\pi(u) = \bigcup_{i\in I} \Delta_i$; let $u_i = \bigvee \Delta_i$; it follows that $\pi(u_i) = \Delta_i$ hence u_i is connected; so $u = \dot{\bigvee}_{i\in I} u_i$ and by lemma 4 there exists $i_0 \in I$ such that $u = u_{i_0}$. The "only if" part follows from lemma 2. ∎

Proposition 1 *The category* **Sstr** *of S-structures is algebraic.*

Proof: Let **Slat** be the category of meet-semilattices with 0 and 1. We have a forgetful functor F : **Sstr** → **Slat**. We define a left adjoint dd-Idl : **Slat** → **Sstr** to F as follows:

- If $A \in$ **Slat**, we define dd-Idl A as the set of non empty dd downwards closed subsets of A, ordered under inclusion. That is, $u \subseteq A$ is in dd-Idl A iff: $0 \in u$, if $a \in u$ and $b \leq a$ then $b \in u$, and last u is dd.
- If $\varphi : A \to B$ is a morphism in **Slat**, and $u \in$ dd-Idl A, then dd-Idl $(\varphi)(u) = \downarrow (\varphi)(u)$.

First, we prove that dd-Idl A is an S-structure. As 0, we take $\{0\}$, and as 1, we take A itself. Let $u, v \in$ dd-Idl A. Then $u \cap v$ is a lower set (clear). Let us check it is dd. Let $a, b \in u \cap v$ be such that $a \wedge b \neq 0$. Since u and v are dd, we can find $c \in u$ and $d \in v$ such that $c \geq a, b$ and $d \geq a, b$. But $c \wedge d \in u \cap v$ and we conclude. Let now $D \subseteq$ dd-Idl A be dd. $\bigcup D$ is clearly a lower set. Let $a, b \in \bigcup D$ be such that $a \wedge b \neq 0$. Let $u, v \in D$ be such that $a \in u$ and $b \in v$. Since $a \wedge b \in u \cap v$, and $a \wedge b \neq 0$, we have $u \cap v \neq 0$. Hence there is a $w \in D$ such that $u, v \subseteq w$. Now $a, b \in w$ and w is dd, so there is a $c \in w$ such that $c \geq a, b$ and we conclude. Distributivity is obvious, so we have defined an S-structure, the free S-structure over A. Checking that dd-Idl (φ) is a an S-morphism is a straightforward verification.

Now we check that dd-Idl is the left adjoint of F. For this, we define naturally an S-morphism s : dd-Idl F$\mathcal{L} \to \mathcal{L}$. Let $I \in$ dd-Idl F$\mathcal{L}$. We set $s(I) = \bigvee I$, which is defined since I is dd. Now we prove that s is an S-morphism. Preservation of 0, 1 and all dd lubs are obvious. Let $I, J \in$ dd-Idl F$\mathcal{L}$. We have

$$\begin{aligned} s(I \cap J) &= \bigvee\{u \wedge v \mid u \in I,\ v \in J\} \\ &= (\bigvee I) \cap (\bigvee J) \end{aligned}$$

by distributivity. Let f : dd-Idl $A \to \mathcal{L}$. To conclude that the required djunction holds, we have to show that there exists a unique $\varphi : A \to$ F$\mathcal{L}$ such that $f = s \circ$ dd-Idl (φ). We set, for $a \in A$, $\varphi(a) = f(\downarrow (a))$ (there is no choice indeed since $s(\downarrow (a)) = a$). That φ preserves 0 and 1 is obvious. Furthermore

$$\varphi(a \wedge b) = f(\downarrow (a \wedge b)) = f(\downarrow (a) \cap \downarrow (b)) = f(\downarrow (a)) \wedge f(\downarrow (b))$$

and we conclude.

It remains to prove that the adjunction is monadic. This is a routine verification. ∎

Definition 7 *A dd-prime filter of an S-structure $\mathcal{L}$ is a subset p of $\mathcal{L}$ which is a filter and such that, for any $D \subseteq \mathcal{L}$ dd, if $\bigvee D \in p$ then $p \cap D \neq \emptyset$. We note* pt($\mathcal{L}$) *the poset of dd-prime filters of $\mathcal{L}$ ordered by inclusion.*

Remark that any dd-prime filter is the inverse image of $1 \in \mathcal{O}$ (the two elements S-structure $\{0,1\}$) under a morphism of S-structure, and conversely.

Let us now define the cpo's on which we shall define S-topologies.

Definition 8 *A cpo $(X, \leq)$ is said to be a* meet-cpo *if it has bounded binary glbs and if these glbs ditribute over directed lubs.*

Now we define stable open subsets of such cpo's. The word "stable" has here to be understood as "conditionally multiplicative", in the terminology introduced by Berry (see [Ber]).

Definition 9 *A subset U of a meet-cpo X is a* stable open *subset if it is Scott-open and closed under bounded binary glbs. We note $\Omega_S^{\wedge} X$ the poset of stable open subsets of X ordered by inclusion.*

Definition 10 *A Scott-continuous function $f : X \to Y$ between two meet-cpos is said to be* stable *iff it preserves binary bounded glbs.*

Last, we recall Berry's stable ordering for stable functions.

Definition 11 *Let $f, g : X \to Y$ be two stable functions. One says that f is stably less than g and write $f \sqsubseteq g$ iff for all $x, x' \in X$ such that $x \leq x'$ one has*

$$f(x) = f(x') \wedge g(x) .$$

2 Duality results

The aim of this section is to establish some connection between categories of S-structures and S-spaces.

2.1 Basic duality

The first result we prove is the existence of an adjunction between the categories **Ssp** and **Sstr^{op}**.

We can always define a map $\varphi : \mathcal{L} \to \mathcal{P}(\mathrm{pt}(\mathcal{L}))$ by

$$\varphi(u) = \{p \in \mathrm{pt}(\mathcal{L}) \mid u \in p\}$$

We will write sometimes $\varphi_{\mathcal{L}}$ instead of φ in order to stress the dependence of φ wrt $\mathcal{L}$.

Conversely, for a S-space X it is possible to define a S-map $\Phi : X \to \mathrm{pt}(\Omega_S(X))$ by

$$\Phi(x) = \{U \mid x \in U\} .$$

For the same purpose we will write sometimes Φ_X instead of Φ.

Lemma 7 *For any S-structure $\mathcal{L}$ there exists an S-space* $(\mathrm{pt}(\mathcal{L}), \Omega^\circ_S\mathcal{L})$ *which is the least (wrt inclusion) among all S-topologies on* $\mathrm{pt}(\mathcal{L})$ *containing* $\varphi(\mathcal{L})$.

Furthermore, $\varphi_{\mathcal{L}}$ *is an S-morphism from* $\mathcal{L}$ *to* $\Omega^\circ_S\mathcal{L}$.

Proof: Simlpy take for $\Omega^\circ_S\mathcal{L}$ the intersection of all S-topologies on $\mathrm{pt}(\mathcal{L})$ which contain $\varphi(\mathcal{L})$. Remark that there is at least one of such S-topologies, namely $\mathcal{P}(\mathrm{pt}(\mathcal{L}))$. However for technical reasons we give an ordinal construction of $\Omega^\circ_S\mathcal{L}$ by setting $L_0 = \varphi(\mathcal{L})$,

$$L_{\eta+1} = L_\eta \cup \{U \cap V \mid U, V \in L_\eta\} \cup \{\bigcup D \mid D \subseteq L_\eta \text{ and } D \text{ is dd}\}$$

$$L_\lambda = \bigcup_{\eta<\lambda} L_\eta \text{ for } \lambda \text{ limit}$$

It is easily verified that this increasing sequence becomes stationary for some ordinal ρ and that $\Omega^\circ_S\mathcal{L} = L_\rho$.

The last statement of the lemma is clear. ∎

Lemma 8 *If* $f : \mathcal{L} \to \mathcal{M}$ *is a S-morphism then* $f_* : \mathrm{pt}(\mathcal{M}) \to \mathrm{pt}(\mathcal{L})$ *is a S-map wrt the S-topologies* $\Omega^\circ_S\mathcal{M}$ *and* $\Omega^\circ_S\mathcal{L}$.

Proof: We have to show that $f_*^{-1}(U) \in \Omega^\circ_S\mathcal{M}$ for all $U \in \Omega^\circ_S\mathcal{L}$. Let (L_η) (resp. (M_η)) denote the ordinal construction of $\Omega^\circ_S\mathcal{L}$ (resp. $\Omega^\circ_S\mathcal{M}$). We prove by induction that for all η, $f_*^{-1}(U) \in M_\eta$ for $U \in L_\eta$.

For $\eta = 0$ we have $f_*^{-1}(\varphi(u)) = \varphi(f(u))$.

Assume that the condition holds at η and let $U \in L_{\eta+1}$. The case $U \in L_\eta$ is trivial. If U is an intersection of elements of L_η then the condition holds by preservation of intersections under inverse image. If U is the union of a dd subset of L_η then it suffices to remark that the inverse image of this subset is a dd subset of M_η and that unions are preserved under inverse image.

The case where η is a limit ordinal is trivial. ∎

So we can define the functor $\mathbf{Pt} : \mathbf{Sstr}^{\mathrm{op}} \to \mathbf{Ssp}$ defined on morphisms by $\mathbf{Pt}f = f_*$. Conversely we have the functor $\mathbf{\Omega_S} : \mathbf{Ssp} \to \mathbf{Sstr}^{\mathrm{op}}$ defined by $\mathbf{\Omega_S}(X, \Omega_S(X)) = \Omega_S(X)$ and $\mathbf{\Omega_S}f = f^{-1}$.

Lemma 9 *Let* X *be an S-space. Then* $\varphi : \Omega_S(X) \to \Omega^\circ_S\Omega_S(X)$ *is an isomorphism, and for any* $U \in \Omega_S(X)$ *we have* $\Phi^{-1}(\varphi(U)) = U$, *so that* Φ *is an S-map from* X *to* $\mathbf{Pt}\mathbf{\Omega_S}X$.

Proof: It suffices to remark that, for $x \in X$ and $U \in \Omega_S(X)$, we have

$$x \in U \quad \text{iff} \quad \Phi(x) \in \varphi(U)\ .$$

The lemma follows easily from this fact. ∎

Proposition 2 *The functor* $\mathbf{\Omega_S}$ *is left adjoint to* $\mathbf{Pt}$.

Proof: It suffices to remark that the family of morphisms $(\varphi_{\mathcal{L}})$ defines the unit of the adjunction and the family (Φ_X) its counit. ∎

Definition 12 *An S-structure $\mathcal{L}$ is* spatial *iff $\varphi_{\mathcal{L}}$ is an isomorphism between $\mathcal{L}$ and $\Omega^{\circ}_{S}\mathcal{L}$.*

Definition 13 *A S-space X is said* sober *if Φ_X is an isomorphism.*

As a consequence of the previous proposition we get the following:

Corollary 1 *the functors* **Pt** *and* Ω_S *define an equivalence between the dual of the category of spatial S-structures and the category of sober S-spaces.*

2.2 Duality in the algebraic case

From now on, we restrict our interest to the special case where the S-spaces are meet-cpo's and the S-topology is the one induced by stability. Actually:

Proposition 3 *For any meet-cpo X, the set $\Omega^{\wedge}_{S}X$ is an S-topology on X called the stable S-topology on X.*

Proof: Let $\mathcal{D} \subseteq \Omega^{\wedge}_{S}X$ be a dd set. We prove that $V = \bigcup\mathcal{D}$ is in $\Omega^{\wedge}_{S}X$. That V is Scott open is standard. Let $x, y \in V$ be bounded. Let $U, U' \in \mathcal{D}$ be such that $x \in U$ and $y \in U'$. Then since x and y are bounded, $U \cap V$ is non empty and thus has an upper bound $W \in \mathcal{D}$. Since $x \wedge y \in W$, we have $x \wedge y \in V$.

It is obvious that $\Omega^{\wedge}_{S}X$ has binary glbs (which are intersections), a 0 (the empty set) and a 1 (the whole space). Distributivity is clear too (since glbs are intersections and lubs unions). ∎

Proposition 4 *Let $f : X \to Y$, where X and Y are two algebraic meet-cpo's. Then f is stable iff it is an S-map with respect to the stable S-topologies of X and Y. .*

Proof: Let f be an S-map and let us prove that it is stable, the converse being trivial. Indeed we do not really need algebraicity of domains, but only a T_0 property for the S-topolgy of Y, insured for instance by spatiality. But there is no place here to go into these developments. The fact that f is Scott-continuous is plain, since any S-open is Scott open, and since the S-open sets of the shape $\uparrow k$ where k is compact form a basis of the Scott-topology. In particular, f is monotonous. Now let $x, y \in X$ be bounded. We have $f(x \wedge y) \leq f(x) \wedge f(y)$. Take any compact k in Y such that $f(x) \wedge f(y) \in \uparrow k$. We have $f(x), f(y) \in \uparrow k$, hence $x, y \in f^{-1}(\uparrow k)$, and thus since $f^{-1}(\uparrow k) \in \Omega^{\wedge}_{S}X$, we have $x \wedge y \in f^{-1}(\uparrow k)$, that is $f(x \wedge y) \geq k$ and therefore $f(x \wedge y) \geq f(x) \wedge f(y)$. ∎

Let now $\mathcal{L}$ be any S-structure.

Proposition 5 *The poset* $(\mathrm{pt}(\mathcal{L}), \subseteq)$ *is a meet-cpo.*

Proof: The proof that $\mathrm{pt}(\mathcal{L})$ is a cpo is straightforward. Let $p, q \in \mathrm{pt}(\mathcal{L})$ be two bounded points. Let $r \in \mathrm{pt}(\mathcal{L})$ be an upper bound of p and q. We prove that $p \cap q$ is a point. That it is upper closed and stable under binary meets is clear. Let $D \subseteq \mathcal{L}$ be a dd set such that $\bigvee D \in p \cap q$. Let $u \in p \cap D$ and $v \in q \cap D$. Then $u, v \in r$ and thus $u \wedge v \in r$, therefore $u \wedge v \neq 0$. Since D is dd, there exists a $w \in D$ greater than u and v. We have thus $w \in p \cap q$ and we conclude.

Furthermore, $\mathrm{pt}(\mathcal{L})$ has a bottom, namely the empty set. ∎

Proposition 6 *If $\mathcal{L}$ is algebraic then it is spatial and* $\mathrm{pt}(\mathcal{L})$ *is an algebraic cpo. Futhermore* $\varphi(\mathcal{L}) \simeq \Omega_S^{\wedge}\mathrm{pt}(\mathcal{L})$ *and* $\mathrm{Kpt}(\mathcal{L}) \simeq \mathrm{Pr}\mathcal{L}^{\mathrm{op}}$.

Therefore, the functor **Pt** : **Sstr** $\rightarrow$ **Ssp** *restricts to a functor* **ASL** $\rightarrow$ **AIC** *that we still note* **Pt**.

Proof: We assume that $\mathcal{L}$ is algebraic.

We prove first that $\mathrm{pt}(\mathcal{L})$ is algebraic and that the compact elements of this cpo are the points of the shape $\uparrow \alpha$. For this we just have to prove that, for any $p \in \mathrm{pt}(\mathcal{L})$, p is the union of the set $D = \{\uparrow \alpha \mid \alpha \in p\}$ and that this set is directed. The fact that $\bigcup D \subseteq p$ is clear. Conversely, let $u \in p$, then by algebraicity $u = \bigvee \pi(u)$ and thus, since $\pi(u)$ is dd, $\pi(u) \cap p \neq \emptyset$. Let α be in this intersection, we have $u \in \uparrow \alpha \in D$. Let us prove now that D is directed. Let $\uparrow \alpha, \uparrow \beta \in D$, we have $\alpha, \beta \in p$, and thus by algebraicity again, there is a $\gamma \leq \alpha, \beta$ such that $\gamma \in p$, so we have $\uparrow \alpha, \uparrow \beta \subseteq \uparrow \gamma \in p$.

Let $U \in \Omega_S(\mathrm{pt}(\mathcal{L}))$. We prove that the set $B_U = \{\alpha \mid \uparrow \alpha \in U\}$ is dd. Let $\alpha, \beta \in B_U$ be such that $\alpha \wedge \beta \neq 0$. Then there exists a γ such that $\gamma \leq \alpha, \beta$ (by algebraicity). Thus $\uparrow \alpha \cap \uparrow \beta \in U$. By algebraicity of $\mathrm{pt}(\mathcal{L})$, we know that $\uparrow \alpha \cap \uparrow \beta$ is the directed union of all cones $\uparrow \gamma$ such that $\gamma \geq \alpha, \beta$, thus since U is Scott open there is such a gamma satisfying $\uparrow \gamma \in U$. Therefore we can define a function $\psi : \Omega_S(\mathrm{pt}(\mathcal{L})) \rightarrow \mathcal{L}$ by $\psi(U) = \bigvee B_U$. Clearly, ψ is monotonous. We conclude by proving that $\varphi \circ \psi = \mathrm{Id}$ and $\psi \circ \varphi = \mathrm{Id}$.

$$\begin{aligned} \psi\varphi(u) &= \bigvee \{\alpha \mid \uparrow \alpha \in \varphi(u)\} \\ &= \bigvee \{\alpha \mid u \in \uparrow \alpha\} \\ &= u \end{aligned}$$

by algebraicity of $\mathcal{L}$. On the other hand:

$$\begin{aligned} \varphi\psi(U) &= \{p \mid \psi(U) \in p\} \\ &= \{p \mid B_U \cap p \neq \emptyset\} \\ &= \{p \mid \exists \alpha \in p \ \uparrow \alpha \in U\} \\ &= U \end{aligned}$$

by algebraicity of $\mathrm{pt}(\mathcal{L})$.

The fact that **Pt** : **Sstr** $\rightarrow$ **Ssp** restricts to a functor **ASL** $\rightarrow$ **AIC** is a consequence of proposition 4. ∎

Proposition 7 *Let X be an algebraic meet-cpo. Then $(X, \Omega_S^{\wedge}X)$ is a sober S-space and $\Omega_S^{\wedge}X$ is an algebraic S-structure. Furthermore,* $\mathrm{Pr}\Omega_S^{\wedge}X \simeq \mathrm{K}X^{\mathrm{op}}$.

Therefore, the functor Ω_S : **Ssp** $\rightarrow$ **Sstr** *restricts to a functor* **AIC** $\rightarrow$ **ASL** *that we still note* Ω_S.

Proof: We first prove that $\Omega_S^{\wedge}X$ is algebraic and that its dd-prime elements are the stable opens of the shape $\uparrow k$ where $k \in X$ is compact. This amounts to proving that for any $U \in \Omega_S^{\wedge}X$, the set $P_U = \{\uparrow k \mid k \in U \cap \mathrm{K}X\}$ is dd. Let $\uparrow k, \uparrow l \in P_U$ having a non empty intersection. This means that k and l are bounded. Thus $k \wedge l \in U$, and since U

is Scott open and X is algebraic, this implies that k and l have a compact lower bound m in U. Then $\uparrow k, \uparrow l \subseteq \uparrow m \in P_U$.

Now we define a map $\Psi : \mathrm{pt}(\Omega_S^{\wedge} X) \to X$ by $\Psi(p) = \bigvee B_p$ where

$$B_p = \{k \in \mathrm{K}X \mid \uparrow k \in p\} .$$

Actually, let us prove that this set is directed. Let $k, l \in B_p$. Then $\uparrow k \cap \uparrow l \in p$ since p is a filter. Since $\Omega_S^{\wedge} X$ is algebraic, $\uparrow k \cap \uparrow l$ is the dd lub of its prime lower bounds in $\Omega_S^{\wedge} X$, which are of the shape $\uparrow m$ with $m \geq k, l$. Since p is a dd-prime filter, there is a compact $m \geq k, l$ such that $\uparrow m \in p$, that is $m \in B_p$. So Ψ is well defined by completeness of X.

We finish the proof by a simple calculation:

$$\begin{aligned} \Phi\Psi(p) &= \{U \mid \exists k \in \mathrm{K}X \ \uparrow k \in p \text{ and } k \in U\} \\ &= \{U \mid U \in p\} = p \end{aligned}$$

and

$$\begin{aligned} \Psi\Phi(x) &= \bigvee\{k \in \mathrm{K}X \mid x \in \uparrow k\} \\ &= \bigvee\{k \in \mathrm{K}X \mid x \geq k\} \\ &= x \end{aligned}$$

by algebraicity of X.

The fact that Ω_S restricts to a functor **AIC** $\to$ **ASL** results, for morphisms, from proposition 4. ∎

To summarize:

Proposition 8 *The functors* **Pt** *and* Ω_S *define an equivalence of categories between the category* **ASL** *of algebraic S-structures with S-morphisms and the dual of the category* **AIC** *of algebraic meet-cpos with stable maps.*

3 Intrinsic stable ordering

In the following, we assume $\mathcal{L}$ to be a fixed algebraic S-structure.

Definition 14 *Let $u, v \in \mathcal{L}$. We say that u is stably less than v and we write $u \sqsubseteq v$ iff $u \leq v$ and*

$$\forall \alpha \ \alpha \leq v \text{ and } u \wedge \alpha \neq 0 \Rightarrow \alpha \leq u .$$

Proposition 9 *i) $\sqsubseteq$ is an order included in $\leq$.*

ii) If $u, v \in \mathcal{L}$ are stably bounded then $u \wedge v$ is the glb of u and v wrt $\sqsubseteq$.

iii) If $D \subseteq \mathcal{L}$ is directed wrt $\sqsubseteq$, then $\bigvee D$ is the lub of D wrt $\sqsubseteq$.

Proof:

i) We have just to check transitivity: $u \sqsubseteq v \sqsubseteq w$ and $\alpha \leq w$ and $\alpha \wedge u \neq 0$ then $\alpha \wedge v \neq 0$ so $\alpha \leq v$ and $\alpha \wedge u \neq 0$, hence $\alpha \leq u$.

ii) Let $u, v \in \mathcal{L}$ be stably bounded; we have to show that $u \wedge v \sqsubseteq u$; so let $\alpha \leq u$ such that $\alpha \wedge u \wedge v \neq 0$; now there exists w such that $u \sqsubseteq w \sqsupseteq v$, because u, v are bounded, so we can state that $\alpha \leq w$ because $u \leq w$ and $\alpha \wedge u \neq 0$ because $\alpha \wedge u \geq \alpha \wedge u \wedge v$; by the same argument $\alpha \wedge v \neq 0$, so we conclude that $\alpha \leq u \wedge v$.

iii) Let $u \in D$; it is easily verified that $u \sqsubseteq \sup D$ iff $\forall \alpha, \forall v \in D\ \alpha \leq v, \alpha \wedge u \neq 0 \Rightarrow \alpha \leq u$; but for $w \in D, u, v \sqsubseteq w$ we can deduce that if $\alpha \leq v, u \wedge \alpha \neq 0$ then $\alpha \leq w \sqsupseteq u$ hence $\alpha \leq u$, which proves that $u \sqsubseteq \bigvee D$.
Let suppose now that exists a w which is another stable upper bound for D and $w \sqsubseteq \bigvee D$; then we can conclude that $w = \bigvee D$ because $\sqsubseteq$ is included in $\leq$.

■

Proposition 10 *Let $a, b, u \in \mathcal{L}$.*

i) If $a \sqsubseteq b$ then $a = b$.

ii) If $a, b \sqsubseteq u$ then either $a = b$ or $a \wedge b = 0$.

Proof:

i) If $\alpha \leq b$ then we can find a $\gamma \leq a$ such that $\alpha \leq \gamma$; this γ exists because by lemma 6 $\pi(a), \pi(b)$ are directed set and $\pi(a) \subseteq \pi(b)$; so for $\beta \leq a$ there exists γ, such that $\alpha, \beta \geq \gamma \in \pi(b)$; it follows that the set $H = \{\gamma \leq b \mid \gamma \wedge a \neq 0\}$ is a directed set included in $\pi(b)$ and moreover $b = \bigvee H$; but $a \sqsubseteq b$ implies that $H \subseteq \pi(a)$ so $b = \bigvee H \leq \bigvee \pi(a) = a$.

ii) Suppose $a \wedge b \neq 0$; then $a \wedge b \sqsubseteq a$; let $\alpha \leq a$, then there exists $\gamma \leq a$ such that $\alpha \leq \gamma$ and $a \wedge b \wedge \gamma \neq 0$ (because $\pi(a)$ is directed); now $a \wedge b \sqsubseteq a$ implies $\gamma \leq a \wedge b$; it follows that $H = \{\gamma \leq a \mid a \wedge b \wedge \alpha \neq 0\}$ is directed and $a = \bigvee H = a \wedge b$; the same holds for b giving the desired equality $a = a \wedge b = b$.

■

Definition 15 *Let $u \in \mathcal{L}$. We call* trace *of u and we note* $\mathrm{tr}(u)$ *the set of connected stable lower bounds of u.*

Remark: A connected stable lower bound of u is just a connected maximal lower bound of u so in the previous definition the reference to definition of stable ordering is not necessary.

Proposition 11 *i) $u \sqsubseteq v$ iff $\mathrm{tr}(u) \subseteq \mathrm{tr}(v)$.*

ii) $u \leq v$ iff for all $a \in \mathrm{tr}(u)$ there exists a $b \in \mathrm{tr}(v)$ such that $a \leq b$ (and then this b is unique).

Proof:

i) If $u \sqsubseteq v$ and $a \in \mathrm{tr}(u)$ then $a \sqsubseteq u \sqsubseteq v$ so $a \sqsubseteq v$ and hence $a \in \mathrm{tr}(v)$. Conversely let $\mathrm{tr}(u) \subseteq \mathrm{tr}(v)$, let $\alpha \leq v$ be such that $\alpha \wedge u \neq 0$. Then there exists an unique $a \in \mathrm{tr}(v)$ such that $\alpha \leq a$ (because α is connected); now $\alpha \wedge u \neq 0$ implies $a \in \mathrm{tr}(u)$ (by disjointness of traces) so we get $\alpha \leq a$ which proves $u \sqsubseteq v$.

ii) If $u \leq v$ and $a \in \mathrm{tr}(u)$ then $a \leq v = \dot{\bigvee} \mathrm{tr}(v)$ so there exists an unique $b \in \mathrm{tr}(v)$ which is greater than a. The converse follows from monotonicity of sup.

■

Proposition 12 *The cpo $(\mathcal{L}, \sqsubseteq)$ is isomorphic to the coherence space $C(\mathcal{L})$ the web of which is the set of connected elements of $\mathcal{L}$ endowed with the coherence relation defined by: $a \frown b$ iff $a \wedge b = 0$. This isomorphism $\Theta : \mathcal{L} \to C(\mathcal{L})$ is given by: $\Theta(u) = \mathrm{tr}(u)$ and $\Theta^{-1}(A) = \bigvee A$.*

Proof: We have already shown that $u = \dot{\bigvee}_{i \in I} (\bigvee \Delta_i)$ where for each i the set Δ_i is directed and $\bigcup_{i \in I}(\Delta_i) = \pi(u)$, so $u = \dot{\bigvee}_{i \in I} a_i$ where $\{a_i \mid i \in I\}$ is the set of (disjont) connected elements of $\mathcal{L}$ such that $\bigvee \Delta_i = a_i \sqsubseteq u$ for any i; this shows that Θ is a bijection. The fact that is an isomorphism follows from proposistion 11.i) . ■

Corollary 2 *The cpo $(\mathcal{L}, \sqsubseteq)$ is algebraic. But in general, the cardinality of its basis is strictly larger than the cardinality of the basis of $(\mathcal{L}, \leq)$.*

Proposition 13 *The evaluation map $\mathrm{Ev} : C(\mathcal{L}) \times \mathrm{pt}(\mathcal{L}) \to \mathbf{O}$ defined by $\mathrm{Ev}(u, p) = \top$ iff $u \in p$ is stable.*

Proof: Let $u, v \sqsubseteq w$ and $p, p' \subseteq q$; we have to show that $\mathrm{Ev}(u \wedge v, p \cap p') = \mathrm{Ev}(u, p) \wedge \mathrm{Ev}(v, p')$, i.e. $u \in p$ and $v \in p' \Rightarrow u \wedge v \in p \cap p'$ the other implication being trivial; now $u \in p$ and $v \in p'$ iff there exists $a \sqsubseteq u$, $b \sqsubseteq v$ such that $a \in p$, $b \in p'$ (because p, p' are dd-prime filters), and $u, v \sqsubseteq w$ forces $a = b$ or $a \wedge b = 0$; but $a \wedge b = 0$ implies $0 \in q$ contradicting the fact that q is a dd-prime filter; hence $u \wedge v \in p \cap p'$. ■

4 Function spaces

In this section, we assume that $\mathcal{L}$ and $\mathcal{M}$ are two algebraic S-structures.

4.1 Stable ordering of S-morphism

In this paragraph we introduce the notions of stable ordering between S-morphism and trace of an S-morphism and then we show that these definitions are the natural ones, that is stable ordering between S-morphism corresponds (by duality) to the Berry' order on stable maps.

Definition 16 *Let $f, g : \mathcal{L} \to \mathcal{M}$ be two S-morphisms.*

- *We say that f is stably less than g and we write $f \sqsubseteq g$ iff for all $\alpha \in \mathcal{L}$ we have $f(\alpha) \sqsubseteq g(\alpha)$.*
- *The trace of f is the set $\mathrm{tr}(f) = \{(\alpha, b) \in \mathrm{Pr}\mathcal{L} \times \mathrm{Conn}(\mathcal{M}) \mid b \sqsubseteq f(\alpha)\}$.*

Proposition 14 *For any α, $f(\alpha) \sqsubseteq g(\alpha)$ iff for any u, $f(u) \sqsubseteq g(u)$.*

Proof: We have just to prove the "left to right" implication. Let $\alpha \le f(u)$ and $\alpha \wedge g(u) \neq 0$. Then there exist $\beta, \beta' \le u$ such that $\alpha \le f(\beta)$ and $\alpha \wedge g(\beta') \neq 0$ (because f and g preserve the dd lubs). It follows that $\beta \wedge \beta' \neq 0$ (because otherwise $0 = g(\beta \wedge \beta') = g(\beta) \wedge g(\beta') \ge \alpha \wedge g(\beta)$) and therefore, by using the fact that $\pi(u)$ is dd, there exists $\gamma \le u$, such that $\beta, \beta' \le \gamma$, $\alpha \le f(\gamma)$, $\alpha \wedge g(\gamma) \neq 0$, so we conclude that $\alpha \le g(u)$. ∎

Remember that we have already defined a notion of stable ordering between stable maps. The following connects the two notions.

Proposition 15 *Let $f, g : \mathcal{L} \to \mathcal{M}$ be two S-morphisms. Then $f \sqsubseteq g$ iff $f_* \sqsubseteq g_*$.*

Proof: Suppose $f_* \sqsubseteq g_*$; this means that $\forall x' \ge x (f_*(x) = g_*(x) \wedge f_*(x'))$; we have to show that for any prime $\beta \in \mathcal{L} \simeq \Omega_S(\mathrm{pt}(\mathcal{L}))$ $f_*^{-1}(\beta) \sqsubseteq g_*^{-1}(\beta)$, i.e. $\forall \alpha (\alpha \subseteq g_*^{-1}(\beta)$ and $\alpha \cap f_*^{-1}(\beta) \neq \emptyset \Rightarrow \alpha \subseteq f_*^{-1}(\beta))$ which is the same (by using the isomorphism between $\mathrm{Pr}\mathcal{L}$ and $\mathrm{Kpt}(\mathcal{L})^{\mathrm{op}}$) that for $y \in \mathrm{Kpt}(\mathcal{M})$ and for any $x \in \mathrm{Kpt}(\mathcal{M})$ $(\uparrow x \subseteq g_*^{-1}(\uparrow y)$ and $\uparrow x \cap f_*^{-1}(\uparrow y) \neq \emptyset \Rightarrow \uparrow x \subseteq f_*^{-1}(\uparrow y))$ which is true whenever, if $g_*(x) \ge y$ and there exists $z \ge x$ such that $f_*(z) \ge y$ then $f_*(x) \ge y$. Now, by hypothesis, we have $f_*(x) = g_*(x) \wedge f_*(z)$ hence the last condition is true.

By definition $f_*(p) = f^{-1}(p)$, so let $p \subseteq p'$ and what we have to show is that

$$f^{-1}(p) = g^{-1}(p) \cap f^{-1}(p')$$

which is true iff for any u $f(u) \in p$ iff $g(u) \in p$ and $f(u) \in p'$. This last equivalence is true because if $f(u) \in p$ then $g(u) \in p$ (because $f(u) \le g(u)$) and $f(u) \in p'$ (because $p \subseteq p'$). On the other side, let $f(u) \in p'$ and $g(u) \in p$; then there exist $\alpha \le f(u), \alpha \in p'$ and $\alpha' \le g(u), \alpha' \in p$; now $\alpha' \le g(u)$ and $\alpha' \wedge f(u) \neq 0$ because $\alpha \wedge \alpha' \in p'$ so $\alpha' \le f(u)$ which allows to us to conclude that $f(u) \in p$. ∎

Proposition 16 *For any $u \in \mathcal{L}$ we have $f(u) = \bigvee \{b \mid \exists \alpha \le u \ (\alpha, b) \in \mathrm{tr}(f)\}$ and $f \sqsubseteq g$ iff $\mathrm{tr}(f) \subseteq \mathrm{tr}(g)$.*

Proof: In fact $f(\alpha) = \dot{\bigvee} \{b \mid b \sqsubseteq f(\alpha)\}$ so

$$\begin{aligned} f(u) &= f(\bigvee \pi(u)) \\ &= \bigvee f(\pi(u)) \\ &= \bigvee \{b \mid \exists \alpha \le u \ (\alpha, b) \in \mathrm{tr}(f)\} \end{aligned}$$

and

$$\begin{aligned} f \sqsubseteq g \quad & \text{iff} \quad \forall \alpha \ f(\alpha) \sqsubseteq g(\alpha) \\ & \text{iff} \quad \forall \alpha \ \mathrm{tr}(f(\alpha)) \subseteq \mathrm{tr}(g(\alpha)) \\ & \text{iff} \quad \mathrm{tr}(f) \subseteq \mathrm{tr}(g) \end{aligned}$$

∎

4.2 Space of function's traces

In the following we define the set of weakly coherent S-structure. For $\mathcal{L}$, $\mathcal{M}$ in this class we can represent the set of S-morphism from $\mathcal{L}$ to $\mathcal{M}$ as a subset of $\mathcal{P}(\mathrm{Pr}\mathcal{L} \times \mathrm{Conn}(\mathcal{M}))$ which we call the set of trace on $\mathcal{L}$, $\mathcal{M}$. It is interesting to remark that, as far we know, when we consider more general domains than dI domains, there is no simple definition of trace of stable map, since there is no more notion of minimal element. Now the definition of connected element idealizes, always by duality, the notion of minimal element, and so allows us to speak about traces in the usual sense. This representation property will be used for proving the cartesian closedness of the category of L-domains with stable maps which is the dual of the category of weakly coherent S-structure with S-morphism.

Definition 17 *An S-structure $\mathcal{L}$ is called* weakly coherent *if for any α, β we have* $\mathrm{tr}(\alpha \wedge \beta) \subseteq \mathrm{Pr}\mathcal{L}$.

Remark: By duality a weakly coherent S-structure is an algebraic meet-cpo such that any finite set A of bounded compact points has a set B of compact upper bounds which is complete, i.e. any two elements of this set are unbounded, and any upper bound of A is greater than an element of B. This is equivalent to say that the set of lower bound of any point is a lattice, and that is the definition of an L-domain.

Proposition 17 *We assume that $\mathcal{L}$ is weakly coherent. Let T be a subset of $\mathrm{Pr}\mathcal{L} \times \mathrm{Conn}(\mathcal{M})$. T is the trace of an S-morphism $\mathcal{L} \to \mathcal{M}$ iff it satisfies the three following conditions:*

i) For any α, the set $T_\alpha = \{b \mid (\alpha, b) \in T\}$ is disjoint.

ii) For any $(\alpha, b) \in T$, for any $\alpha' \geq \alpha$, there exists a b' such that $(\alpha', b') \in T$.

iii) For any $(\alpha, b), (\alpha', b') \in T$, for any $c \sqsubseteq b \wedge b'$, there is a $\beta \sqsubseteq \alpha \wedge \alpha'$ such that $(\beta, c) \in T$.

Furthermore, when these conditions hold, the b' of ii) is unique, and the β of iii) too.

Any set satisfying these conditions will be called a trace.

Proof: Let $f : \mathcal{L} \to \mathcal{M}$ be an S-morphism. Let us prove that $T = \mathrm{tr}(f)$ satisfies conditions i), ii) and iii).

i) It simply results from the fact that $T_\alpha = \mathrm{tr}(f(\alpha))$.

ii) It results from proposition 11 ii).

iii) Let $(\alpha, b), (\alpha', b') \in T$. We have

$$\begin{aligned} f(\alpha \wedge \alpha') &= \dot{\bigvee} \{f(\beta) \mid \beta \sqsubseteq \alpha \wedge \alpha'\} \\ &= \dot{\bigvee} \{c \mid \exists \beta \sqsubseteq \alpha \wedge \alpha' \ (\beta, c) \in T\} \end{aligned}$$

and

$$\begin{aligned} f(\alpha) \wedge (\alpha') &= \dot{\bigvee} \{b \wedge b' \mid (\alpha, b), (\alpha', b') \in T\} \\ &= \dot{\bigvee} \{c \mid \exists (\alpha, b), (\alpha', b') \in T \; c \sqsubseteq b \wedge b'\} . \end{aligned}$$

Since $f(\alpha) \wedge (\alpha') \leq f(\alpha \wedge \alpha')$ and since the lubs written above are both disjoint lubs of connected elements, we conclude that *iii*) holds for T.

Let now $T \subseteq \mathrm{Pr}\mathcal{L} \times \mathrm{Conn}(\mathcal{M})$ be a set satisfying conditions i), ii) and iii). Remark first that, for any $u \in \mathcal{L}$, the set $F_T(u) = \{b \mid \exists \alpha \leq u \; (\alpha, b) \in T\}$ is dd, so that by setting $f(u) = \bigvee F_T(u)$ we define a function from $\mathcal{L}$ to $\mathcal{M}$. Actually, let $b, b' \in F_T(u)$ be such that $b \wedge b' \neq 0$. Let $\alpha, \alpha' \leq u$ be such that $(\alpha, b), (\alpha', b') \in T$. Then $\alpha \wedge \alpha' \neq 0$ because of condition iii). Since $\pi(u)$ is dd, there exists $\beta \geq \alpha, \alpha'$ such that $\beta \leq u$. By condition ii), we can find $c \geq b$ and $c' \geq b'$ such that $(\beta, c), (\beta, c')$. But $c \wedge c' \geq b \wedge b' \neq 0$, so by condition i), $c = c'$. So we have found in $F_T(u)$ a c such that $c \geq b, b'$, and thus $F_T(u)$ is dd.

It is obvious that $f(0) = 0$ and that $f(1) = 1$. Let us prove that f preserves binary glbs. Let $u, v \in \mathcal{L}$. On one hand,

$$f(u \wedge v) = \bigvee \{b \mid \exists \alpha \leq u \wedge v \; (\alpha, b) \in T\} .$$

On the other hand,

$$f(u) \wedge f(v) = \bigvee \{b \wedge b' \mid \exists \alpha \leq u, \alpha' \leq v \; (\alpha, b), (\alpha', b') \in T\}$$

and we have to prove that $f(u) \wedge f(v) \leq f(u \wedge v)$ since f is clearly monotonous, by condition ii). So let $(\alpha, b), (\alpha', b') \in T$ be such that $\alpha \leq u$ and $\alpha' \leq v$. Let c be such that $c \sqsubseteq b \wedge b'$. By property iii), there exists a $\beta \sqsubseteq \alpha \wedge \alpha'$ such that $(\beta, c) \in T$. Since $\beta \leq \alpha \wedge \alpha' \leq u \wedge v$, we have $c \leq f(u \wedge v)$, and we conclude.

Last, let $D \subseteq \mathcal{L}$ be dd. We have

$$\begin{aligned} f(\bigvee D) &= \bigvee \{b \mid \exists \alpha \leq \bigvee D \; (\alpha, b) \in T\} \\ &= \bigvee \{b \mid \exists \alpha \; \exists u \in D \; \alpha \leq u \text{ and } (\alpha, b) \in T\} \\ &= \bigvee_{u \in D} f(u) \end{aligned}$$

hence f preserves dd lubs.

To conclude, let us check the second unicity statement, for any T satisfying the three conditions (the first one is easy too and results from i)). Let $(\alpha, b), (\alpha', b') \in T$ and let $c \sqsubseteq b \wedge b'$. Let $\beta, \beta' \sqsubseteq \alpha \wedge \alpha'$ be such that $(\beta, c), (\beta', c) \in T$. There must exist (by iii)) a $\gamma \leq \beta, \beta'$ such that $(\gamma, c) \in T$, so $\beta \wedge \beta' \neq 0$, so $\beta = \beta'$ since they both are in $\mathrm{tr}(\alpha \wedge \alpha')$. ∎

Proposition 18 *Let T be a trace. Let E be any subset of T. Then there exists a trace $[E]_T$ such that $E \subseteq [E]_T \subseteq T$ and which is minimal among the traces having this property.*

Proof: We define $[E]_T = \bigcap A$ where $A = \{T' \mid E \subseteq T' \subseteq T \text{ and } T' \text{ is a trace}\}$. We have to show that $[E]_T$ is a trace. Condition *i*) holds for any element of A, so it holds for their intersection. If $(\alpha, b) \in T'$ for any $T' \in A$ then for any $\alpha' \geq \alpha$, there exists a b' such that $(\alpha', b') \in T'$ and since this b' is unique in T, it must be the same for any T', hence condition *ii*) holds. Finally, by using the same unicity argument, condition *iii*) is verified. ∎

As a corollary, we retrieve an already known result:

Proposition 19 *The poset of all traces ordered by inclusion is an L-domain.*

Proof: Checking that the union of a directed family of traces is a trace is straightforward.

Let T be a fixed trace. If S, S' are two traces included in T then by proposition18 $S \cap S'$ is a trace and so it is their glb

We can now prove algebraicity. It simply results from the fact that, for any finite subset E of T, the trace $[E]_T$ is compact. Actually, let $\mathcal{D}$ be a directed family of traces such that $[E]_T \subseteq \bigcup \mathcal{D}$. Since E is finite, there exists a trace $S \in \mathcal{D}$ such that $E \subseteq S$. Now $[E]_T, S \subseteq \bigcup \mathcal{D}$ so $[E]_T \cap S$ is a subtrace of T containing E; hence $[E]_T \subseteq [E]_T \cap S \subseteq S$.

We finally show that the set of lower bounds of a point is a lattice. It suffices to show the existence of finite lubs in $\downarrow(T)$ where T is a trace. The lub of S, S' is $S'' = [S \cup S']_T$; by definition, S'' is a trace greater than S and S' and less than T. Now S'' is the lub of S and S' in $\downarrow(T)$, because of proposition 18.

∎

From this result, it is easy to prove that the category of L-domains and stable functions is cartesian closed. Indeed what remains to be proved is that evaluation and abstraction are stable and this is a routine exercise, since stable ordering has been choosen in order to make evaluation stable.

References

[Ab] S. Abramsky. *Domain theory in logical form.* Proceedings of the second annual symposium on Logic in Computer Science, 1987.

[Ber] G. Berry. *Modèles complètement adéquats et stables des lambda-calculs typés.* Thèse d'Etat.

[John] P. Johnstone *Stone Spaces.* Cambridge University Press 1982.

[Jung] A. Jung. *Cartesian closed categories of domains.* Dissertation, FB Mathematik der TH Darmstadt 1988.

[McL] S. MacLane. *Categories for the working mathematician.* Springer Verlag 1971.

[Zha] G.Q. Zhang. *Logics of domains.* Technical report n° 185. University of Cambridge, Computer Laboratory, dec 1989.

Bifinite Domains: Stable Case

Roberto M. Amadio[1]
Labo. Informatique, Ecole Normale Supérieure,
45 rue d'Ulm, F-75230, Paris. e-mail: amadio@dmi.ens.fr

Abstract

Let $\mathbf{Cpo}_\wedge$ be the ccc of complete partial orders with "continuous glbs" of compatible pairs, and stable maps. We introduce the full subccc $\mathbf{Bif}_\wedge$ of stable bifinites (or equivalently sfp) over $\mathbf{Cpo}_\wedge$. Its objects are characterized as ω-algebraic $\text{Cpos}_\wedge$ satisfying a "combination" of property M, as in Smyth theorem, and property I, as introduced by Berry, that we call property (MI)*.

We "test" the category $\mathbf{Bif}_\wedge$ via a series of classical constructions in domain theory. Inter alia we show that: (1) $\mathbf{Bif}_\wedge$ is an ω-algebroidal category and it has a universal, homogeneous object. (2) The image of a stable retraction over a stable bifinite is again a stable bifinite. (3) If $D \in \text{Bif}_\wedge$, and Prj(D) denotes the collection of projections over D then $\text{Prj(D)} \in \text{Bif}_\wedge$.

Next we investigate which full, cartesian closed, sub-categories of ω-algebraic $\text{Cpo}_\wedge$ and stable maps are contained in $\mathbf{Bif}_\wedge$. It is shown that property M and "2/3" of property I are necessary for preserving the ω-algebraicity of the functional space. The remaining "1/3" of property I is also necessary under rather mild hypothesis. As a fall out we show that a stable, countably based, version of L-domains, introduced by Coquand, is contained in $\mathbf{Bif}_\wedge$ and that such L-domains are the "largest ccc" under the assumption that principal ideals of domains are distributive.

Contents: 1. Introduction, 2. Stable Bifinites, 3. Characterization Stable Bifinites, 4. Some Domain Theoretic Constructions, 5. Property M, Property I, and ω-Algebraicity, 6. Conclusion, References, Acknowledgments, Appendix 1: Trace properties, Appendix 2: Retractions, Appendix 3: L-domains in the Stable Case.

1. Introduction

Conventions

A *preorder* is a binary relation on a set that is reflexive and transitive. If the relation is also anti-symmetric then we speak of *partial order* (poset). If (P, ≤) is a pre-order and $X \subseteq P$ then X is *directed* if $\forall x, y \in X.\ \exists z \in X.(x \le z \ \wedge\ y \le z)$. If $d, d' \in P$ then $\downarrow d$ ($\uparrow d$) denotes $\{e \in P \mid e \le d\}$ ($\{e \in P \mid d \le e\}$). Also $d \uparrow d'$ iff $\exists e.(d \le e \wedge d' \le e)$. An *ideal* I is a non-empty, directed, downward closed subset of P. ∧, ∨ denote respectively glbs and lubs of a pair of elements. ∩, and ∪ denote respectively glbs and lubs of a family of elements.

The category **Cpo** of complete partial orders and continuous maps has: (i) as *objects* posets that are complete in the sense that they have a least element and every directed set has a lub (least upper bound). (ii) as *morphisms* monotone maps that preserve the lubs of directed sets, that is: X directed $\Rightarrow f(\cup X) = \cup_{x \in X} f(x)$; after Scott topology we also call such maps *continuous*. Given a cpo D and $d \in D$ we say that d is *compact* if for each X directed: $d \le \cup X \Rightarrow \exists x \in X.(d \le x)$. We denote with D_0 the collection of compact elements. A cpo D is *algebraic* if $\forall d \in D$, $\{d' \in D_0 \mid d' \le d\}$ is directed and $d = \cup\{d' \in D_0 \mid d' \le d\}$. It is *ω-algebraic* if moreover D_0 is countable. A cpo is *bounded complete* if every bounded set of elements has a lub.

More Conventions

A category **C** is an *O-category* if (1) every hom-set is a partial order with lubs of every ω-chain; (2) composition of morphisms is a continuous operation with respect to the orders of the hom-sets.

[1] Research supported by "Consiglio Nazionale delle Ricerche", Joint Collaboration Contract "Typed Lambda Calculus" of the EEC, and Basic Research Action "Categorical Logic in Computer Science" of the EEC.

Let **C** be an O-category. The category $\mathbf{C}^{ep}$ has the same objects as C and as morphisms embedding-projection pairs:

$$\mathbf{C}^{ep}[A, B] \triangleq \{(i, j) \mid i: A \to B, j: B \to A, j \circ i = id_A, i \circ j \le id_B\}.$$

Let **C** be an O-category, A, B∈C. We write:

$A \lhd B$ if $\exists$ (i, j), i: A→B, j: B→A, $j \circ i = id_A$. *(Retraction)*

$A \sqsubseteq B$ if $\exists$ (i, j), i: A→B, j: B→A, $j \circ i = id_A$, $i \circ j \le id_B$. *(Embedding-Projection)*

1.1 Bifinite Domains

The category of *bifinite domains*[2] was introduced in Plotkin[76] under the name of SFP (Sequences of Finite inductive Partial orders). There, bifinite domains are characterized as ω-algebraic cpos satisfying three additional conditions. Plotkin also conjectured that these conditions are *necessary* for getting cartesian closure. This result was finally proved in Smyth[83]. Jung[90] generalizes Smyth's result to various cases including the one in which the assumption on the *countability* of the compacts is removed. In this case a new subcategory of **Cpo** has to be considered, namely the category of *L-domains.*

The "bifinite approach" to the construction of domains is appealing in its simplicity and generality. Roughly: (1) Start with some idea of how your maps should behave on finite cpos and so define a (cartesian closed) category $\mathbf{C}_0$. (2) Fully embed $\mathbf{C}_0$ into a much larger (cartesian closed) category **C** including infinite complete partial orders, without ever worrying about ω-algebraicity. (3) Consider the related category, $\mathbf{C}^{ep}$, of embedding-projection pairs. (4) Define **Bif(C)** as the full subcategory of **C** whose objects can be obtained as ω-colimits of ω-diagrams in $\mathbf{C}_0{}^{ep}$.

In other words the bifinite objects are those objects of the larger category **C** that can be obtained as ω-colimits (in the related category of embedding-projections pairs) of ω-diagrams of finite objects. Such objects turn out to be almost automatically ω-algebraic; moreover they are closed w.r.t. those domain theoretic constructions that preserve ω-colimits in $\mathbf{C}^{ep}$, and finiteness. In particular cartesian closure is obtained by an elementary argument.

1.2 Stable Maps

The notion of *stable map,* as that of a continuous map that preserves glbs of compatible pairs, has two origins related to λ-calculus and proof theory. It was introduced by Berry as a semantic approximation of the sequential behavior of λ-calculus evaluation[3] and by Girard in the theory of dilators. It appears that similar ideas were also developed independently and with purely algebraic motivations by Diers (see Taylor[90] for an up-to-date account oriented towards computer science).

Once the notion of stable transformation is accepted and we start working with categories of cpos having glbs of compatible elements two basic original ideas appear. The first one is a *new order on the functional space* that refines the pointwise order. This order is naturally discovered by requiring the stability of the evaluation. The second one is *property I.* It says that the principal ideal generated by a compact element is finite. This property has to do with the ω-algebraicity of the functional space. It also appears in Kahn and Plotkin "concrete domains", and in Winskel's "event structures". It is worth observing that property I is *not* preserved by the pointwise ordering, so it also provides an a posteriori justification for the use of the stable ordering. More importantly under property I it is possible to recover the original *computational intuition* of stable map as that of a continuous map over domains, say f: D→E, with the property that:

$$\forall d, e.\ e \le f(d) \Rightarrow \exists \min\{d' \le d \mid e \le f(d')\}.$$

[2] A terminology proposed by P.Taylor and adopted in Gunter and Scott[91].

[3] The stable maps considered here are in Berry's terminology "multiplicative under condition".

Roughly, it is always possible to find "locally" the minimum amount of information needed to generate a given output.

In the following we are particularly indebted to Berry[79]. There will be however a notable variation: we will consider a "bifinite approach to the stable case". On the way of a comparison let us recall that Berry's dI-domains are bounded complete algebraic cpos satisfying property I and distributivity. We already discussed the reasons for the introduction of property I. *Distributivity* arises when trying to prove that the functional space is *bounded complete*. On the other hand in our approach we will drop the requirement of bounded completeness and substitute it with the weaker assumption that compatible pairs have glbs. Then it will turn out that the condition of distributivity can be dropped as well.

An *overview* of the paper goes as follows: in section 2 we introduce the ccc of stable bifinites and stable maps, $\mathrm{Bif}_\wedge$; in section 3 we characterize the objects in $\mathrm{Bif}_\wedge$; in section 4 we study some more closure properties of the category; in section 5 we investigate which of the properties characterizing objects in $\mathrm{Bif}_\wedge$ are necessary in order to preserve the ω-algebraicity of the functional space. All omitted proofs and some more remarks can be found in Amadio[91].

2. Stable Bifinites

$\mathbf{Cpo}_\wedge$ is a category of cpos with continuous glbs of compatible pairs and continuous maps preserving such glbs. We first show that $\mathbf{Cpo}_\wedge$ is a ccc. The main novelty w.r.t. the continuous case lies in the introduction of a *stable ordering* on the functional space that is practically forced by the requirement that the evaluation map is stable.

2.1 Definition *($Cpo_\wedge$)*

$D \in \mathrm{Cpo}_\wedge$ if (1) D is a cpo, (2) $\forall d, e \in D.(d \uparrow e \Rightarrow \exists\, d \wedge e)$,
(3) $\forall X \subseteq D$, directed $(d \uparrow \cup X \Rightarrow d \wedge \cup X = \cup_{x \in X}(d \wedge x))$.[4]

2.2 Definition *(category of $Cpo_\wedge$ and stable maps)*

Let D, $E \in \mathrm{Cpo}_\wedge$. A map f: D→E is stable if
(1) it is Scott continuous, and (2) $\forall d, e \in D.(d \uparrow e \Rightarrow f(d \wedge e) = f(d) \wedge f(e))$.
We denote with $\mathbf{Cpo}_\wedge$ the category of $\mathrm{Cpo}_\wedge$ and stable maps.

2.3 Proposition *($Cpo_\wedge$ is a ccc)*

The category $\mathbf{Cpo}_\wedge$ is cartesian closed. In particular given D, $E \in \mathrm{Cpo}_\wedge$ the exponent D→E is given by the collection of stable maps ordered as follows:

$$f \leq_s g \Leftrightarrow \forall d, d' \in D.(d \leq d' \Rightarrow f(d) = f(d') \wedge g(d)).^5$$

Proof

We assume that the reader is familiar with the proof that cpos and continuous maps form a ccc. We mainly refer to the additional checks that are needed.

$\mathbf{Cpo}_\wedge$ is *cartesian*: take the terminal object and the product as in **Cpo**. Check: (a) terminal object and product are in $\mathrm{Cpo}_\wedge$. (b) constant functions and projections are stable. (c) if f: C→D and G: C→E are stable then the pairing $\langle f, g\rangle \triangleq \lambda c.(fc, gc)$: C→D×E is stable. Let us give the details of this. Suppose $c \uparrow c'$, then:

$$\langle f, g\rangle(c \wedge c') = (f(c \wedge c'), g(c \wedge c')) = (f(c) \wedge f(c'), g(c) \wedge g(c')) = (f(c), g(c)) \wedge (f(c'), g(c')).$$

$\mathbf{Cpo}_\wedge$ is *cartesian closed*. Consider the relation $\leq_s$ on the collection of stable maps D→E defined as above. First let us control that $(D \to E, \leq_s) \in \mathrm{Cpo}_\wedge$.

---

[4] Note that (3) makes sense. If D is algebraic then (2) implies (3). We use (3) many times in showing that D→E is a $\mathrm{Cpo}_\wedge$. Another way to state (3) is to say that whenever we look at principal ideals the $\wedge$ distributes w.r.t. directed sets.

[5] Henceforth we denote with $\leq_s$ the stable ordering on the functional space.

• $(D \to E, \preceq)$ is a partial order with least element.

Reflexivity: $f \preceq f$. $d \leq d' \Rightarrow f(d) = f(d') \wedge f(d)$, by monotonicity. *Symmetry:* $f \preceq g \preceq f \Rightarrow f = g$. $d \leq d \Rightarrow f(d) = f(d) \wedge g(d) \leq g(d)$ and symmetrically $d \leq d \Rightarrow g(d) = f(d) \wedge g(d) \leq f(d)$. *Transitivity:* $f \preceq g \preceq h \Rightarrow f \preceq h$. $d \leq d' \Rightarrow f(d) = f(d') \wedge g(d)$, $g(d) = g(d') \wedge h(d) \Rightarrow$ $f(d) = f(d') \wedge g(d') \wedge h(d) = f(d') \wedge h(d)$. *Least Element:* $\lambda d.\bot \preceq f$ as $d \leq d' \Rightarrow \bot = \bot \wedge g(d)$.

• Compatible elements have a glb. Given $f, g \preceq h$ define $(f \wedge g)(d) \triangleq f(d) \wedge g(d)$. Let us verify $f \wedge g$ is well-defined and stable. Well definedness: observe $f(d), g(d) \leq h(d)$. Let X directed in D. Then: $(f \wedge g)(\cup X) = f(\cup X) \wedge g(\cup X) = \cup f(X) \wedge \cup g(X) = \cup_{x \in X} (f(x) \wedge g(x))$, by the assumption that $\wedge$ is continuous. Moreover: $\cup_{x \in X} (f(x) \wedge g(x)) = \cup_{x \in X} (f \wedge g(x))$. Suppose $d \uparrow d'$. Then:

$$(f \wedge g)(d \wedge d') = f(d \wedge d') \wedge g(d \wedge d') = f(d) \wedge f(d') \wedge g(d) \wedge g(d') = (f \wedge g)(d) \wedge (f \wedge g)(d').$$

• $f \wedge g = \mathrm{glb}\{f, g\}$. Recall: $f, g \preceq h$. Hence $d \leq d'$ implies $f(d) = f(d') \wedge h(d)$ and $g(d) = g(d') \wedge h(d)$. We first show $f \wedge g \preceq f$. $d \leq d' \Rightarrow f \wedge g(d) = f(d) \wedge g(d) = f(d') \wedge g(d') \wedge h(d) = f(d') \wedge g(d') \wedge f(d') \wedge h(d) =$ $f \wedge g(d') \wedge f(d)$. And symmetrically $f \wedge g \preceq g$.

Suppose $k \preceq f, g$. This means: $d \leq d' \Rightarrow k(d) = k(d') \wedge f(d)$ and $k(d) = k(d') \wedge g(d)$.

Now $k \preceq f \wedge g$ iff $d \leq d' \Rightarrow k(d) = k(d') \wedge f(d) \wedge g(d)$. Observe:

$$k(d) = k(d) \wedge k(d) = k(d') \wedge f(d) \wedge k(d') \wedge g(d) = k(d') \wedge f(d) \wedge g(d).$$

• Given F directed in $D \to E$ define $(\cup F)(d) \triangleq \cup_{f \in F} f(d)$. Check: $\cup F$ is well-defined and stable. Remark that if F is directed w.r.t. $\preceq$ then it is also directed w.r.t. the pointwise ordering. In particular $\{f(d)\}_{f \in F}$ is directed in E and therefore $(\cup F)$ is well-defined. $\cup F$ is a continuous map by the standard proof. Let us check stability: suppose $d \uparrow d'$. Then $(\cup F)(d \wedge d')$ $= \cup_{f \in F} f(d \wedge d') = \cup_{f \in F}(f(d) \wedge f(d')) = \cup_{f \in F} f(d) \wedge \cup_{f \in F} f(d')$, since $\cup_{f \in F} f(d) \uparrow \cup_{f \in F} f(d')$ and $\wedge$ is continuous. Now: $\cup_{f \in F} f(d) \wedge \cup_{f \in F} f(d') = (\cup F)(d) \wedge (\cup F)(d')$.

• $\cup F = \mathrm{lub}\ F$. First we show $f \in F \Rightarrow f \preceq \cup F$. Observe:

$\cup F(d) = \cup_{g \in F} g(d) = \cup\{g(d) \mid f \preceq g, g \in F\}$. Now $f \preceq g$, $d \leq d' \Rightarrow f(d) = f(d') \wedge g(d)$. So:

$$f(d) = \cup\{f(d') \wedge g(d) \mid f \preceq g, g \in F\} = f(d') \wedge \cup_{g \in F} g(d) = f(d') \wedge \cup F(d).$$

Suppose $\forall g \in F.(g \preceq h)$. Then $d \leq d' \Rightarrow g(d) = g(d') \wedge h(d)$. Hence:

$$\cup F(d) = \cup_{g \in F} g(d) = \cup_{g \in F} g(d') \wedge h(d) = \cup F(d) \wedge h(d), \text{ and } \cup F \preceq h.$$

• $\wedge$ is continuous. Suppose $f \uparrow \cup F$, F directed.

$(f \wedge \cup F)(d) = f(d) \wedge \cup_{g \in F} g(d) = \cup_{g \in F} f(d) \wedge g(d) = \cup_{g \in F}(f \wedge g)(d) = (\cup_{g \in F}(f \wedge g))(d)$.

Finally verify that: $\mathrm{eval}_{D,E} \triangleq \lambda f.\lambda d.f(d, e)$: $(D \to E) \times D \to E$ and $\Lambda_{D,D',E} \triangleq \lambda g.\lambda d'.\lambda e.g(d, e)$: $(D' \times E \to D) \to (D' \to (E \to D))$ are stable. □

Notes

(1) Here are some examples to keep in mind:

$cpo_\wedge$, $\neg$bound. compl.

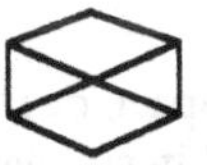

$\neg\ cpo_\wedge$

$\neg\omega$-algebraic, $\neg$cont. inf

(2) The stable ordering arises naturally when requiring the stability of the evaluation map.

(3) Let D, E $\in Cpo_\wedge$ f, g: $D \to E$, f continuous, g stable and $f \preceq g$. Then f is stable.

Next we apply the "bifinite approach" to obtain a ccc of ω-algebraic cpos and stable maps denoted with $\mathbf{Bif}_\wedge$.

2.4 Proposition *(Stable Projections)*

Let D be a $cpo_\wedge$ and $p, q \preceq id_D$. Then:

(1) $p \circ q = p \wedge q$. (2) p is a projection. (3) im(p) is downward closed.

Proof

(1) Remark first that since p and q are bounded their glb exists. Next observe:
$qd \le d \Rightarrow p(qd) = pd \wedge qd = (p \wedge q)(d)$.
(2) For p=q we obtain from (1): $p(pd) = pd \wedge pd = pd$.
(3) $d \le pd' \Rightarrow pd = p(pd') \wedge d = pd' \wedge d = d$. □

2.5 Definition *(Stable Bifinite)*

Let $D \in Cpo_\wedge$. D is a stable bifinite, and we will write $D \in Bif_\wedge$, if there is a chain of (stable) projections $\{p_n\}_{n\in\omega}$ such that $im(p_n)$ is finite and $\cup_{n\in\omega} p_n = id_D$.

2.6 Proposition *($Bif_\wedge$ is a ccc)*

The category $\mathbf{Bif}_\wedge$ of stable bifinites and stable maps is cartesian closed.

Proof

Given 2.3 we only need to check that if $D, E \in Bif_\wedge$ then $D \times E$, $D \to E \in Bif_\wedge$.
Let $\{p_n\}_{n\in\omega}$ and $\{q_n\}_{n\in\omega}$ be the sequences on D and E respectively. One verifies that: $\{\lambda(d, e).(p_n(d), q_n(e))\}_{n\in\omega}$ and $\{\lambda f.\lambda d.q_n(f(p_n(d)))\}_{n\in\omega}$ define the needed sequences of stable projections on D×E and D→E respectively. E.g. for the function space after expansion we have to show: $f \le g \Rightarrow \alpha(d) \equiv q_n(f(p_n(d))) = q_n(g(p_n(d))) \wedge f(d) \equiv \beta(d)$. Observe:

$$f(p_n(d)) \le f(d),\ q_n \le id \Rightarrow q_n(f(p_n(d))) = q_n(f(d)) \wedge f(p_n(d)) \text{ and}$$

$f(d) \le g(d),\ q_n \le id \Rightarrow q_n(f(d)) = q_n(g(d)) \wedge f(d).\ p_n(d) \le d,\ f \le g \Rightarrow f(p_n(d)) = f(d) \wedge g(p_n(d))$.
Therefore: $\alpha(d) = q_n(g(d)) \wedge f(d) \wedge g(p_n(d))$. On the other hand:

$$g(p_n(d)) \le g(d),\ q_n \le id \Rightarrow \beta(d) = q_n(g(d)) \wedge g(p_n(d)) \wedge f(d). \text{ So } \alpha(d) = \beta(d).$$

For the finiteness of the images observe:
$im(\lambda(d, e).(p_n(d), q_n(e))) \cong p_n(D) \times q_n(E)$, $im(\lambda f.\lambda d.q_n(f(p_n(d)))) \cong p_n(D) \to q_n(E)$. □

3. Characterization Stable Bifinites

The main result here is a characterization of the objects in $\mathbf{Bif}_\wedge$. Roughly they are ω-algebraic $cpos_\wedge$ satisfying a combination of properties M and I that we call (MI)*, where property M guarantees that each finite collection of compacts has a finite number of minimal upper bounds (mubs), and property I requires that the principal ideal generated by a compact element is finite. Then, in first approximation, the combined property (MI)* consists in asking that the iteration of the operator that computes the mubs and the operator that computes the principal ideals on a finite collection of compacts returns a finite collection.

3.1 Definition *(property I)*

A cpo D has *property I* if for each compact, $d \in D_o$, the principal ideal $\downarrow d \triangleq \{e \mid e \le d\}$ is finite. It is convenient to decompose property I in simpler properties.

3.2 Definition *(properties I_1, I_2, I_3)*

Let D be a cpo. We say that it has:

- *property I_1* if every "non increasing" sequence of compacts is finite:
$$(\{x_n\}_{n\in\omega} \subseteq D_o \wedge \forall n \in \omega. x_n \ge x_{n+1}) \Rightarrow \{x_n\}_{n\in\omega} \text{ finite.}$$
- *property I_2* if every "non decreasing sequence" of compacts under a compact is finite:
$$(x \in D_o \wedge \{x_n\}_{n\in\omega} \subseteq D_o \wedge \forall n \in \omega.\ x_n \le x_{n+1} \le x) \Rightarrow \{x_n\}_{n\in\omega} \text{ finite.}$$
- *property I_3* if the "immediate predecessors" of a compact are finite:
$$x \in D_o \Rightarrow Pred(x) \triangleq \{y \in D \mid y < x \wedge y \le y' < x \Rightarrow y = y'\} \text{ is finite.}$$

Example Three basic situations in which property I can fail:

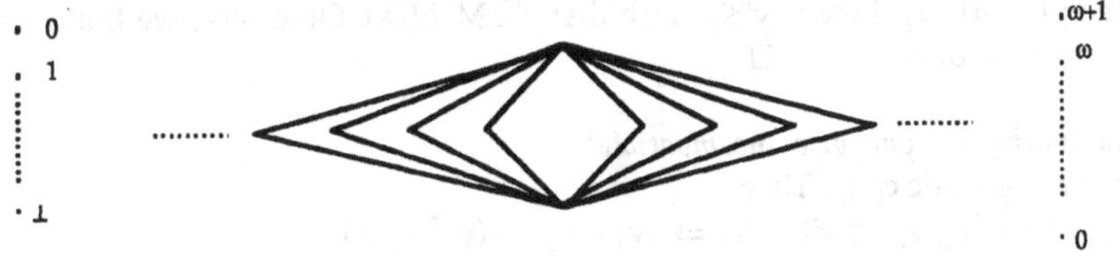

Decreasing chain,$\neg I_1$ ω-branching,$\neg I_3$ Increasing chain,$\neg I_2$

3.3 Proposition *(decomposition property I)*

Let D be an algebraic cpo. Then it has property I iff it has properties I_1, I_2, I_3.

Proof

($\Rightarrow$) Immediate.

($\Leftarrow$) Let $d \in D_0$. First observe that $\downarrow d \subseteq D_0$. If there is a non compact element, say x, under d then since D is an algebraic cpo $\downarrow x \cap D_o$ is directed and $\cup \downarrow x \cap D_o = x$. So we can build an ascending chain under d contradicting I_2.

Again from property I_2 follows that Pred(d) is complete in the sense that: $e < d \Rightarrow \exists e' \in \text{Pred}(d).\ (e \leq e' < d)$. Otherwise we can again build a growing chain under d. From property I_1 we conclude that by iterating the Pred operator every given element $e \in \downarrow d$ can be reached in a finite number of steps. If not using the completeness of Pred we build a decreasing chain included in $\uparrow e \cap \downarrow d$.

Now define: $X_0 \triangleq \{d\}$, $X_{n+1} \triangleq \cup\{\text{Pred}(x) \mid x \in X_n\} \cup X_n$. Then: (a) $\cup_{n\in\omega} X_n = \downarrow d$. (b) $\exists n \in \omega.\ X_{n+1} = X_n$, o.w. we contradict property I_1. (c) By property I_3, $\forall x \in D_o$. Pred(x) is finite. Hence: $\forall n \in \omega.\ X_n$ is finite. That implies: $\downarrow d$ is finite.[6] □

3.4 Property M

Given (P, ≤) poset, and $X \subseteq P$ subset, let (M)UB(X) denote the collection of (minimal) upper bounds of X in P. We say that MUB(X) is *complete* if each upper bound of X is minorated by some element of MUB(X). We say that X *has property M* if MUB(X) is finite and complete. Next define the operator U as follows:

$$U(X) \triangleq \cup\{\text{MUB}(Y) \mid Y \subseteq_{fin} X\}$$

and denote with U*(X) the least set containing X and closed w.r.t. the U operator.
If D is an algebraic cpo then we say that *D has property M* if

$$\forall X \subseteq_{fin} D_o.\ \text{MUB}(X) \text{ has property M.}$$

The following fact is proved by induction on the cardinality of X.

3.5 Fact *(see Smyth[83])*

If $\forall x,y \in P$.MUB({x,y}) has property M then $\forall X \subseteq_{fin} P$.MUB(X) has property M.

3.6 Lemma *($Bif_\wedge \Rightarrow$ property I, property $I_1 \Rightarrow$ mub-completeness)*

(1) If $D \in Bif_\wedge$ then D is an ω-algebraic $cpo_\wedge$ satisfying property I.

(2) If D is an algebraic cpo satisfying property I_1 then for each X, finite collection of compacts in D, MUB(X) is complete.

Proof

(1) Let $\{p_n\}_{n\in\omega}$ be the sequence related to D. Observe that $d \in D$ is compact iff $\exists n.\ p_n(d) = d$. This gives ω-algebraicity. As for property I observe: $d' \leq d = p_n(d) \Rightarrow d' \in im(p_n)$ that is finite (2.4.(3)).

6 In case you wonder this is the contrapositive of König lemma adapted to directed acyclic graphs.

(2) Given any y, upper bound for X, there is a compact $y' \leq y$ that is also an upper bound for X. By the property I_1 there is $y'' \leq y'$ such that $y'' \in MUB(X)$. Otherwise we build an infinite decreasing chain under y'. □

3.7 Lemma *(mubs are pairwise incompatible)*

Let D be an algebraic $cpo_\wedge$. Then

$$\forall X \subseteq_{fin} D_o.(\{y_1, y_2\} \subseteq MUB(X) \Rightarrow (y_1 = y_2 \vee \neg(y_1 \uparrow y_2))).$$

Proof

If $y_1 \uparrow y_2$ then $\exists\, y_1 \wedge y_2 \in MUB(X)$. This forces $y_1 = y_2$. □

3.8 Lemma *(the U operator collapses at the second level)*

Let D be an algebraic $cpo_\wedge$ satisfying property I_1. Then $\forall X \subseteq D_o.\ U(U(X)) = U(X)$.

Proof

By taking the mub of singleton subsets of U(X) one proves that $U(X) \subseteq U(U(X))$. Vice versa observe that $z \in U(U(X))$ iff $\exists\, Y \subseteq_{fin} U(X).(z \in MUB(Y))$. Say $Y \triangleq \{y_1, \ldots, y_n\}$, $(n \geq 0)$. By definition of U(X) we have:

$$y_i \in MUB(X_i) \text{ for } X_i \subseteq_{fin} X \text{ and } i=1, \ldots, n.$$

We want to show that $z \in MUB(X_1 \cup \ldots \cup X_n)$. Then by definition $z \in U(X)$. Certainly $z \in UB(X_1 \cup \ldots \cup X_n)$. Suppose $\exists z' \in UB(X_1 \cup \ldots \cup X_n).\ (z' < z)$. Then $z' \notin UB(Y)$ as $z \in MUB(Y)$. Thus $\exists j \in \{1, \ldots, n\}.\ \neg(y_j \leq z')$ then, by completeness of $MUB(X_j)$ (induced by property I_1), $\exists y_j' \in MUB(X_j).\ (y_j' \leq z')$. So $y_j' \neq y_j$. But $y_j' \leq z' \leq z \geq y_j$, i.e $y_j' \uparrow y_j$. Then by lemma 3.7 $y_j' = y_j$. Contradiction. □

Suppose $D \in Bif_\wedge$. What additional information can we gather about the structure of compact elements? Clearly they are countable. Also let X be a finite collection of compacts. From 3.6 we know that there is some stable projection p_n with finite image such that $X \subseteq im(p_n)$. The image of stable projections are downward closed so the principal ideals generated by elements in $im(p_n)$ are contained in $im(p_n)$. Moreover, as it will be formally shown below, $im(p_n)$ is closed w.r.t. the U operator. Towards a characterization of $Bif_\wedge$ objects we are therefore led to the following

3.9 Definition *((MI)* property)*[7]

Given $(P, \leq)$ poset, and $X \subseteq P$ subset, set $\downarrow(X) \triangleq \cup\{\downarrow x \mid x \in X\}$ and $U\downarrow(X) \equiv U(\downarrow(X))$. Let $(U\downarrow)^*(X)$ be the least set containing X and closed w.r.t. the $U\downarrow$ operator. Then we say that X has property $(MI)^*$ if $(U\downarrow)^*(X)$ is finite. If D is an algebraic $cpo_\wedge$ then we say that D has property $(MI)^*$ if $\forall X \subseteq_{fin} D_o$. X has property $(MI)^*$.

Example *((MI)* strictly implies M and I)*

Let D be an algebraic $cpo_\wedge$. If D has property $(MI)^*$ then it also has property I and property M. On the other hand the following domain Δ provides an example of ω-algebraic $cpo_\wedge$ with properties I and M but without property $(MI)^*$.

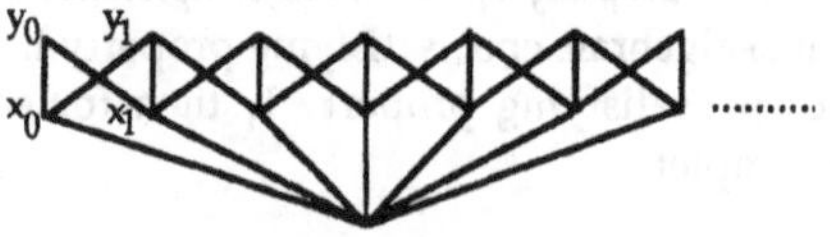

Observe that $(U\downarrow)^*(\{y_i\}) = \Delta$ and y_i is compact. Such domain does not belong to $Bif_\wedge$ (o.w.

[7] When this work was finished M. Droste (Droste[91]) pointed out to me that R. Göbel and himself had already observed the connection between stable projections and property $(MI)^*$ in Droste&Göbel[90]. In particular the category $Bif_\wedge{}^{eP}$ coincides with their category "$\omega BL_{st}{}^{sepp}$" as finite finite $cpos_\wedge$ are the same as finite L-domains, and theorem 4.1.2 is a consequence of their theorem 1.3.

there should be a stable projection p_n with finite image and $p_n(y_i) = y_i$).[8]

3.10 Theorem *(Characterizing Bif$_\wedge$)*

$D \in Bif_\wedge$ iff D is an ω-algebraic cpo$_\wedge$ with property (MI)*.

Proof

(⇒) Let $\{p_n\}_{n\in\omega}$ be the chain of projections associated to D. From 3.6 D is an ω-algebraic cpo$_\wedge$ with property I. Let $X \subseteq_{fin} D_o$ and n such that $X \subseteq im(p_n)$ ($\equiv D_n$). We observe $MUB_D(X) = MUB_{Dn}(X) \subseteq D_n$. This follows from:

$$x \in UB_D(X) \Rightarrow p_n(x) \in UB_{Dn}(X) \cap UB_D(X).$$

That forces: $x \in MUB_D(X) \Rightarrow p_n(x) = x$. Now $U(X) = \cup\{MUB(Y) \mid Y \subseteq_{fin} X\}$. If Y is empty then $\cup Y = \bot$ and $p_n(\bot) = \bot$. Therefore we can infer:

$$X \subseteq im(p_n) \Rightarrow U(X) \subseteq im(p_n).$$

Next recall that by 2.4.(3) $im(p_n)$ is downward closed, hence:

$$X \subseteq im(p_n) \Rightarrow \downarrow(X) \subseteq im(p_n).$$

Since $im(p_n)$ contains X and is closed w.r.t the $U\downarrow$ operator we conclude $(U\downarrow)^*(X) \subseteq im(p_n)$, that is finite.

(⇐) Let $\{d_i\}_{i\in\omega}$ be an enumeration of D_o. Define: $X_n \triangleq \{d_0, ..., d_n\}$, and $D_n \triangleq (U\downarrow)^*(X_n)$, that is finite by assumption. Observe that D_n is downward closed, moreover given $d \in D$ $\downarrow d \cap D_n$ is directed by the definition of the U operator and the mub completeness. Therefore we can set:

$$p_n(d) \triangleq \max(\downarrow d \cap D_n) \equiv d_n$$

Let us check that $\{p_n\}_{n\in\omega}$ provides the needed chain:

• p_n is continuous. Let X directed in D. Since D_n is finite $\exists x \in X.(\cup X)_n = x_n$.

• p_n is a projection with a finite image. Clearly $im(p_n) = D_n$ that is finite. We prove $p_n \leq_s id$. Let $x \leq y$. We check $x_n = y_n \wedge x$. Of course $x_n \leq y_n \wedge x$. If $z \leq y_n$ and $z \leq x$ then $z \leq x$ and $z \in D_n$ and this implies $z \leq x_n$. This also shows that p_n is stable being continuous and less than the identity (note (3), §2). Finally by proposition 2.4 p_n is a projection.

• $p_n \leq_s p_{n+1}$. Let $x \leq y$. We check $x_n = y_n \wedge x_{n+1}$. Of course $x_n \leq y_n \wedge x_{n+1}$. If $z \leq y_n$ and $z \leq x_{n+1}$ then $z \leq x$ and $z \in D_n$ and this implies $z \leq x_n$.

• $\cup\{p_n\}_{n\in\omega} = id$. Observe: $\forall d \in D_o.\exists n.p_n(d) = d$. □

4. Some Domain Theoretic Constructions

In this section we want to support the view that categories of domains and stable maps provide *a theory of finite approximation* that can stand a comparison with the standard theory based on the notion of continuous map as summarized for example in Gunter&Scott[90].

4.1 Bif$_\wedge$^{ep} as an ω-Algebroidal Category

In this section we verify that **Bif$_\wedge$^{ep}** is an *ω-algebroidal category* with the *amalgamation property*. Therefore it has a *universal homogeneous object*.

Conventions (see e.g. Banaschewski&Herrlich[76], Smyth& Plotkin[82])

A category **K** is a *category of monos* if $\forall f, g, h. (f \circ g = f \circ h \Rightarrow g = h)$.

Let **K** be a category of monos. An object $A \in K$ is *compact (or finite)* iff for each ω-chain $\{B_n, f_n\}_{n\in\omega}$ with colimit (B, g_n) the following holds: given any morphism $h: A \to B$ for any sufficiently large n there is a (unique) morphism $k_n: A \to B_n$ such that $h = g_n \circ k_n$. We denote with K_o the collection of compact objects.

[8] It is shown in Amadio[91] that: (1) id_Δ is compact, (2) $\Delta \to \Delta$ has not property I_2, (3) $(\Delta\to\Delta)\to(\Delta\to\Delta)$ is not ω-algebraic.

A category of monos **K** is *algebroidal* if (1) K has an *initial object*. (2) Every object is the colimit of an ω-chain of compact objects. (3) Every ω-chain of compact objects has a colimit. An algebroidal category is *ω-algebroidal* if the collection of compact objects, K_o, is countable (up to isomorphism) and so is the hom-set between any two compact objects.[9]

Let U be an object in a category K of monos, and K* a full subcategory. Then:
(1) U is *K*-universal* if $\forall A\in K^*.\exists f: A\to U$
(2) U is *K*-homogeneous* if $\forall A\in K^*.\ \forall f, g: A\to U.\ \exists h: U\to U.\ (h\circ g = f)$
(3) K* has the *amalgamation property* if
$\forall A, B, B'\in K^*.\forall f: A\to B.\forall f': A\to B'.\exists C\in K^*.\exists g: B\to C.\exists g': B'\to C.\ (g\circ f = g'\circ f')$.

4.1.1 Fact *(Droste&Göbel[90])*

Let K be an ω-algebroidal category of monos. The following are equivalent:
(1) There is a K-universal, K_o-homogeneous object. (2) K_o has the amalgamation property. Moreover a K-universal, K_o-homogeneous object is uniquely determined up to isomorphism.

4.1.2 Theorem*($Bif_\wedge{}^{ep}$ has a Universal Homogeneous Object)*

Let $\mathbf{Bif_\wedge{}^{ep}}$ be the category of stable bifinite domains and stable embedding projection pairs.
(1) $\mathbf{Bif_\wedge{}^{ep}}$ is a category of monos with initial object.
(2) Let $D\in Bif_\wedge$. Then D is compact in $\mathbf{Bif_\wedge{}^{ep}}$ iff the cardinality of D is finite.
(3) Each object in $\mathbf{Bif_\wedge{}^{ep}}$ is an ω-colimit of compact objects.
(4) Each ω-diagram of (compact) objects in $\mathbf{Bif_\wedge{}^{ep}}$ has a colimit.
(5) $\mathbf{Bif_\wedge{}^{ep}}$ is an ω-algebroidal category and the collection of compact objects has the amalgamation property.
(6) $\mathbf{Bif_\wedge{}^{ep}}$ has a universal homogeneous object.

Proof

The proof, although long, does not require original techniques. □

4.2. Retractions, Projections, and Operators Representation

In this section we are interested in the problem of representing "subdomains" of a domain D as certain operators over D. In particular we will concentrate on *retractions* and *projections,* the idea being that subdomains are represented by the image of such maps. In the continuous case not every retraction (or projection) corresponds to a domain (i.e. an algebraic cpo).[10] In the stable case the development of the theory is simpler. Making an essential use of property I it can be shown that the image of a stable retraction over $Bif_\wedge$ still belongs to $Bif_\wedge$. Therefore all stable retractions (and projections) turn out to be *finitary*.

Generally proofs are simpler when dealing with projections. In particular it will not be difficult to show that the collection of projections over a stable bifinite D is again a stable bifinite. Standard results on the representation of functors over a universal domain smoothly extend.

4.2.1 Proposition *(the structure of fixpoints)*

Let D be a $cpo_\wedge$ and f: D→D a stable map. Then $Fix(f) \triangleq \{ d\in D \mid f(d) = d\}$ is a $cpo_\wedge$.

---

[9] Terminology: following Banaschewski and Herrlich we call ω-algebroidal what is called algebroidal by Smyth and Plotkin.

[10] In general it can only be shown that the image of a retraction is a *continuous* cpo. When considering the collection of retractions things get even worse. The collection of retractions over Pω is not a continuous lattice, hence a fortiori not the image of a retraction (Ershov, see exercise 18.4.10 in Barendregt's book).

Convention

Given D $cpo_\wedge$ we denote with $Ret_\wedge(D)$ the collection of stable retractions over D. If r is a retraction then r(D) denotes its image. To improve readability we sometime write fx for the functional application f(x).

4.2.2 Corollary

Let D be a $cpo_\wedge$. Then:
(1) $Ret_\wedge(D) = Fix(\lambda f{:}D\to D.\ f\cdot f)$ is a $cpo_\wedge$. (2) If $r \in Ret_\wedge(D)$ then $r(D) = Fix(r)$ is a $cpo_\wedge$.

4.2.3 Fact *(compacts in r(D))*[11]

Let D be an algebraic cpo and r: D→D a continuous retraction. Then
(1) $r(D)_0$, the collection of compacts in r(D), can be characterized as follows:

$r(D)_0 = \{rd \mid d \in D_0 \wedge d \leq rd\}$.

(2) If p is a projection then $p(D)_0 = p(D) \cap D_0$.

Berardi[88] was apparently the first to observe that when working over dI-domains the image of a retraction is still a dI-domain. It was then possible to adapt a technique of Scott[80] to show that Ret(D) ◅ D→D if D is a dI-domain. Here is a partial counterpart of this result for $Bif_\wedge$ whose proof we delay to A.2 as it needs some technical lemmas.

4.2.4 Theorem *($D \in Bif_\wedge \Rightarrow r(D) \in Bif_\wedge$)*

Let D be a $Bif_\wedge$ and r: D→D be a stable retraction then r(D) is a $Bif_\wedge$.

4.2.5 Proposition *($D \in Bif_\wedge \Rightarrow Prj(D) \in Bif_\wedge$)*

(1) Let $D \in Cpo_\wedge$ and suppose p is a projection, p ≤ id. If D is an (ω-)algebraic $cpo_\wedge$ ($Bif_\wedge$) then im(p) is an (ω-)algebraic $cpo_\wedge$ ($Bif_\wedge$).

(2) If $D \in Bif_\wedge$ then $Prj(D) = \downarrow(id_D)$ is a stable bifinite and a lattice.[12]

Proof

(1) Since im(p) is downward closed (2.4) and $p(D)_0 = p(D) \cap D_0$ we can infer that im(p) is an (ω-)algebraic $cpo_\wedge$. If $D \in Bif_\wedge$ and $\{q_n\}_{n\in\omega}$ is the corresponding chain of projections then we define $\{p \wedge q_n\}_{n\in\omega}$ as the chain related to im(p).

(2) Use $Prj(D) = \downarrow(id_D)$. □

Let U be a universal (homogeneous) domain for $Bif_\wedge$. Then each domain can be identified with a projection over U. Since by 4.2.5 Prj(U) is bifinite we can conclude by 4.1.2.(6) that Prj(U) ⊑ U. It can also be shown that the basic operators over domains can be adequately represented as stable transformations over Prj(U). As a fall-out one gets a technique to solve domain equations via the standard least-fixed point theorem (all this was already observed by D. S. Scott for the continuous case).

5. Property M, Property I, and ω-Algebraicity

We are now going to present a series of technical lemmas in order to establish if property M and property I are necessary to enforce the *ω-algebraicity* of functional spaces. It turns out that properties M, I_1, and I_2 are necessary (5.2, 5.3). We can also show that property I_3 is necessary under rather mild hypothesis (5.5, 5.9). As a fall out we show that a stable, countably based version of L-domains introduced in Coquand[89] is "the largest ccc" whenever principal ideals of domains are distributive (5.8). The necessity of property (MI)*

[11] This is attributed to G.D. Plotkin in Scott[80].

[12] Let D be the example of finite $cpo_\wedge$ not bounded complete given in note (1), §2. Then it is not the case that Prj(D) ⊑ D→D.

is still open.

In the first place we want to guarantee that in any full subcategory of algebraic $Cpo_\wedge$ if the terminal object, the product, and the exponent exist then they coincide up to isomorphism with the ones defined in $\mathbf{Cpo}_\wedge$.

5.1 Proposition *(terminal object, product, and exponent are pre-determined)*

Let **C** be a full subcategory of algebraic $Cpo_\wedge$ and stable maps. Then:

(1) If **C** has terminal object T then T is a one point cpo.

(2) If **C** has terminal object T and products $D \overset{\pi D}{\leftarrow} D\otimes E \overset{\pi E}{\rightarrow} E$ then $D\otimes E \cong D\times E$, the standard product in $\mathbf{Cpo}_\wedge$.

(3) If **C** has terminal object T, finite products $D \overset{\pi D}{\leftarrow} D\otimes E \overset{\pi E}{\rightarrow} E$, and exponent $E^D\otimes D \overset{ev}{\rightarrow}{}_C E$ then $E^D \cong D\Rightarrow E$, the standard exponent in $\mathbf{Cpo}_\wedge$.

Proof

This proof consists of a slight variation of the one in Smyth[83] so that the arguments apply also to the stable case. □

5.2 Lemma *(D, D→D ω-algebraic ⇒ D has property M)*[13]

If D and D→D are ω-algebraic $cpo_\wedge$ then $\forall x, x' \in D_0.\ MUB(\{x, x'\})$ is finite.

Proof

The following is a (non-trivial) adaptation of lemma 3 in Smyth[83]. We show that if MUB({x, x'}) is infinite then D→D has uncountably many compacts (by fact 3.5 it is enough to consider two elements). Let $Y \equiv MUB(\{x, x'\})$ infinite, and $y, y' \in Y$, $y \neq y'$. By lemma 3.7 we know $\neg(y\uparrow y')$. Let $S \subseteq Y$ such that $y \in S$ and $y' \notin S$. There are uncountably many of such S. Define a map f_S: D→D as follows:[14]

$$f_S(d) \triangleq \begin{cases} y & \text{if } \exists e\in S.(e\leq d) & \equiv (1)\\ y' & \text{if } \exists e\in Y\backslash S.(e\leq d) & \equiv (2)\\ y & \text{if } d\in UB(x)\cap UB(x') \wedge \neg(1) \wedge \neg(2) & \equiv (3)\\ x & \text{if } d\in UB(x)\backslash UB(x') & \equiv (4)\\ x' & \text{if } d\in UB(x')\backslash UB(x) & \equiv (5)\\ x\wedge x' & \text{if } d\notin UB(x)\cup UB(x') & \equiv (6) \end{cases}$$

It is rather easy to check that f_S is continuous. Let us verify that f_S is *stable*. Let $d\uparrow d'$. The problem being symmetric we have to check $f_S(d\wedge d') = f_S(d)\wedge f_S(d')$ for the $6(6+1)/2 = 21$ possible situations in which d satisfies (i), d' satisfies (j), and $1\leq i\leq j\leq 6$.

1-1: if $d\geq e_d$, $d'\geq e_{d'}$, $e_d, e_{d'}\in S$ then $e_d = e_{d'} \leq d\wedge d'$. 1-2: Impossible, distinct mubs are incompatible. 1-3 Impossible, contradicts minimality of ub. 1-4: then also $d\wedge d' \in UB(x)\backslash UB(x')$. 1-5: like 1-4. 1-6: then also $d\wedge d' \notin UB(x)\cup UB(x')$.

2-2: like 1-1. 2-3: Impossible, like 1-3. 2-4: like 1-4. 2-5: like 1-4. 2-6: like 1-6.

3-3: then also $d\wedge d' \in UB(x)\cap UB(x') \wedge \neg(1) \wedge \neg(2)$. 3-4: : then also $d\wedge d' \in UB(x)\backslash UB(x')$. 3-5: like 3-4. 3-6: like 1-6.

4-4: then also $d\wedge d' \in UB(x)\backslash UB(x')$. 4-5: then $d\wedge d' \notin UB(x)\cup UB(x')$ and $f_S(d\wedge d') = x\wedge x' = f_S(d)\wedge f_S(d')$. 4-6: like 1-6. 5-5: like 4-4. 5-6: like 1-6. 6-6: like 1-6.

Observe $\forall d\in \{x, x'\}.\ f_S(d) = d$. If D→D is ω-algebraic then there is a compact f'_S such that $f'_S \leq_s f_S$ and $\forall d\in \{x, x'\}.\ f'_S(d) = d$. By monotonicity we have:

$$\forall e\in Y.\ \forall d\in \{x, x'\}.\ f'_S(d) \leq f'_S(e) \leq f_S(e) \in MUB(\{x, x'\})$$

This forces: $\forall e\in Y.\ f'_S(e) = f_S(e)$. But then $S\neq S' \Rightarrow f'_S \neq f'_{S'}$. Hence we have an uncountable number of compacts. □

[13] It will be shown later that property M is not necessary if we do not ask for the countability of compact elements.

[14] Next lemma 5.3 shows that also property I_1 is necessary. If we assume I_1 then we have mub completeness and case (3) in f_S definition does not arise anymore.

5.3 Lemma *(D, D→D ω-algebraic ⇒ D has properties I_1, I_2)*

Suppose D and D→D are ω-algebraic $cpo_\wedge$ then D has properties I_1 and I_2.

Proof

Preliminary remarks:

(A) Suppose f, g: D→D, $g \leq_s f$. Given $d \in D$ we have: $f(d) = g(d) \Rightarrow f_{|\downarrow d} = g_{|\downarrow d}$. As:

$e \leq d \Rightarrow g(e) = g(d) \wedge f(e) = f(d) \wedge f(e) = f(e)$.

(B) Suppose f: D→D, and $f(d) \in D_o$ for $d \in D$. Let $\{g_i\}_{i \in I}$ be directed in D→D such that $f = \cup_{i \in I}\{g_i\}$. Then $\exists i \in I.\ f_{|\downarrow d} = g_{i|\downarrow d}$. For $f(d) = \cup\{g_i\}_{i \in I}(d)$ implies $\exists i \in I.\ f(d) = g_i(d)$ and we can apply (A).

• *Property I_1*. Suppose there is an infinite decreasing chain of compact elements:

$\{x_n\}_{n \in \omega} \subseteq D_o \wedge \forall n \in \omega.\ x_n > x_{n+1}$

Let σ: ω→ω be any strictly increasing map: $i<j \Rightarrow \sigma(i) < \sigma(j)$. There are uncountably many of them. Define f_σ as follows:

$$f_\sigma(x) \triangleq \begin{cases} \bot & \text{if } \forall i \in \omega.\ \neg(x_i \leq x) \\ x_{\sigma(i)} & \text{if } i = \min_\omega\{j \mid x_j \leq x\} \end{cases}$$

Observe $\forall i \in \omega.\ f_\sigma(x_i) = x_{\sigma(i)}$, therefore: $\sigma \neq \sigma' \Rightarrow f_\sigma \neq f_{\sigma'}$. Let us verify that f_σ is *stable*. Let X be a directed set in D. If $\forall i \in \omega.\ \neg(x_i \leq \cup X)$ then $f_\sigma(\cup X) = \bot = \cup_{x \in X} f_\sigma(x)$.
If $i = \min_\omega\{j \mid x_j \leq \cup X\}$ then, by compactness, $\exists y \in X.\ (x_i \leq y)$ and therefore:

$f_\sigma(\cup X) = x_{\sigma(i)} = f_\sigma(y) = \cup_{x \in X} f_\sigma(x)$.

Let $x, y \in D$, $x \uparrow y$. If $\forall i \in \omega.\ \neg(x_i \leq x)$ or $\forall i \in \omega.\ \neg(x_i \leq y)$ then $f_\sigma(x) \wedge f_\sigma(y) = \bot$, and we are done. Otherwise let $i = \min_\omega\{j \mid x_j \leq x\}$, $k = \min_\omega\{j \mid x_j \leq y\}$, $i \leq k$. Then

$f_\sigma(x) \wedge f_\sigma(y) = x_{\sigma(i)} \wedge x_{\sigma(k)} = x_{\sigma(k)} = f_\sigma(x \wedge y)$.

Next we show that f_σ is *compact* in D→D, hence contradicting the assumption that D→D is ω-algebraic. Since D→D is algebraic the compacts below f form a directed set, say $\{g_i\}_{i \in I}$, whose lub is f_σ. We can apply remark (B): $\exists i \in I.\ f_{\sigma|\downarrow d} = g_{i|\downarrow d}$, for $d = x_0$. We show that this implies $f_\sigma = g_i$. Consider $x \in D$ and let $k = \min_\omega\{j \mid x_j \leq x\}$ (if this set is empty then $f_\sigma(x) = \bot$ and we are done). Then:

$g(x_k) \leq g(x) \leq f_\sigma(x) = f_\sigma(x_k) = x_{\sigma(k)} = g(x_k) \Rightarrow f_\sigma(x) = g(x)$.

• *Property I_2*. The argument is very similar. Let $x_{\omega+1}$ be a compact with an infinite growing chain of compacts below whose lub is x_ω and such that $x_0 = \bot$. That is:

$x_{\omega+1} \in D_o \wedge \{x_n\}_{n \in \omega} \subseteq D_o \wedge x_0 = \bot \wedge \forall n \in \omega.\ x_n < x_{n+1} < x_\omega < x_{\omega+1} \wedge \cup\{x_n\}_{n \in \omega} = x_\omega$.

Given σ: ω→ω increasing map as above extend it to ω+2 so that : $\sigma(\alpha) = \alpha$ for $\alpha=\omega$, $\alpha=\omega+1$. Define:

$f_\sigma(x) \triangleq x_{\sigma(\alpha)}$ where $\alpha = \cup\{i \in \omega+2 \mid x_i \leq x\}$.

Again $\forall i \in \omega.\ f_\sigma(x_i) = x_{\sigma(i)}$ and $\sigma \neq \sigma' \Rightarrow f_\sigma \neq f_{\sigma'}$. Let us verify that f_σ is *stable*. Let X be a directed set in D and $\alpha = \cup\{i \in \omega+2 \mid x_i \leq \cup X\}$. If $\alpha = \omega+1$ or $\alpha < \omega$ then by compactness $\exists y \in X.\ (x_\alpha \leq y)$. Then $f_\sigma(\cup X) = x_{\sigma(\alpha)} = f_\sigma(y) = \cup_{x \in X} f_\sigma(x)$. If $\alpha = \omega$ then $\forall i \in \omega.\ x_i \leq \cup X \wedge \forall i \in \omega.\ \exists y_i \in X.\ (x_i \leq y_i)$. Therefore:

$x_\omega = f_\sigma(\cup X) \geq \cup_{i \in \omega} f_\sigma(y_i) \geq \cup_{i \in \omega} x_{\sigma(i)} = x_\omega$.

Let $x, y \in D$, $x \uparrow y$, $\alpha = \cup\{i \in \omega+2 \mid x_i \leq x\}$, $\beta = \cup\{i \in \omega+2 \mid x_i \leq y\}$, $\beta = \min\{\alpha, \beta\}$. Then

$f_\sigma(x) \wedge f_\sigma(y) = x_{\sigma(\alpha)} \wedge x_{\sigma(\beta)} = x_{\sigma(\beta)} = f_\sigma(x \wedge y)$. As $x_i \leq x, y \Rightarrow x_i \leq x \wedge y$.

Again by remark (B) $\exists g$ compact.$(f_{\sigma|\downarrow x_{\omega+1}} = g_{|\downarrow x_{\omega+1}})$. This implies $f_\sigma = g$ as follows. Let $x \in D$ and let $\alpha = \cup\{i \in \omega+2 \mid x_i \leq x\}$. Then:

$g(x_\alpha) \leq g(x) \leq f_\sigma(x) = f_\sigma(x_\alpha) = x_{\sigma(\alpha)} = g(x_\alpha) \Rightarrow f_\sigma(x) = g(x)$. □

5.4 Definition *(distributivity)*

Let D be a bounded complete algebraic cpo.[15] D is *distributive* if

$$\forall d, d'.\ d\uparrow d' \Rightarrow (d \vee d') \wedge e = (d \wedge e) \vee (d' \wedge e).$$

5.5 Lemma *(D,D→D ω-alg. + distributivity principal ideals ⇒ D has property I_3)*

Suppose D and D→D are ω-algebraic $cpo_\wedge$. Then

(1) If $d \in D$ then $\downarrow d$ is an ω-algebraic lattice.

(2) If $d \in D_0$ and $\downarrow d$ is distributive then $\downarrow d$ is finite.

(3) If for each $d \in D_0$ $\downarrow d$ is distributive then D has property I.

Proof

(1) Consider the $cpo_\wedge$ $\downarrow d$ with the order induced by D. Show: $(\downarrow d)_0 = \downarrow d \cap D_0$. Next we show that any two compact elements have a lub in $\downarrow d$. This is enough to guarantee the existence of arbitrary lubs and therefore of arbitrary glbs. Given $x, y \in \downarrow d$ clearly $d \in UB(\{x, y\})$. By lemma 5.3 property I_1 holds and by lemma 3.6 we know that $MUB(\{x, y\})$ is complete. So $\exists z \leq d.\ z \in MUB(\{x, y\})$. Moreover such z is the unique mub bounded by d as by lemma 3.7 mubs are pairwise incompatible. Therefore $z = lub_{\downarrow d}\{x, y\}$.

(2) Let $e \in D_0$ and suppose $\downarrow e$ is distributive and infinite. From property I_1 we can find a minimal d such that $\downarrow d$ is infinite (and distributive since $\downarrow e$ is). Now consider its infinite immediate predecessors (o.w. we contradict minimality), say $\{x_0, x_1, x_2, ...\}$. Observe that, for glb computed in $\downarrow d$: $\forall i, j \in \omega.\ (i \neq j \Rightarrow x_i \vee x_j = d)$, $\forall i, k \in \omega.\ (i \neq k \Rightarrow x_i \wedge x_k < x_k)$.
Select x_k, from the hypothesis on $\downarrow d$ we know $\downarrow x_k$ is finite. Therefore:

$$\exists i, j \in \omega.\ (i \neq j \wedge i \neq k \wedge j \neq k \wedge x_i \wedge x_k = x_j \wedge x_k).$$

Hence: $(x_i \vee x_j) \wedge x_k = d \wedge x_k = x_k > x_j \wedge x_k = (x_i \wedge x_k) \vee (x_j \wedge x_k)$, contradicting the distributivity of $\downarrow e$.

(3) From (2) it has property I_3. From lemma 5.3 it has properties I_1, I_2. From Proposition 3.3 it has property I. □

We now recall some basic definitions and facts about a stable version of L-domains. The line of the proofs is given in A.3.

5.6 Definition *(L, $L_\wedge$, and $L_{0\wedge}$ domains)*

D is an *L-domain* if it is a cpo and every principal ideal is an algebraic lattice. D is an *$L_\wedge$-domain* if it is a cpo and every principal ideal is a dI-domain. If moreover it is countably based and it has property M then it is a *$L_{0\wedge}$-domain*. We denote, respectively, with $\mathbf{L}_\wedge$ and $\mathbf{L}_{0\wedge}$ the category of $L_\wedge$-domains and stable maps and its full subcategory of $L_{0\wedge}$-domains.

5.7 Fact *(Coquand[89])*

The categories $\mathbf{L}_\wedge$ and $\mathbf{L}_{0\wedge}$ are cartesian closed.

5.8 Theorem *(the largest ccc for the locally distributive case)*

(1) The collection of $L_{0\wedge}$-domains is composed of those domains that are both $L_\wedge$-domains and $Bif_\wedge$-domains.

(2) The category $\mathbf{L}_{0\wedge}$ is the largest full subccc of ω-algebraic $cpos_\wedge$ and stable maps with the property that principal ideals of its objects are distributive lattices.

Proof

(1) If $D \in L_\wedge \cap Bif_\wedge$ then $D \in L_{0\wedge}$, by the characterization of $Bif_\wedge$-domains. Vice versa given $D \in L_{0\wedge}$ let $\{d_i\}_{i \in \omega}$ be an enumeration of the compacts. Define the chain $\{p_n\}_{n \in \omega}$ as: $p_n(x) \triangleq lub_{\downarrow x}\{z \mid z \leq x \wedge z \in \downarrow d_1 \cup ... \cup \downarrow d_n\}$.
Observe: (a) $\{z \mid z \leq x \wedge z \in \downarrow d_1 \cup ... \cup \downarrow d_n\}$ is finite. (b) p_n is continuous. (c) $\forall i \leq n.\ p_n(d_i) = d_i$.

[15] See also note (1), §3. Note that a bounded complete algebraic cpo has glb of every non empty set.

(d) $p_n \leq id$. (e) $p_n \leq p_{n+1}$. Distributivity plays a crucial role in (d), (e).

(2) Let **C** be a full sub-ccc of ω-algebraic cpos$_\wedge$ and stable maps. From 5.1 the exponent is determined up to isomorphism. Let D∈**C**, then by 5.2 and 5.3 D satisfies properties M, I_1, and I_2. From 5.5.(1) all principal ideals are algebraic lattices, moreover they are distributive by assumption. By 5.5.(3) D has property I_3. Hence D satisfies all properties characterizing $L_{0\wedge}$-domains. □

5.9 Lemma *(D, D→D, and ↓d→↓d ω-algebraic ⇒ D has property I_3)*[16]

Suppose D and D→D are ω-algebraic cpo$_\wedge$ and for each $d \in D_o$ ↓d→↓d is an ω-algebraic cpo$_\wedge$. Then D has property I .

Proof

We will use lemma 5.3 and just check property I_3. We show that if $d \in D_o$ and ↓d does not satisfy property I_3 then we can find $d_\top \leq d$ such that $\downarrow d_\top \to \downarrow d_\top$ is not ω-algebraic. We set:

$Suc_{\downarrow e}(x) \triangleq \{y \leq e \mid x < y \wedge x < y' \leq y \Rightarrow y = y'\}$

Given d there is a (minimal) element $d_\top \in \downarrow d$ such that: (a) $Pred(d_\top)$ is infinite; (b) $\forall y \in Pred(d_\top).\ \downarrow y$ is finite. If not we contradict property I_1. Given $d_\top$ there is a (maximal) element $d_\bot \in \downarrow d_\top$ such that: (a) $Suc_{\downarrow d_\top}(d_\bot)$ is infinite; (b) $\forall y \in Suc(d_\bot).\ \downarrow d_\top \cap \uparrow y$ is finite. If not we contradict property I_2.

Now we define certain operators of upward and downward closure.

Set for $]d_\bot, d_\top[\triangleq (\downarrow d_\top \cap \uparrow d_\bot) \backslash \{d_\top, d_\bot\}$, $X \subseteq]d_\bot, d_\top[$:

$\sqcup(X) \triangleq (\cup\{\uparrow x \mid x \in X\} \cap \downarrow d_\top) \backslash \{d_\top\}$ $\qquad \sqcap(X) \triangleq (\cup\{\downarrow x \mid x \in X\} \cap \uparrow d_\bot) \backslash \{d_\bot\}$

It is crucial to observe that if X is finite then $\sqcup(X)$ and $\sqcap(X)$ are finite. Moreover, if $x \in X$ and $y \in]d_\bot, d_\top[\backslash \sqcup(\sqcap(\sqcup(X)))$ then $x \wedge y = d_\bot$ and $x \vee y = d_\top$. Otherwise, suppose, e.g., $x \wedge y \in]d_\bot, d_\top[$ then $x \in X \Rightarrow x \in \sqcup(X) \Rightarrow x \wedge y \in \sqcap(\sqcup(X)) \Rightarrow y \in \sqcup(\sqcap(\sqcup(X)))$.

We now build an infinite sequence $\{x_i\}_{i \in \omega} \subseteq Suc_{\downarrow d_\top}(d_\bot)$ with the property that: $\forall i, j.(i \neq j \Rightarrow x_i \vee x_j = d_\top \wedge x_i \wedge x_j = d_\bot)$. Let x_0 be arbitrarily chosen in $Suc_{\downarrow d_\top}(d_\bot)$. Suppose to have built the sequence up to x_n then select x_{n+1} in:

$Suc_{\downarrow d_\top}(d_\bot) \backslash \sqcup(\sqcap(\sqcup(\{x_0, ..., x_n\})))$

By what has been observed above this is well defined and gives the desired x_{n+1}.

We are now ready to code a set $S \subseteq \omega$ via a map f_S: $\downarrow d_\top \to \downarrow d_\top$ with the property that $f_S(x_i) = x_i$ iff $i \in S$. Define:

$$f_S(x) \triangleq \begin{cases} d_\top & \text{if } x = d_\top & \equiv (1) \\ x_i & \text{if } x \in \uparrow x_i \wedge i \in S \wedge \neg(1) & \equiv (2) \\ d_\bot & \text{if } d_\bot \leq x \wedge \neg(1) \wedge \neg(2) & \equiv (3) \\ \bot & \text{if } \neg(d_\bot \leq x) & \equiv (4) \end{cases}$$

First verify that f_S is a stable map using the properties of the sequence $\{x_i\}_{i \in \omega}$. Next remark that $d_\top \in (\downarrow d_\top)_o$, so we can show by the usual technique that f_S is a compact in $\downarrow d_\top \to \downarrow d_\top$. Hence one concludes that there are uncountably many compacts in $\downarrow d_\top \to \downarrow d_\top$. Contradiction. □

Problem One would like to remove the assumption in the previous lemma 5.9 that the functional space of the principal ideals generated by compact elements is still ω-algebraic. Towards this goal it would be enough to prove an "extension lemma" that goes as follows. Suppose D and D→D are ω-algebraic cpo$_\wedge$. Then: $\forall d \in D_o.\ \forall f \in (\downarrow d \to \downarrow d)_o .\exists f' \in (D \to D)_o.\ f'_{\downarrow d} = f$. This can be proved under the additional hypothesis that the principal ideal is distributive[17] but we were unable to tackle the general case.

[16] So, roughly speaking, every full subccc of ω-algebraic cpo$_\wedge$ whose objects are closed by principal ideals has property I_3.

[17] We omit this proof as lemma 5.5 provides a much simpler argument for the distributive case.

In view of the connections with classical computability theory and the problem of full abstraction it seems to us that it is particularly important to consider categories of domains where the countability of the compacts of the functional space is preserved. However an interesting and natural mathematical problem is that of classifying the largest ccc of algebraic domains and stable maps. It turns out that L-domains are the simple answer to this question (Taylor[91], see Taylor[91(a)] and Ehrhard&Malacaria[91] for proofs that L-domains and stable functions form a ccc). Since L-domains can be characterized as mub-complete algebraic cpos$_\wedge$ one has to show that mub-completeness is a necessary condition for the algebraicity of the functional space (in the continuous case this is the first lemma of Smyth[83]).

Conclusion

We have applied the "bifinite approach" to the stable case obtaining a ccc, Bif$_\wedge$, whose objects can be characterized via property (MI)*. We have checked that some basic domain constructions can be applied to Bif$_\wedge$; for example it has a universal homogeneous object, and it is possible to represent domains as projections. All ccc of ω-algebraic cpos$_\wedge$ and stable maps known by the author are contained in Bif$_\wedge$. The problem remains if Bif$_\wedge$ is the "largest" ccc of ω-algebraic cpos$_\wedge$ and stable maps. For the time being it has been shown that the category $L_{0\wedge}$ gives the "largest" ccc if we restrict our attention to the "locally distributive" case.

Some more problems concern the closure under "dependent products" (see Jung[91] for a negative result in the continuous case) and "powerdomains constructions". As a matter of fact there are finite posets in $L_{0\wedge}$ for which the convex and upper powerdomain constructions return a poset outside Cpo$_\wedge$ (see Amadio[91]). Unfortunately, in the absence of glbs of compatible pairs, it is not clear what a stable map is. For example one can take Berry's original definition of stability (see 1.2, or A.1) but then it is possible to see in very simple cases that the pairing does not need to be a stable function (take the example of finite cpo not cpo$_\wedge$ in note (1), §2). It would be unfair, however, to conclude that "powerdomains do not work in the stable case", one has probably to look for more refined powerdomains constructions that take stability into account.

Acknowledgments

This work is a fall-out of an attempt at writing joint lecture notes on domain theory with Pierre-Louis Curien. Thanks to him for critically reading a preliminary draft. The paper was actually finished while visiting Prof. Engeler group at ETH, Zürich. I am grateful to the "Forschungsinstitut für Mathematik" for the warm and kind hospitality and to Prof. Engeler for making this possible.

References

Amadio R. [1991] "Bifinite Domains: Stable Case", LIENS-TR 91-3, March 1991.

Berardi S. [1988] "Retractions on dI-domains as a model for Type:Type", preprint.

Berry G. [1979] "Modèles complètement adéquats et stables des lambda-calculs typés", Thèse d' Etat, Université Paris VII.

Coquand T. [1989] "Categories of embeddings", TCS, 68, (221-237).

Droste M., Göbel R. [1990] "Universal domains in the theory of denotational semantics of programming languages", IEEE-LICS 90, Philadelphia.

Droste M. [1991] Personal Communication dated 22.5.91.

Ehrhard T., Malacaria P. [1991] "Stone Duality for Stable Functions", in Proc. Category Theory in Computer Science 91, Paris.

Gierz G., Hofmann K.H., Keimel K., Lawson J.D., Mislove M., Scott D.S. [1980] "A compendium of continuous lattices", Springer-Verlag.

Gunter C., Scott D. [1990] "Semantic Domains", in Handbook of Theoretical Computer Science, North Holland.

Jung A. [1990] "Cartesian closed categories of algebraic cpos", TCS, 70, (233-250).

Jung A. [1991] "The dependent product construction in various categories of domains", TCS, 79, (359-363).

Mac Lane S. [1971] "Categories for the working mathematician", Springer-Verlag, New York.

Plotkin G. [1976] "A powerdomain construction", SIAM J. of Computing, 5, 3, (452-487).

Scott D. [1972] "Continuous lattices", Toposes, Algebraic Geometry and Logic, (Lawvere ed.), SLNM 274, (97-136).

Scott D. [1976] "Data types as lattices", SIAM J. of Computing, 5 (522-587).

Scott D. [1980] "A space of retracts", Manuscript, Bremen.

Smyth M. [1983] The largest cartesian closed category of domains, TCS, 27, (109-119).

Smyth M., Plotkin G. [1982] "The category-theoretic solution of recursive domain equations", SIAM J. of Computing, 11, (761-783).

Taylor P. [1990] "An algebraic approach to stable domains", J. of Pure and Apl. Algebra, 64, (171-203).

Taylor P. [1991] Personal Communication dated 19.3.91.

Taylor P. [1991(a)] "Notes on Stable Domain Theory", collection of manuscripts dated from 1988 to 1991, Imperial College, London, April 1991.

Appendix 1: Trace properties

Let us start by observing that under the hypothesis that domains satisfy property I it is possible to give an alternative characterization of stable maps that follows Berry's original intuition. The trace of a stable map over algebraic cpos$_\wedge$ with property I plays a bit the role of the graph of a continuous function over algebraic cpos. Most of the proofs although lengthy are routine, this is why they are omitted.

1. Lemma *(stability and minimal points)*

Let D, E$\in$Cpo$_\wedge$ and f: D$\to$E continuous. Then:

(1) If $\forall d\in D.\ \forall e\in E.\ (e\le f(d) \Rightarrow \exists \min(\downarrow d \cap f^{-1}(\uparrow e))$ then f is stable.

(2) If D, E are algebraic cpo$_\wedge$ with property I and f: D$\to$E is stable then:

$\forall d\in D.\ \forall e\in E.\ (e\le f(d) \Rightarrow \exists \min(\downarrow d \cap f^{-1}(\uparrow e))$.[18]

2. Definition *(trace)*

Let D, E be algebraic cpos$_\wedge$ with property I and f: D$\to$E stable. Then we define the *trace* of f as: $tr(f) \triangleq \{(\min(\downarrow d \cap f^{-1}(\uparrow e)), e) \mid e \le f(d), (d, e)\in D_0 \times E_0\}$.

3. Lemma *(some trace properties)*

Let C, D, E be algebraic cpo$_\wedge$ with property I. Then:

(1) If f: D$\to$E is a stable map then tr(f) is well-defined and $tr(f) \subseteq graph(f)$, where $graph(f) \triangleq \{(d, e)\in D_0\times E_0 \mid e\le f(d)\}$.

(2) If f, g: D$\to$E are stable maps then $f \le_s g$ iff $tr(f) \subseteq tr(g)$.

(3) If f, g: D$\to$E are stable maps and $f\uparrow g$ then $tr(f\wedge g) = tr(f)\cap tr(g)$.

(4) If F is a directed set in D$\to$E and g is stable then $g = \cup F$ iff $tr(g) = \cup\{tr(f) \mid f\in F\}$.

(5) If f: D$\to$E is a stable map and tr(f) is finite then f is a compact in the functional space. In particular every step function is stable and compact.[19]

(6) If f: C$\to$D, g: D$\to$E are stable and $(c, d)\in tr(f)$, $(d, e)\in tr(g)$ then $(c, e)\in tr(g\circ f)$.

---

[18] We sometime refer to $\min(\downarrow d \cap f^{-1}(\uparrow e))$ as local minimum.

[19] Can we characterize compact stable maps as those having finite trace? The answer is yes for stable bifinites (exercise!), and in a "locally distributive case" presented in lemma 4, A.3. The answer is no in general, as shown by the domain Δ defined in §3.

Appendix 2: Retractions

This appendix is dedicated to the study of retractions in the stable case. We are indebted to Berardi[88] for the following lemma

1. Lemma

Let D be an (ω-)algebraic $cpo_\wedge$ with property I and r: $D \to D$ a stable retraction.

(1) $\forall x, y \in D_0.\ \exists z \in D_0.\ (x \le ry \Rightarrow x \le rz \wedge z \le rz \le ry)$. That is:

(2) r(D) is an (ω-)algebraic $cpo_\wedge$ with property I.

Proof

(1) We exploit the existence of local minima (see lemma 1, A.1).

$x \le ry = r(ry) \Rightarrow \exists z \in D_0.\ z = \min(\downarrow ry \cap r^{-1}(\uparrow x))$. Hence $z \le ry$ and $x \le rz$.

$z \le ry = r(ry) \Rightarrow \exists z' \in D_0.\ z' = \min(\downarrow ry \cap r^{-1}(\uparrow z))$. Hence $z' \le ry$ and $z \le rz'$.

Now z, $z' \le ry$ are compatible. Also $(z, x) \in tr(r) \wedge (z', z) \in tr(r) \Rightarrow (z', x) \in tr(r \circ r) = tr(r)$, by 3.(6), A.1. Hence z'=z.

(2) We already know that r(D) is a $cpo_\wedge$, moreover we have a characterization of $r(D)_0$ as $\{rd \mid d \in D_0 \wedge d \le rd\}$. Part (1) implies the following:

$\forall x \in D_0.\ \forall w \in r(D).\ (x \le w \Rightarrow \exists u \in r(D)_0.\ (x \le u \le w))$. (A)

In fact if $x \le w = rd$ then for some $y \in D_0$ we have $x \le ry \le w = rd$. Now use (1) to find z such that $x \le rz \wedge z \le rz \le ry$ and set u = rz.

• *(ω-)algebraicity*. From the characterization of $r(D)_0$ immediately follows that the cardinality of $r(D)_0$ is bounded by the cardinality of D_0. Next let w = r(w) and consider the set $r(D)_0 \cap \downarrow w$. Let us show that it is directed. Suppose $u_1 \le ru_1$, $u_2 \le ru_2$, $ru_1 \le w$, $ru_2 \le w$. We know from 5.5 that $\downarrow w$ is an algebraic lattice. So consider $u_1 \vee u_2 \in D_0$ (lub relative to $\downarrow w$). Since $u_1 \vee u_2 \le w = rw$ by (A) follows $\exists u \in r(D)_0.\ (u_1 \vee u_2 \le u \le w)$. Then $ru_1, ru_2 \le r(u_1 \vee u_2) \le ru = u \le w$. So $r(D)_0 \cap \downarrow w$ is directed moreover from (A) we know that it is cofinal with $D_0 \cap \downarrow w$ whose lub is w. Therefore: $\exists \cup(r(D)_0 \cap \downarrow w) = \cup(D_0 \cap \downarrow w) = w$.

• *Property I*. Suppose $d \le r(d)$ and $e \le r(e) \le r(d)$, where d, $e \in D_0$. We show using the existence of local minima that $\exists\ e' \le d.\ (r(e') = r(e))$. Since $\downarrow d$ is finite we can then conclude that $\downarrow r(d) \cap r(D)_0$ is finite. Observe: $\exists e' \in D_0.\ e' = \min(\downarrow rd \cap r^{-1}(\uparrow re))$. Now $e' \le d$ as $d \le r(d)$ and $r(e) \le r(d)$. Also $e' \le r(e)$ as $r(e) \le r(d)$ and $r(e) \le r(re)$. Finally $r(e) \le r(e')$ by definition of e'. Hence $e' \le r(e') = r(e)$, and $e' \in \downarrow d$. □

2. Theorem *($D \in Bif_\wedge \Rightarrow r(D) \in Bif_\wedge$)*[20]

Let D be a $Bif_\wedge$ and r: $D \to D$ be a stable retraction then r(D) is a $Bif_\wedge$.

Proof

First let us show that $\forall x, y \in r(D)_0.\ MUB_{r(D)}\{x, y\}$ is finite. Suppose:

$x_1 \le rx_1$, $x_2 \le rx_2$, $z \le rz$ and $rz = MUB_{r(D)}\{rx_1, rx_2\}$.

Let x_1', x_2', z' be the minima under rz generating respectively $rx_1' = rx_1$, $rx_2' = rx_2$, $rz' = rz$. Consider $x_{12} = lub_{\downarrow rz}\{x_1', x_2'\}$. We have:

$x_1' \le rx_1$, $x_2' \le rx_2$, $rx_i \le rx_{12} \le rz$ (i=1, 2). So $x_{12} \le rx_{12} \in r(D)_0$.

Now observe: $x_1' \le z'$ and $x_2' \le z'$ as $z' \le rz$, $rx_1 \le rz'$, and $rx_2 \le rz'$. Therefore $rx_{12} \le rz' = rz$. But rz is a mub so $rx_{12} = rz$. Conclusion: given any $rz \in MUB_{r(D)}\{rx_1, rx_2\}$ we can find $x_{12} \in MUB_D\{x_1', x_2'\}$ with $rx_{12} = rz$, where $x_1' \in \downarrow x_1$, $x_2' \in \downarrow x_2$, and all these sets are finite.

It remains to show that r(D) enjoys property (MI)*. We write:

C(R, X) iff $R \subseteq r(D)_0 \wedge X \subseteq D_0 \wedge \forall w \in R.\exists x \in X.\ (x \le rx = w)$.

[20] It remains to check if $D \in Bif_\wedge$ implies $Ret_\wedge(D) \in Bif_\wedge$.

Observe that: (a) $C(R, X) \wedge X \subseteq Y \subseteq D_0 \Rightarrow C(R, Y)$, (b) $C(R, X) \Rightarrow C(\downarrow_{r(D)}R, \downarrow X)$.

Next we want to prove $C(R, X) \Rightarrow C(U_{r(D)}(R), (U\downarrow)^*(X))$. This seems to require some intermediate step. Define a new operator U_2 as follows:

$$U_2(X) \triangleq \cup\{MUB(Y) \mid \#Y \leq 2, Y \subseteq X\}.$$

As usual $U_2^*(X)$ is the least set closed w.r.t. the U_2 operator. If D enjoys mub-completeness (e.g. if D is an algebraic cpo with property I_1) then it is not difficult to show that $\forall X \subseteq D_0$. $U_2^*(X) = U^*(X)$.

Suppose C(R, X). From what we have observed above about mub-finiteness we have: $C(U_{2\ r(D)}(R), U_2(\downarrow X))$. If we iterate we get $C(U_{2\ r(D)}{}^*(R), (U_2\downarrow)^*(X))$. But in algebraic cpo$_\wedge$ with property I_1 $U_2^*(X) = U^*(X) = U(X)$, so we have $C(U_{r(D)}(R), (U_2\downarrow)^*(X))$ that by (a) implies $C(U_{r(D)}(R), (U\downarrow)^*(X))$. Now we can combine with (b) and iterate to get:

$C((U\downarrow)^*_{r(D)}(R), (U\downarrow)^*(X))$. □

Appendix 3: L-domains: stable case

We introduce the main ideas about the construction of a stable version of L domains. The categories $L_\wedge$ and $L_{o\wedge}$ that we define in the following appear in Coquand[89] as the "poset case" of a more general categorical construction.

1. Definition *(Lcpo$_\wedge$)*

D is a *Lcpo$_\wedge$* if it is a cpo and every principal ideal is a completely distributive lattice.[21] We denote with **Lcpo$_\wedge$** the category of Lcpos$_\wedge$ and stable maps.[22]

2. Proposition *(Lcpo$_\wedge$ is a ccc)*

The category **Lcpo$_\wedge$** is cartesian closed.

We now wish to develop a category of *algebraic* Lcpos$_\wedge$. For technical reasons we will use the trace characterization of the stable ordering. Then property I is also needed.

3. Definition *($L_\wedge$-domains)*

D is an *$L_\wedge$-domain* if it is a cpo and every principal ideal is a dI-domain.[23] An *$L_{o\wedge}$-domain* is an $L_\wedge$-domain that is ω-algebraic and has property M.
We denote with **L$_\wedge$** (**L$_{o\wedge}$**) the category of $L_\wedge$-domains ($L_{o\wedge}$-domains) and stable maps.

4. Lemma

Let D, E be $L_\wedge$-domains and f: D→E be a stable map.

(1) If $T \subseteq tr(f)$ then there is a stable map g such that $g \leq_s f$ and

$$tr(g) = \{(d', e') \mid (d', e') \leq (d, e), (d', e') \in tr(f), (d, e) \in T\}$$

Moreover if T is finite then tr(g) is finite and g is compact.

(2) If f is compact then tr(f) is finite.

5. Proposition *($L_\wedge$ is a ccc)*

The categories **L$_\wedge$** and **L$_{o\wedge}$** are cartesian closed.

---

[21] A completely distributive lattice satisfies: $\cup X \wedge y = \cup\{x \wedge y \mid x \in X\}$.

[22] Verify that Lcpo$_\wedge$ is a subcategory of Cpo$_\wedge$.

[23] From this it follows that every principal ideal is an algebraic distributive lattice with property I.

Local Variables and Non-Interference in Algol-like Languages

(Summary of Invited Talk)

R. D. Tennent*
Queen's University
Kingston, Canada

Finding a suitably abstract semantics for the stack-implementable local variables in Algol-like languages has proved to be a difficult problem [?]. In the traditional denotational-semantic approach, each state records which storage variables are in use. This approach is operationally adequate, but does not capture many of the abstract properties of the local-storage discipline and is inadequate for reasoning about programs with local-variable declarations.

To address this problem, researchers [?,?] have shown how to define the "storage support" of semantical entities by induction on types. Then the storage variable chosen to be the meaning of a locally declared variable identifier can be *any* variable outside the support of the block body, and the meaning of the block is independent of which such variable is chosen. However, this approach is technically very intricate.

J. C. Reynolds [?] and F. J. Oles [?,?] proposed the use of a category-theoretic form of possible-world semantics to capture the local-storage discipline more directly: types and type assignments are interpreted as *functors* from a suitable category of "worlds" (e.g., a world is a set of states) to a category of domains, and phrases are then interpreted as *natural transformations* of these functors. Local-variable declarations are easily treated in this framework by allowing changes of world that "expand" the set of states; e.g., a change from some set S to a set of the form $S \times V$, where V is the set of values that a new variable might contain.

The author and P. O'Hearn [?,?,?] have since used this functor-category framework to interpret the concept of *non-interference* used by Reynolds [?, ?] in his "specification logic," a version of Hoare's logic that allows modular reasoning about programs in Algol-like languages with higher-order procedures.

*Supported by research grants from the Natural Sciences and Engineering Research Council of Canada and the Information Technology Research Centre of Ontario.

An essential technique is to allow "changes" of world whose only effect is to constrain state transformations to be compatible with some pre-ordering on the set of states.

The resulting treatment of local variables is quite successful: virtually all of the examples that have been suggested as tests of the abstractness of the interpretation are correctly treated; often this can be shown using specification logic alone.

The functor-category approach to the semantics of local variables seems cleaner and more systematic than other treatments; however, there remain *ad hoc* aspects of current models, and some counter-examples have been discovered. We are exploring ways of addressing these problems.

References

[1] J. Y. Halpern, A. R. Meyer, and B. A. Trakhtenbrot. The semantics of local storage, or what makes the free-list free? In *Conf. Record 11th ACM Symp. on Principles of Programming Languages*, pages 245–257, Austin, Texas, 1984. ACM, New York.

[2] A. R. Meyer and K. Sieber. Towards fully abstract semantics for local variables: preliminary report. In *Conf. Record 15th ACM Symp. on Principles of Programming Languages*, pages 191–203, San Diego, California, 1988. ACM, New York.

[3] P. W. O'Hearn. *The Semantics of Non-Interference: a Natural Approach.* Ph.D. thesis, Queen's University, Kingston, Canada, 1990.

[4] P. W. O'Hearn and R. D. Tennent. Semantical analysis of specification logic, part 2. Technical Report 91-304, Dept. of Computing and Information Science, Queen's University, 1991.

[5] F. J. Oles. *A Category-Theoretic Approach to the Semantics of Programming Languages.* Ph.D. thesis, Syracuse University, Syracuse, 1982.

[6] F. J. Oles. Type algebras, functor categories and block structure. In M. Nivat and J. C. Reynolds, editors, *Algebraic Methods in Semantics*, pages 543–573. Cambridge University Press, Cambridge, 1985.

[7] J. C. Reynolds. *The Craft of Programming.* Prentice-Hall International, London, 1981.

[8] J. C. Reynolds. The essence of Algol. In J. W. de Bakker and J. C. van Vliet, editors, *Algorithmic Languages*, pages 345–372, Amsterdam, 1981. North-Holland, Amsterdam.

[9] J. C. Reynolds. Idealized Algol and its specification logic. In D. Néel, editor, *Tools and Notions for Program Construction*, pages 121–161. Cambridge University Press, Cambridge, 1982.

[10] R. D. Tennent. Semantical analysis of specification logic. *Information and Computation*, 85(2):135–162, 1990.

Categories of Information Systems

(Extended Abstract)

Abbas Edalat
(email: ae@doc.ic.ac.uk)
and
Michael B. Smyth

Department of Computing, Imperial College of Science, Technology and Medicine,
180 Queen's Gate, London SW7 2BZ, England.

Abstract

An abstract notion of "information category" (I-category) is introduced as a generalization of Scott's well-known category of information systems. The proposed axioms introduce a global partial order on the morphisms of the category, making them an ω-algebraic cpo. An initial algebra theorem for a class of endofunctors continuous on the cpo of morphisms is proved, thus giving canonical solution of domain equations. An effective version of these results, in the general setting, is also provided. Some basic examples of categories of information systems are dealt with.

1 Introduction

A distinctive feature of information systems representing Scott domains, as expressed in [Sco82, LW84], is that the collection of all information systems itself has an "information ordering". This has the consequence that domain equations of the form

$$D \cong F(D)$$

for the recursive specification of types, can be handled by ordinary (cpo) fixed point techniques. The ordering of information systems is, typically, defined as follows. Let I, J be information systems, with token sets L, L'. Then $I \trianglelefteq J$ iff $L \subseteq L'$ and the operations and relations of L are the restrictions to L of those of L'. If we further ask that the tokens of information systems all be drawn from a common pool of tokens and that the very *same* token, Δ, appear as the distinguished member in all information systems that we consider (in a given context), we find (typically) that the collection of information systems under $\trianglelefteq$ is a cpo (even, for example, an ω-algebraic cpo) with the trivial information system with token set $\{\Delta\}$ as its least element; see Section 5.2. The result is an approach to domain equations which is appreciably simpler than other standard approaches in terms of inverse limits, universal objects, or O-categories. But it must be pointed out that the simple approach, in terms of the $\trianglelefteq$ -ordering of information systems, does not give us everything that the more elaborate category-theoretic methods yield: Specifically, we do not get an initial algebra theorem, characterising the intended solutions of the above domain equation.

Our enquiry really starts with the question: What needs to be added to the information system method, to get something fully equivalent to the other standard methods for solving a domain equation?

The answer, we suggest, is that a global information ordering of morphisms, $\trianglelefteq^m$, is needed, in addition to (or, including) that for objects. Concretely, we can define:

$$(f : I_1 \to I_2) \trianglelefteq^m (g : J_1 \to J_2) \text{ if } I_n \trianglelefteq J_n \ \ (n = 1,2) \text{ and } f \subseteq g$$

(note that f, g are sets of pairs of tokens). For the general theory, however, we do not work concretely in this way: rather, we abstract the required properties of the global ordering of objects and morphisms, arriving at an *axiomatic* notion of "category of information systems", or "I-category". All the required theory of domain equations, at least as far as the (effective) initial algebra theorem, can be developed at the abstract level, and is then available to be applied routinely to various concrete situations, such as Scott domains, stable domains, Stone spaces, and even metric spaces. Various constructions, such as the functor category of two categories of information systems, can also be handled.

A category of information systems, in the narrow sense (see below in Section 2.3 under ω-algebraic I-category), has an ω-algebraic cpo as its class of morphisms with (continuous) partial operations representing domain, codomain, and composition of morphisms. The methodological principle involved is that the approximation of infinite objects (morphisms, etc.) by finite ones is always accomplished via an information *ordering*; categorical notions, such as colimit, are not required for approximation.

2 Basic Definitions and Axioms

In this section we give the basic definitions of a category of information systems, which we simply call an I-category, and the related notions. Our notations are fairly standard. For a category P, we denote its class of objects by Obj_P and its class of morphisms by Mor_P, and we often delete the subscript P. The mappings dom and cod : Mor $\to$ Obj denote the domain and the codomain mappings associated with P, and $\mathrm{Id} : \mathrm{Obj} \to \mathrm{Mor}$ denotes the mapping which takes an object to the identity morphism of that object, so that the identity morphism on A is denoted by $\mathrm{Id}(A)$. We denote the morphism f with domain A and codomain B by $f : A \to B$. For $f, g \in \mathrm{Mor}$ with $\mathrm{cod}(f) = \mathrm{dom}(g)$, we denote the composition of f and g by $f; g$ (instead of the common notation $g \circ f$), which should be read as f followed by g. Given a functor $F : P_1 \to P_2$ between two categories P_1 and P_2 we write $F_o : \mathrm{Obj}_{P_1} \to \mathrm{Obj}_{P_2}$ and $F_m : \mathrm{Mor}_{P_1} \to \mathrm{Mor}_{P_2}$ for the induced mappings on objects and morphisms respectively. The identity endofunctor is denoted by ID. We say P is a *countable category* if Mor_P is countable. $\mathcal{P}_f$ denotes the *finite* power set constructor, i.e. $\mathcal{P}_f(A)$ is the set of all finite subsets of the set A. Given a partial order we write $a \uparrow b$ if the elements a and b are bounded above. We denote the class of compact elements of an algebraic cpo A by $\mathcal{K}_A$.

2.1 I-categories

An *I-category* is, intuitively speaking, a category with ordered hom-sets and with a distinguished class of morphisms, called *inclusion morphisms*, which induces a partial order on the class of morphisms as well as on the class of objects. Here is the precise definition.

Definition 2.1 An *I-category* is a four tuple $(P, \mathrm{Inc}, \sqsubseteq, \Delta)$ where:

- P is a category,
- $\mathrm{Inc} \subseteq \mathrm{Mor}$ is the subclass of *inclusion morphisms* of P such that in each hom-set, $\mathrm{hom}(A, B)$, there is at most one inclusion morphism which we denote by $\mathrm{in}(A, B)$ or $A \rightarrowtail B$,

- $\sqsubseteq^{A,B}$ is a partial order on hom(A, B), for all $A, B \in$ Obj (the superscripts in $\sqsubseteq^{A,B}$ will be often deleted),
- $\Delta \in$ Obj is a distinguished object,

which satisfy the following two axioms:

Axiom 1 (i) *The class of objects Obj and the inclusion morphisms Inc form a partial order represented as a category.*

(ii) *in*(Δ, A) *exists, for all* $A \in$ *Obj and in*$(\Delta, A) \sqsubseteq f$ *for all morphisms* $f \in \mathrm{hom}(\Delta, A)$.

(iii)
$$f; in(A,B) \sqsubseteq g; in(A,B) \;\Rightarrow\; f \sqsubseteq g,$$
for all $f, g \in$ *Mor and in*$(A, B) \in$ *Inc, such that the compositions are defined.*

Axiom 2 *Composition of morphisms is monotone with respect to the partial order on hom-sets, i.e.*
$$f_1 \sqsubseteq f_2 \;\&\; g_1 \sqsubseteq g_2 \;\Rightarrow\; f_1; g_1 \sqsubseteq f_2; g_2$$
whenever the compositions are defined.

For convenience, we often denote an I-category by its carrier and write P for $(P, \mathrm{Inc}, \sqsubseteq, \Delta)$. We now define partial orders $\trianglelefteq$ on Obj_P and $\trianglelefteq^m$ on Mor_P of an I-category $(P, \mathrm{Inc}, \sqsubseteq, \Delta)$ as follows:

- $A \trianglelefteq B$ if in(A, B) exists.
- $f \trianglelefteq^m g$ if
  - (i) dom$(f) \trianglelefteq$ dom(g)
  - (ii) cod$(f) \trianglelefteq$ cod(g)
  - (iii) f; in(cod(f), cod(g)) $\sqsubseteq$ in(dom(f), dom(g)); g.

Note that $f \trianglelefteq^m g$ iff the diagram

$$\begin{array}{ccc} \mathrm{dom}(g) & \xrightarrow{\;g\;} & \mathrm{cod}(g) \\ \uparrow & \sqsupseteq & \uparrow \\ \mathrm{dom}(f) & \xrightarrow{\;f\;} & \mathrm{cod}(f) \end{array}$$

weakly commutes. It is easily checked that the relation $\trianglelefteq^m$ is in fact a partial order. Note that if in Axiom 1(i) we require to have a pre-order rather than a partial order, we obtain the notion of a *pre-I-category* , in which $\trianglelefteq$ and $\trianglelefteq^m$ will be just pre-orders.

Proposition 2.2 (i) *Id*(Δ) *is the least element of* $(Mor, \trianglelefteq^m)$.

(ii) *Inclusion morphisms are monomorphisms.*

(iii) *Composition of morphisms is monotonic with respect to* $\trianglelefteq^m$.

Example 2.3 PARTIAL ORDERS
Any partial order with least element, $(Q, \sqsubseteq)$, considered as a category in the standard way, gives rise to an I-category $(Q, \mathrm{Inc}, =, \perp)$, where Inc = Mor, "=" is the discrete partial order on hom-sets and $\Delta = \perp$. Similarly, pre-orders are examples of pre-I-categories.

Example 2.4 SETS
Consider the category of sets which we denote by **Set**. It can be easily checked that $(\mathbf{Set}, \mathrm{Inc}, =, \emptyset)$ is a large I-category, where Inc is just the set inclusions, "=" is the discrete partial order on hom-sets and $\emptyset$ is the empty set. Here, we have

$$(f : A \to B) \trianglelefteq^m (g : C \to D)$$

iff $A \subseteq B, C \subseteq D$ and $f = g \restriction_A$.

2.2 Complete I-categories

In the same way that we define a cpo as a partial order having lubs of increasing chains, we can define a complete I-category as an I-category with some completion properties:

Definition 2.5 A *complete I-category* $(P, \mathrm{Inc}, \sqsubseteq, \Delta)$ is an I-category which satisfies the following three axioms:

Axiom 3 *$(Mor, \trianglelefteq^m)$ is a cpo.*

Axiom 4 *$(Inc, \trianglelefteq^m)$ is a subcpo of $(Mor, \trianglelefteq^m)$.*

Axiom 5 *Composition of morphisms is a continuous operation with respect to $\trianglelefteq^m$, i.e.*

$$\bigsqcup_i (f_i; g_i) = (\bigsqcup_i f_i); (\bigsqcup_i g_i),$$

whenever $\langle f_i \rangle_{i \geq 0}$ and $\langle g_i \rangle_{i \geq 0}$ are increasing chains in $(Mor, \trianglelefteq^m)$ with $cod(f_i) = dom(f_i)$, for all $i \geq 0$.

Proposition 2.6 *In a complete I-category, $(Obj, \trianglelefteq)$ is a cpo and the mappings*

$$Id : (Obj, \trianglelefteq) \to (Mor, \trianglelefteq^m),$$

and

$$dom,\ cod : (Mor, \trianglelefteq^m) \to (Obj, \trianglelefteq)$$

are continuous.

Example 2.7 COMPLETE PARTIAL ORDERS
Any cpo, considered as a category, is a complete I-category. See Example 2.3. It is straightforward to check the axioms.

Example 2.8 SETS
Observe that $(\mathbf{Set}, \mathrm{Inc}, \subseteq, \emptyset)$ of Example 2.4 is a complete large I-category. An increasing chain of objects

$$A_0 \subseteq A_2 \subseteq A_3 \subseteq \cdots$$

has lub $\bigcup_i A_i$; whereas an increasing chain of morphisms $\langle f_i : A_i \to B_i \rangle_{i \geq 0}$ has lub

$$\bigcup_i f_i : (\bigcup_i A_i) \to (\bigcup_i B_i),$$

where $(\bigcup_i f_i)(x) = f_j(x)$ if $x \in A_j$. It is routine to verify the axioms.

2.3 ω-algebraic I-categories

An *ω-algebraic I-category* is a complete I-category in which $(\text{Mor}, \trianglelefteq^m)$ is ω-algebraic and compact morphisms and objects are related in a desirable way as follows.

Definition 2.9 An *ω-algebraic I-category* is a complete I-category which satisfies the following two axioms:

Axiom 6 *$(Mor, \trianglelefteq^m)$ is ω-algebraic.*

Axiom 7 (i) *$f \in (Mor, \trianglelefteq^m)$ is compact $\Rightarrow$ $cod(f) \in (Obj, \trianglelefteq)$ is compact.*

(ii) *$A, B \in (Obj, \trianglelefteq)$ are compact with $A \trianglelefteq B$ $\Rightarrow$ $in(A, B) \in (Mor, \trianglelefteq^m)$ is compact.*

(iii) *The composition of compact morphisms is compact.*

See Proposition 2.10 below about Axiom 7(i) and the remarks after Proposition 2.18 about Axiom 7(ii) and Axiom 7(iii) .

Proposition 2.10 *In an ω-algebraic I-category, $(Obj, \trianglelefteq)$ is an ω-algebraic cpo and compact morphisms have compact domains.*

Example 2.11 ω-ALGEBRAIC CPO'S
Any ω-algebraic cpo, considered as a category, is an ω-algebraic I-category. See Example 2.3. The compact objects are precisely the compact points and the compact morphisms are precisely the morphisms between the compact points. It is routine to check all the axioms.

Example 2.12 SETS WITH ELEMENTS FROM A COUNTABLE ALPHABET
Consider the category of sets with elements from a given countable alphabet. Denoting this category by **Set-ISys**, we can easily check that $(\textbf{Set-ISys}, \text{Inc}, =, \emptyset)$ is an ω-algebraic I-category. In fact, given a morphism $f : A \to B$ we can construct an increasing chain of morphisms between finite sets with lub f as follows. Choose finite sets $A_i, B_i, i \geq 0$ with $A = \bigcup_i A_i, B = \bigcup_i B_i$ and for each $i \geq 0$ let n_i be the least integer with $n_i \geq i$ and $f(A_i) \subseteq B_{n_i}$. Define $f_i : A_i \to B_{n_i}$ by $f_i = f \restriction_{A_i}$. Then we have $f = \bigcup_i f_i$ and $B = \bigcup_i B_{n_i}$.

Finally in this section, we note that the product $P_1 \times P_2$ of two I-categories P_1 and P_2 is an I-category, with $\Delta_{P_1 \times P_2} = (\Delta_{P_1}, \Delta_{P_2})$ and the partial order on hom-sets and inclusion morphisms defined co-ordinatewise in the obvious way. Furthermore $P_1 \times P_2$ is complete (ω-algebraic) if P_1 and P_2 are complete (ω-algebraic).

2.4 Completion of an I-category

Recall that the ideal completion of a countable poset, $(L, \sqsubseteq)$, gives rise to an ω-algebraic cpo, $(\overline{L}, \subseteq)$, of directed ideals of L ordered by inclusion. In the same way we can take a certain completion of a countable I-category P to obtain an ω-algebraic I-category $\overline{P}$, whose objects and morphisms will be the directed ideals of $(\text{Obj}_P, \trianglelefteq)$ and $(\text{Mor}_P, \trianglelefteq^m)$ respectively. Here is the precise definition.

Definition 2.13 Let $P = (P, \text{Inc}, \sqsubseteq, \Delta)$ be a countable I-category; its *I-completion* denoted by $\overline{P} = (\overline{P}, \overline{\text{Inc}}, \subseteq, \overline{\Delta})$ is given by:

- $\mathrm{Obj}_{\overline{P}} = \overline{\mathrm{Obj}_P}$;
- $\mathrm{Mor}_{\overline{P}} = \overline{\mathrm{Mor}_P}$, with

$$\begin{array}{ll} \mathrm{dom}(\mathcal{F}) = \{\mathrm{dom}(f) \mid f \in \mathcal{F}\}, & \mathrm{cod}(\mathcal{F}) = \downarrow\{\mathrm{cod}(f) \mid f \in \mathcal{F}\}, \\ \mathrm{Id}(\mathcal{A}) = \downarrow\{\mathrm{Id}(A) \mid A \in \mathcal{A}\}, & \mathcal{F};\mathcal{G} = \downarrow\{f;g \mid f \in \mathcal{F}, g \in \mathcal{G}\}; \end{array}$$

(assuming in the latter equation that $\mathrm{cod}(\mathcal{F}) = \mathrm{dom}(\mathcal{G})$, and denoting the downward closure of S by $\downarrow S$)

- $\mathrm{in}(\mathcal{A}, \mathcal{B})$ exists iff $\mathcal{A} \subseteq \mathcal{B}$, and when it exists we have:

$$\mathrm{in}(\mathcal{A}, \mathcal{B}) = \downarrow\{\mathrm{in}(A, B) \mid A \in \mathcal{A}, B \in \mathcal{B}, A \trianglelefteq B\};$$

- the inclusion $\subseteq$ as the partial order on hom-sets;
- $\overline{\Delta} = \{\Delta\}$.

In order to show that $\overline{P}$ is a well defined category, it is convenient to use the following simple lemma.

Lemma 2.14 *Suppose A, B are posets with A directed and $h : A \to B$ a monotonic surjection. Then B is also directed and if S, T are finite subsets of A, B respectively, there exists $a \in A$ such that a is an upperbound of S in A and $h(a)$ is an upper bound of T in B.*

We will now have:

Proposition 2.15 *Let P be an I-category. Then $\overline{P}$ is a well-defined category.*

Finally, we get the expected result:

Proposition 2.16 *Let P be a countable I-category. Then $\overline{P}$ is an ω-algebraic I-category.*

Conversely, any ω-algebraic I-category is isomorphic to the completion of a countable I-category. To make this more precise we need :

Definition 2.17 The *base* of an ω-algebraic I-category, P, is the subcategory P^b of compact objects and compact morphisms of P, i.e.

$$\mathrm{Obj}(P^b) = \mathcal{K}_{\mathrm{Obj}_P}, \qquad \mathrm{Mor}(P^b) = \mathcal{K}_{\mathrm{Mor}_P}.$$

Clearly, P^b is a countable category. We also have:

Proposition 2.18 *If P is an ω-algebraic I-category, then P is isomorphic to $\overline{P^b}$.*

It is this completion result which explains the form of Axiom 7 above, in particular Axiom 7(ii). Had we, for example, chosen to postulate (as might be reasonable) that in an ω-algebraic I-category, instead of Axiom 7(ii), we have:

$$A, B \text{ are compact } \Rightarrow \mathrm{hom}(A, B) \text{ is finite},$$

we would have obtained the stronger result that any ω-algebraic I-category is the completion of an I-category with finite hom-sets.

3 The Initial Algebra Theorem

In this section, we present one of our main results, i.e. the initial algebra theorem for a suitable class of endofunctors on a complete I-category. We begin with some definitions.

3.1 Strict morphisms

Definition 3.1 A morphism, $f : A \to B$, of an I-category P is *strict* if the diagram

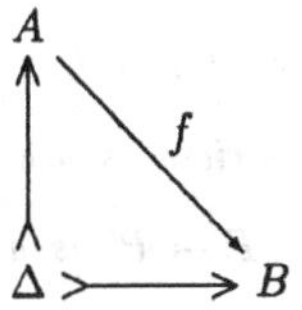

commutes.

In the case of Scott information systems, strict morphisms represent the usual notion of a strict function, i.e. one which preserves the least element.

Proposition 3.2 *Let P be an I-category.*

(i) *If $f \in Mor_P$ is strict and $g \trianglelefteq^m f$, then g is strict as well.*

(ii) *Inclusion morphisms of P are strict.*

(iii) *The composition of strict morphisms of P is strict.*

(iv) *If P is complete, then the lub of an increasing chain of strict morphisms of P is strict.*

We now define, for an I-category P, the *subcategory of strict morphisms*, P^s, by:

$$\mathrm{Obj}_{P^s} = \mathrm{Obj}_P, \qquad \mathrm{Mor}_{P^s} = \{f \in \mathrm{Mor}_P \mid f \text{ is strict}\}.$$

Using the above proposition, the following can now be shown.

Proposition 3.3 (i) *If P is an I-category, so is P^s.*

(ii) *If P is an (ω-algebraic) complete I-category, so is P^s.*

3.2 Standard functors

In dealing with functors between I-categories, it is natural to require that the functors preserve the class of inclusion morphisms.

Definition 3.4 Let P, P' be I-categories.

(i) An endofunctor $F : P \to P'$ is *standard* if it preserves inclusion morphisms; it is *strictness preserving* if it preserves strict morphisms.

(ii) Suppose P, P' are complete. A functor $F : P \to P'$ is *continuous* if the induced map on morphisms $F_m : (\mathrm{Mor}_P, \trianglelefteq^m) \to (\mathrm{Mor}_{P'}, \trianglelefteq^m)$ is a continuous (cpo) function.

We will work with standard and continuous endofunctors on I-categories. Here is a list of their basic properties:

Proposition 3.5 *Let P, P' be I-categories.*

(i) *A functor $F : P \to P'$ is standard iff*

$$F(in(A,B)) = in(F(A), F(B)),$$

for all $in(A,B)) \in Inc_P$.

(ii) *Composition of standard functors is standard.*

(iii) *A standard functor $F : P \to P'$ is strictness preserving.*

(iv) *Suppose P, P' are complete. If $F : P \to P'$ is a continuous endofunctor, then the induced map on objects*

$$F_o : (Obj_P, \trianglelefteq) \to (Obj_{P'}, \trianglelefteq)$$

is continuous.

Since we will only be dealing with standard functors, *we assume from now on that all functors between I-categories are standard.* We now proceed to state our results for I-categories. Recall that we denote the inclusion morphism in(A,B) by $A \rightarrowtail B$.

Lemma 3.6 *The lub of any increasing chain of objects in a complete I-category P is a colimit of the chain in P. It is also a colimit of the chain in P^s.*

Corollary 3.7 *If $F : P \to P$ is a continuous functor on a complete I-category P, then the chain*

$$T = \Delta \rightarrowtail F(\Delta) \rightarrowtail F^2(\Delta) \rightarrowtail \cdots$$

has a colimit, D, in P with $F(D) = D$. D is also a colimit of the chain in P^s.

3.3 Initial algebras

Recall that given an endofunctor $F : C \to C$ on a category C, the category of F-algebras has as objects the pairs (A, f), with $A \in \mathrm{Obj}_C$ and $f \in \mathrm{hom}(F(A), A)$, and as morphisms, between objects (A, f) and (B, g), those $h \in \mathrm{hom}(A, B)$ for which the following diagram:

$$\begin{array}{ccc} A & \xleftarrow{\ f\ } & F(A) \\ {\scriptstyle h}\downarrow & & \downarrow{\scriptstyle F(h)} \\ B & \xleftarrow{\ g\ } & F(B) \end{array}$$

commutes. The *initial algebra* of F is defined to be the initial object, if it exists, of the category of F-algebras; and we then say that F has an initial algebra, or a least fixed point, in C. Initial algebras give a canonical solution of domain equations and therefore play a fundamental role in computing science (see, for example, [MA86] for details). We now state and prove our main result. Remember that all functors between I-categories are assumed to be standard.

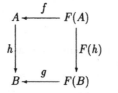

Theorem 3.8 *A continuous endofunctor on a complete I-category P has an initial algebra in P^s.*

Proof Let $D = \bigsqcup F^i(\Delta)$ as in Corollary 3.7. Then $F(D) = D$ and, therefore, $(D, \mathrm{Id}(D))$ is an F-algebra in P^s, since $\mathrm{Id}(D)$ is strict by Proposition 3.2(ii). Suppose (E, k) is any F-algebra in P^s i.e. $\mathrm{dom}(k) = F(E)$, $\mathrm{cod}(k) = E$ and k is strict. We must prove the existence of a unique strict morphism h with $h = F(h); k$ i.e. for which the diagram.

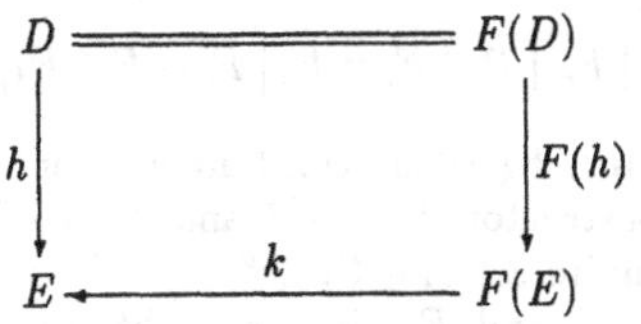

commutes.

Existence of h: We inductively define the morphisms g_i, $i \geq 0$, as follows:

$$g_0 \;=\; \mathrm{in}(\Delta, E), \qquad\qquad g_i \;=\; F(g_{i-1}); k\,.$$

We will show that $\langle g_i \rangle_{i \geq 0}$ is an increasing chain of strict morphisms, whose lub h is strict and satisfies $h = F(h); k$. To show that the g_i's are increasing, first note that, by Axiom 1(ii), and Proposition 2.2(iii) we have:

$$g_0 \trianglelefteq^m \mathrm{in}(\Delta, F(\Delta)); g_1 \trianglelefteq^m g_1.$$

Assume inductively that $g_i \trianglelefteq^m g_{i+1}$. Since F is monotonic on $(\mathrm{Mor}, \trianglelefteq^m)$ and composition of morphisms is a monotonic operation with respect to $\trianglelefteq^m$, it follows that $F(g_i); k \trianglelefteq^m F(g_{i+1}); k$, i.e. $g_{i+1} \trianglelefteq^m g_{i+2}$, proving the inductive step. To show that the g_i's are strict, note that g_0 is strict as it is an inclusion morphism. Assume that g_i is strict. Then $F(g_i)$ is strict since F is a standard functor (Proposition 3.5(iii)), and hence $g_{i+1} = F(g_i); k$ is strict, since the composition of strict morphisms is strict. It now follows by Proposition 3.2(iv) that h is strict as it is the lub of an increasing chain of strict morphisms. Furthermore, the continuity of F implies:

$$h \;=\; \bigsqcup g_i \;=\; \bigsqcup g_{i+1} \;=\; \bigsqcup (F(g_i); k) \;=\; (\bigsqcup F(g_i)); k \;=\; (F(\bigsqcup g_i)); k \;=\; F(h); k.$$

Therefore $(h, F(h))$ is a morphism of the category of F-algebras of P^s mediating between $(D, \mathrm{Id}(D))$ and (E, k).

Uniqueness of h: Let f be another strict morphism also satisfying $h = F(h); k$. We show, by induction, that $\mathrm{in}(F^i(\Delta), D); f = g_i$ for all $i \geq 0$. For $i = 0$ this equation is just the strictness condition for f. Assume that the equation holds for $i \geq 0$, we have:

$$\begin{array}{lll}
\mathrm{in}(F^i(\Delta), D); f = g_i & \Rightarrow & F(\mathrm{in}(F^i(\Delta), D); f) = F(g_i) \;\Rightarrow \\
\mathrm{in}(F^{i+1}(\Delta), F(D)); F(f) = F(g_i) & \Rightarrow & \mathrm{in}(F^{i+1}(\Delta), D); F(f) = F(g_i) \;\Rightarrow \\
\mathrm{in}(F^{i+1}(\Delta), D); F(f); k = F(g_i); k & \Rightarrow & \mathrm{in}(F^{i+1}(\Delta), D); f = g_{i+1}.
\end{array}$$

This completes the inductive proof. It now follows that:

$$\begin{aligned}
h \;&=\; \bigsqcup g_i \;=\; \bigsqcup(\mathrm{in}(F^i(\Delta), D); f) \;=\; (\bigsqcup(\mathrm{in}(F^i(\Delta), D)); f \;=\; (\mathrm{in}(\bigsqcup F^i(\Delta), D)); f \\
&=\; \mathrm{in}(D, D); f \;=\; f.
\end{aligned}$$

This completes the proof. □

We have therefore succeeded in obtaining an initial algebra theorem by essentially a cpo construction without the usual heavy category-theoretic machinery based on global cocontinuity of functors. Finally, we note that a continuous endofunctor on P may not have an initial algebra in P itself. In fact, although the existence of h, in the proof of Theorem 3.8, is always guaranteed, its uniqueness will fail, in general, when we allow non-strict morphisms.

Example 3.9 Let $(Q, \sqsubseteq)$ be a cpo considered as a complete I-category. (See Example 2.7.) All continuous functions $f : Q \to Q$ induce continuous functors on Q, and the initial algebra theorem reduces to the statement that the least fixed point $D = \bigsqcup f^i(\bot)$ is the least pre-fixed point (where E is a pre-fixed point of f if $E \sqsupseteq f(E)$).

Example 3.10 Consider **Set-ISys** which we have seen to be an ω-algebraic I-category. (See Example 2.12.) Using the BNF notation, we can construct a set of endofunctors as follows:

$$F ::= \mathrm{ID} \mid F_A \mid \mathcal{P}_f \mid F_1 \times F_2 \mid F_1 + F_2 \mid F_{A_f} \to F \mid F_1 \circ F_2$$

where ID is the identity functor, F_A is the constant functor which maps all objects to $A \in \mathrm{Obj}$, $\mathcal{P}_f$ is the finite power set constructor, "$- \times -$" and "$- + -$" are the product and the disjoint sum functors, A_f denotes a finite set $A \in \mathrm{Obj}$, "$- \to -$" is the function space constructor and $F_1 \circ F_2$ is the composition of F_1 and F_2. It is straightforward to check that this defines a set of continuous endofunctors. Note, however, that in $F_{A_f} \to F$ the finiteness of the set A is necessary, as for example the endofunctor $\mathbb{N} \to \mathrm{ID}$ is not continuous: $\mathbb{N} \to \mathbb{N}$ is in fact strictly larger than $\bigcup_i(\mathbb{N} \to N_i)$, where N_i is the set of the first i natural numbers. As an example, we solve the domain equation

$$X = 1 + X,$$

in **Set-ISys**, where $1 = \{\mathrm{nil}\}$ is a singleton set. Writing $F_{1+}(X) = 1 + X$ we readily find that $D = \bigsqcup F^i_{1+}(\emptyset)$ is the least solution or colimit. To appreciate this, let's fix our notation for the disjoint sum of two sets and write:

$$A + B = \{(0, a) | a \in A\} \cup \{(1, b) | b \in B\}.$$

Therefore, we have:

$$\begin{aligned} &F^0_{1+}(\emptyset) = \emptyset \\ &F_{1+}(\emptyset) = 1 + \emptyset = \{(0, \mathrm{nil})\} \\ &F^2_{1+}(\emptyset) = 1 + (1 + \emptyset) = \{(0, \mathrm{nil}), (1, (0, \mathrm{nil}))\} \\ &\cdots\cdots \\ &\cdots\cdots \end{aligned}$$

We then find, after dropping all brackets, that

$$D = \{0\,\mathrm{nil}, 10\,\mathrm{nil}, 110\,\mathrm{nil}, 1110\,\mathrm{nil}, \cdots\}$$

which can be identified with $\mathbb{N}$.

4 Effectiveness

In this section, we present an effective theory for ω-algebraic I-categories. The idea is to postulate a suitable recursive structure on the base P^b of the category P and define the computable objects and morphisms of P to be the lubs of effective chains of objects and morphisms of P^b respectively. A computable functor will then be defined as a standard functor which is computable as a cpo function on $(\mathrm{Mor}_P, \trianglelefteq^m)$. Our aim is to obtain an effective version of the initial algebra theorem. We will use the standard notions from classical recursion theory as, for example, in [Cut80]. In particular, ϕ_n, is the nth partial recursive function in the standard enumeration and $\langle m_1, \cdots, m_n \rangle$ is the n-tupling function from $\mathbb{N}^n$ to $\mathbb{N}$.

4.1 Effective cpo's

We briefly recall the theory of computability of ω-algebraic cpo's. See [Plo81] for details. Let $(Q, \sqsubseteq)$ be any algebraic cpo with least element $\perp$ and let $e : \mathsf{N} \rightarrow \mathcal{K}_Q$ be any surjection representing an *enumeration* of its basis $\mathcal{K}_Q$. Write e_n for $e(n)$ and $x \uparrow y$ if $x, y \in Q$ have an upper bound. We say that Q is *effectively given with respect to* e with index $\langle a, r, s\rangle$ if the following three conditions hold:

(i) $e_a = \perp$

(ii) $e_m \uparrow e_n$ is recursive in m and n with index r.

(iii) $e_m \sqsubseteq e_n$ is recursive in m and n with index s.

An *effective chain* (of compact elements) of Q with index j with respect to e is an increasing chain $e_{\phi_j(1)} \sqsubseteq e_{\phi_j(2)} \sqsubseteq e_{\phi_j(3)} \sqsubseteq \cdots$. An element $d \in Q$ is said to be *computable* if there exists an effective chain with lub d. The index of d with respect to e is then defined to be the index of this effective chain. Denoting the partial order of the computable elements of Q by C_Q, we can then obtain an enumeration $e' : \mathsf{N} \rightarrow C_Q$ of C_Q with respect to e. Given two ω-algebraic cpo's Q_1 and Q_2 with e_1, e'_1 and e_2, e'_2 as enumeration of their compact and computable elements respectively, we say that a continuous function $f : Q_1 \rightarrow Q_2$ is *computable* if there exists a recursive function $l : \mathsf{N} \rightarrow \mathsf{N}$ such that the diagram:

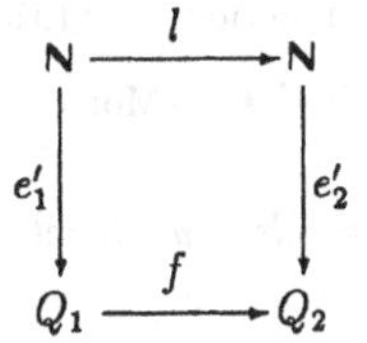

commutes. Any index for l is, then, said to be an index for f. Note that the composition of computable functions is computable, since the composition of recursive functions is recursive. We need the following simple results which can be found in [Plo81].

Lemma 4.1 *The lub of an increasing effective chain of computable elements is computable and an index for it can be effectively obtained from an index for the effective chain.*

Corollary 4.2 *The least fixed point of a computable function on an ω-algebraic cpo is computable and an index for it is effectively obtainable from an index for the function.*

4.2 Effective I-categories

We are now ready to define the notion of an effectively given ω-algebraic I-category.

Definition 4.3 Let P be an ω-algebraic I-category. An *enumeration* (A, f) of the base of P is given by an enumeration $A : \mathsf{N} \rightarrow \mathrm{Obj}_{P^b}$ and an enumeration $f : \mathsf{N} \rightarrow \mathrm{Mor}_{P^b}$ of its compact objects and morphisms (i.e. objects and morphisms of P^b) respectively. P is said to be *effectively given* with respect to (A, f) by index $\langle a, r, s, t, u, v, w\rangle$ if the following five conditions hold:

R1 $(\mathrm{Mor}_P, \trianglelefteq^m)$ is effectively given with respect to f with index $\langle a, r, s\rangle$;

R2 $\mathrm{dom}(f_n) = A_l$ is recursive in n and l with index t;

R3 $\mathrm{cod}(f_n) = A_l$ is recursive in n and l with index u;

R4 $f_n = \mathrm{in}(A_l, A_p)$ is recursive in n, l and p with index v;

R5 $f_n; f_l = f_p$ is recursive in n, l and p with index w.

Proposition 4.4 *Let P be an ω-algebraic I-category effectively given with respect to (A, f).*

(i) *$(Obj_P, \trianglelefteq)$ is effectively given with respect to A with an index which can be effectively obtained from an index for P with respect to (A, f).*

(ii) *The mappings dom,cod: $(Mor, \trianglelefteq^m) \to (Obj, \trianglelefteq)$ and the mapping $Id : (Obj, \trianglelefteq) \to (Mor, , \trianglelefteq^m)$ are computable and an index for each can be effectively obtained from an index for P with respect to (A, f).*

(iii) *If $B, C \in Obj_P$ are computable with indices i, j respectively and $B \trianglelefteq C$, then $in(B, C)$ is computable and an index for it can be effectively obtained from i, j.*

(iv) *If $g, h \in Mor_P$ are computable with $cod(g) = dom(h)$, then $g; h$ is also computable and an index for it can be effectively obtained from one for g and one for h.*

Definition 4.5 (i) A continuous endofunctor F on an effectively given ω-algebraic I-category is *computable* if the induced cpo function $F_m : (\mathrm{Mor}_P, \trianglelefteq^m) \to (\mathrm{Mor}_P, \trianglelefteq^m)$ is computable.

(ii) An F-algebra (E, k) is *computable* if $k \in (\mathrm{Mor}_P, \trianglelefteq^m)$ is computable.

Proposition 4.6 *If F is a computable endofunctor on P, then the induced function $F_o : Obj_P \to Obj_P$ is computable.*

We can now present the effective version of the initial algebra theorem which first appeared in [Eda89].

Theorem 4.7 *Let F be a computable endofunctor on P. Then P has a computable initial algebra $(D, Id(D))$, an index for which is effectively obtainable from one for F. Moreover if (E, k) is a computable F-algebra, the unique morphism h satisfying $h = F(h); k$ is computable and an index for h can be effectively obtained from one for F and one for (E, k).*

We have therefore an effective initial algebra theorem and a satisfactory theory of computability for ω-algebraic I-categories.

5 Basic examples

5.1 Information categories

In practice, I-categories are often *concrete* in the sense that their objects are sets with some internal structure given by operations and predicates defined on the elements of the sets or on their finite subsets, i.e. they are weak second order structures in the terminology of [Bar77]; the partial order on objects corresponds to the substructure relation between objects; and morphisms are relations between elements or finite subsets of the carrier sets of objects. We will call these concrete I-categories *information categories*, which like the abstract I-categories can be complete or ω-algebraic. In information categories the partial order on objects $A \trianglelefteq B$ corresponds to the notion that A is a substructure of B; whereas the partial order on morphisms $f \trianglelefteq^m g$ simply reduces to $f \subseteq g$ i.e. the inclusion of relations. Similarly in complete information categories the lub of a chain of objects will be the union of the chain of structures and the lub of a chain of morphisms will simply be the set union of the relations representing the morphisms. Finally in ω-algebraic information categories compact objects will be precisely the finite objects i.e. objects with a finite carrier set and the compact morphisms will be precisely the relations between finite objects. These features make information categories conceptually simple, and easy to handle; the task of verifying that a certain category is an information category and therefore an I-category becomes more straightforward. Using the notion of information categories, a number of substantial examples of complete (or ω-algebraic) I-categories— including categories of information systems for SFP domains, dI-domains and continuous domains—have already been constructed in detail in [ES91]; and other examples, including a category of information systems for metric spaces, are in preparation. However, for reasons of space, we will not give the full details of information categories here and will briefly treat only some further simple examples of concrete I-categories in this paper.

5.2 Bounded complete information systems

A bounded complete information system is a structure $A = (|A|, \vdash, \wedge, \Delta)$, where $|A|$ is a countable set of tokens, $\vdash$ is a pre-order over $|A|$, and $\wedge$ is a partial binary operation over $|A|$ and Δ is a distinguished token satisfying:

(i) $a \vdash \Delta$ for all $a \in |A|$.

(ii) $a \wedge b$ exists iff we have $c \vdash a$ & $c \vdash b$ for some $c \in |A|$ and in that case $c \vdash a \wedge b$.

Note that this is the same as "propositional languages" of Fourman and Grayson in [FG82]. We require that, in any given context, bounded complete information systems have their tokens from a common pool and have the *same* distinguished token Δ. A morphism or an approximable mapping $f : A \to B$ between two bounded complete information systems $A = (|A|, \vdash_A, \wedge_A)$ and $B = (|B|, \vdash_B, \wedge_B)$ is a relation $f \subseteq |A| \times |B|$ satisfying:

(i) $\Delta f \Delta$

(ii) afb & $afb' \Rightarrow af(b \wedge_B b')$

(iii) $a \vdash_A a'$ & $a'fb'$ & $b' \vdash_B b \Rightarrow afb$

The identity morphism on an object A is given by the approximable mapping $\vdash_A$. The category of bounded complete information systems with approximable mappings, denoted by **BC-ISys**, is equivalent to the category of bounded complete (Scott) domains and continuous functions. (Compare this with Gunter's category of pre-orders and approximable mappings in [Gun87].) We define the set of inclusion morphisms Inc as follows. The inclusion morphism from A to B exists and is equal to $\vdash_B \cap (|A| \times |B|)$ iff A is a substructure of B, i.e. $|A| \subseteq |B|$ and $\vdash_A$ and $\wedge_A$ are respectively the restrictions of $\vdash_B$ and $\wedge_B$ to $|A|$. It can now be easily verified that $(\mathbf{BC\text{-}ISsy}, \mathrm{Inc}, \subseteq, \{\Delta\})$ is an I-category. It is also complete; given an increasing chain of objects $\langle A_k \rangle_{k \geq 0}$, its lub is the object

$$\bigcup A_k = (\bigcup |A_k|, \bigcup \vdash_k, \bigcup \wedge_k),$$

and an increasing chain of morphisms $\langle (f_k : A_k \to B_k) \rangle_{k \geq 0}$ has lub

$$\bigcup f_k : \bigcup A_k \to \bigcup B_k,$$

where $\bigcup f_k$ is just the union of f_k's.

Consider the full subcategory, **BC-ISys***, of objects in which $\vdash$ is a partial order. Since every object of **BC-ISys** is isomorphic to its Lindenbaum algebra (i.e. the quotient of the structure under the equivalence induced by $\vdash$), which is an object of **BC-ISys***, it follows that these two categories are in fact equivalent. It can be easily shown that **BC-ISys*** is an ω-algebraic I-category with the nice feature that an object is compact if and only if its carrier set is finite, and compact morphisms are precisely the morphisms between compact objects. It is therefore convenient to work with **BC-ISys***; all the usual constructors on Scott domains have their counterparts as continuous endofunctors on this category.

5.3 Continuous bounded complete posets

We can use **BC-ISys** to construct a complete I-category for continuous bounded complete posets. It is well known that any continuous bounded complete poset is the projection of a bounded complete (Scott) domain. This means that the category of continuous bounded complete posets and (Scott) continuous functions is equivalent to a full subcategory, **CBC-ISys**, of the *Karoubi envelope* (see for example [LS86, Page 100]) of **BC-ISys**. In more detail, **CBC-ISys** has as objects pairs of the form (A, r), where $A = (|A|, \vdash_A, \wedge_A)$ is an object of **BC-ISys** and $r : A \to A$ is an approximable mapping with $r; r = r$ and $r \subseteq \vdash_A$, i.e. r is a projection. A morphism $f : (A, r) \to (B, s)$ of **CBC-ISys** is an approximable mapping $f : A \to B$ satisfying $r; f; s = f$. The identity morphism on (A, r) is given by r and hom-sets of **CBC-ISys** are ordered by the order induced from **BC-ISys** i.e. by inclusion. For convenience, we denote **CBC-ISys** by P in the rest of this paragraph. In order to define Inc_P, the class of inclusion morphisms of P, we put $(A, r) \trianglelefteq (B, s)$ iff $A \trianglelefteq B$ and $r \trianglelefteq^m s$, i.e. iff the diagram

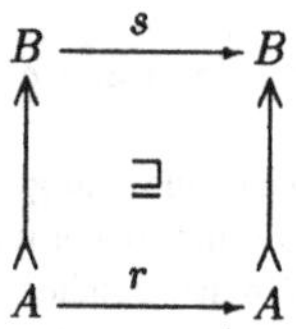

weakly commutes. When $(A, r) \trianglelefteq (B, s)$ holds, we define

$$\mathrm{in}((A, r), (B, s)) = r; \mathrm{in}(A, B); s,$$

which is clearly a morphism of P. It can be shown that $(P, \mathrm{Inc}_P, \subseteq, \Delta_P)$ is a complete I-category, where $\Delta_P = (\Delta, \mathrm{Id}(\Delta))$.

Any standard endofunctor F on **BC-ISys**, which preserves the order in each hom-set, gives rise to a standard endofunctor F' on **CBC-ISys** defined as follows. For an object (A, r), we have $F'(A, r) = (F(A), F(r))$ and for a morphism $f : (A, r) \to (B, s)$ we have $F'(f) = F(f)$. It can also be checked that F' is continuous if F is continuous. This means that all the usual functors on **BC-ISys** give rise to standard and continuous functors on **CBC-ISys** and that the initial algebra theorem can be used to solve domain equations in **CBC-ISys**.

5.4 Boolean algebras

Consider the category of Boolean algebras, $A = (|A|, \vee, \wedge, \neg)$, and Boolean homomorphisms, such that the elements of the algebras are drawn from a countable alphabet, and that all the algebras have the *same* greatest and least elements, 0 and 1 repectively. We denote this category by **Bool-ISys**, and let 2 denote the trivial Boolean algebra with elements 0 and 1. We find that $(\textbf{Bool-ISys}, \mathrm{Inc}, =, 2)$ is a concrete ω-algebraic I-category, where Inc is the set of Boolean inclusions and $=$ is the discrete partial order on hom-sets. Note that, as in the case of **BC-ISys**, the partial order $\trianglelefteq$ on objects induced by Boolean inclusions coincides with the model-theoretic notion of substructure, i.e.

$$(A, \vee_A, \wedge_A, \neg_A) \trianglelefteq (B, \vee_B, \wedge_B, \neg_B)$$

iff $|A| \subseteq |B|$ and the the operations of A are the restrictions of the operations of B to $|A|$.

In [Eda89] it is shown that a set of continuous endofunctors on **Bool-ISys** is given by:

$$F ::= ID \mid F_A \mid F_1 \times F_2 \mid F_1 + F_2 \mid F_V \mid F_1 \circ F_2$$

where ID, F_A, and $F_1 \circ F_2$ are the identity functor, the constant functor, and the composition of F_1 and F_2 respectively; whereas $- \times -$, $- + -$ and F_V are the product, the disjoint sum and the Vietoris functors (see [Smy83, Vic88]).

Note that, by Stone duality, the category **Bool-ISys** is dual to a category of Stone spaces and therefore solving a domain equation in either of these categories is equivalent to solving the dual equation in the other. These domain equations can be quite interesting; for example Abramsky has an example of such equations in his unpublished notes on non-well founded sets (see also [Abr88]). We can find the initial algebra of any endofunctor of the above type and all our construction can be made effective. As an example, we solve the domain equation

$$X = F(X) = 2 \times X$$

in **Bool-ISys**. We readily find that $D = \bigcup F^i(2)$ is the least solution (colimit). Writing the first few terms in the above sum, we get:

$F^0(2) = 2:$ 

$F(2) = 2 \times 2:$ 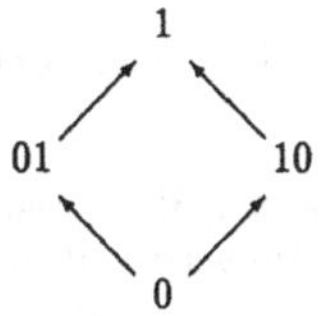

$F^2(2) = 2 \times (2 \times 2):$ 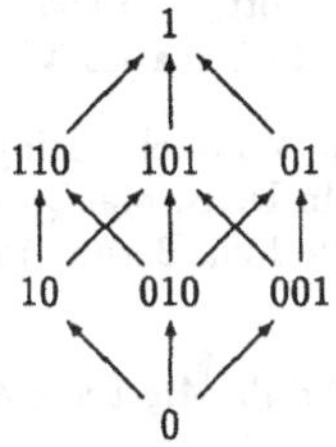

D will therefore consist of 0, 1 and all finite strings ending in 01 or 10. Atoms of D are precisely the strings of 0's ending in 10. These can therefore be identified with $\mathbb{N}$. However unlike $F^i(2), i \geq 0$, which can be identified with the nonnegative integers less than or equal to i, D is not atomic. The ultrafilter containing all the strings of 0's ending in 1 corresponds to the point at ∞, since every such string dominates all the atoms (integers) which have at least the same number of initial 0's. D is in fact the Boolean algebra associated with the final (greatest) solution of the dual equation $Y = 1 + Y$ in the category of Stone spaces, which is just the one point compactification of $\mathbb{N}$.

Acknowledgements

This work is one of the results of the SERC "Foundational Structures for Computer Science" at Imperial College, London; the first author was funded by that project. The first author would like to thank Samson Abramksy and Achim Jung for discussions on information systems and domain theory and for answering numerous questions on these subjects. Thanks are also due to Achim Jung and Mark Ryan for their technical assistance in producing this paper in LaTeX. Paul Taylor's diagram macros have been used in this paper.

References

[Abr88] S. Abramsky. A cooks tour of the finitary non-well founded sets (abstract). *EATCS Bulletin*, 36:233–234, 1988.

[Bar77] K. J. Barwise. An introduction to first order logic. In K. J. Barwise, editor, *The Handbook of Mathematical Logic*, Studies in Logic and Foundations of Mathematics, pages 5–46. North Holland, 1977.

[Cut80] N. J. Cutland. *Computability: An Introduction to Recursive function theory.* Cambridge University Press, 1980.

[Eda89] A. Edalat. Categories of information systems. Master's thesis, Imperial College, University of London, 1989.

[ES91] A. Edalat and M. B. Smyth. Categories of Information Systems. Technical Report Doc-91-21, Imperial College, London, 1991.

[FG82] M. P. Fourman and R. J. Grayson. Formal spaces. In A. S. Trolstra and D. van Dalen, editors, *The L.E.J.Brouwer Centenary Symposium*, pages 107–121. North Holland, 1982.

[Gun87] C. Gunter. Universal profinite domains. *Information and Computation*, 72(1):1–30, 1987.

[Joh82] P. T. Johnstone. *Stone Spaces*, volume 3 of *Cambridge Studies in Advanced Mathematics.* Cambridge University Press, Cambridge, 1982.

[LS86] J. Lambek and P. J. Scott. *Introduction to Higher Order Categorical Logic.* Cambridge Studies in Advanced Mathematics Vol. 7. Cambridge University Press, 1986.

[LW84] K. G. Larsen and G. Winskel. Using information systems to solve recursive domain equations effectively. In D. B. MacQueen G. Kahn and G. Plotkin, editors, *Semantics of Data Types*, pages 109–130, Berlin, 1984. Springer-Verlag. Lecture Notes in Computer Science Vol. 173.

[MA86] E. Manes and M. A. Arbib. *Algebraic Approaches to Program Semantics.* Springer-Verlag, 1986.

[Plo81] G. D. Plotkin. Post-graduate lecture notes in advanced domain theory (incorporating the "Pisa Notes"). Dept. of Computer Science, Univ. of Edinburgh, 1981.

[Sco82] D. S. Scott. Domains for denotational semantics. In M. Nielson and E. M. Schmidt, editors, *Automata, Languages and Programming: Proceedings 1982.* Springer-Verlag, Berlin, 1982. Lecture Notes in Computer Science **140**.

[Smy83] M. B. Smyth. Powerdomains and predicate transformers: a topological view. In J. Diaz, editor, *Automata, Languages and Programming*, pages 662–675, Berlin, 1983. Springer-Verlag. Lecture Notes in Computer Science Vol. 154.

[Vic88] S. J. Vickers. *Topology Via Logic.* Cambridge Tracts in Theoretical Computer Science. Cambridge University Press, 1988.

Collapsing graph models by preorders

Raymond Hoofman* & Harold Schellinx†

–

Department of Computer Science
Utrecht University,
Department of Mathematics and Computer Science
University of Amsterdam

Abstract

We present a strategy for obtaining extensional (partial) combinatory algebras by slightly modifying the well-known construction of graph models for the untyped lambda calculus. Using the notion of weak cartesian closed category an elegant interpretation of our construction in a category theoretical setting is given.

1 Introduction

A lattice L is called *reflexive* if it contains a copy of its own function space (i.e. the lattice $[L \to L]$ of (Scott-)continuous mappings $L \to L$). To be more precise, a lattice L is reflexive if there exist continuous mappings $F : L \longrightarrow [L \to L]$ and $G : [L \to L] \longrightarrow L$ such that $F \circ G = id_{[L \to L]}$.

For any infinite set X, the lattice $(\mathcal{P}(X), \subseteq)$ is reflexive: given some embedding $\langle\cdot,\cdot\rangle : X^{<\omega} \times X \hookrightarrow X$ (where $X^{<\omega}$ denotes the collection of *finite* subsets of X) one easily checks that the mappings F and G defined by $F(x)(y) := \{b \mid \exists \beta \subseteq y.\langle\beta, b\rangle \in x\}$ and $G(f) := \{\langle\beta, b\rangle \mid b \in f(\beta)\}$ are continuous and witness reflexivity. Structures $(\mathcal{P}(X), F, G)$ are called *graph models.* Well-known canonical examples are Engeler's $\mathbf{D}_A$ and the Scott/Plotkin-model $\mathcal{P}\omega$ (see also Schellinx(1991)).

Reflexive lattices L are natural models for the untyped lambda-calculus. In particular, through the mapping F, they define applicative structures $(L, \bullet)$ that are *combinatory algebras*: just define $a \bullet b$ (*'application of a to b'*) := $F(a)(b)$ (see e.g. chapter 5 of Barendregt(1984)).

*raymond@ruuinf.cs.ruu.nl

†harold@fwi.uva.nl

An applicative structure $(L, \bullet)$ is said to be *extensional* if, for all $a, b \in L$, we have that $\forall x.a \bullet x = b \bullet x$ holds if and only if $a = b$: i.e., each element of L *uniquely* represents a mapping $L \to L$. It is easy to show that reflexive lattices (L, F, G) induce extensional combinatory algebras $(L, \bullet)$ iff they are *strict reflexive*; i.e., the mappings F, G additionally have to satisfy $G \circ F = id_L$. Clearly the applicative structure $(\mathcal{P}(X), \bullet)$, obtained from a graph model is *not* extensional: we have $\{\langle\emptyset, b\rangle, \langle\{b\}, b\rangle\} \bullet x = \{\langle\emptyset, b\rangle\} \bullet x$ for all $x \in \mathcal{P}(X)$, while obviously $\{\langle\emptyset, b\rangle, \langle\{b\}, b\rangle\} \neq \{\langle\emptyset, b\rangle\}$. Therefore a graph model $(\mathcal{P}(X), F, G)$, though reflexive, never is *strict*: we can not have $G \circ F = id_{\mathcal{P}(X)}$.

In this paper we will study the process of extensionalising graph models at three increasing levels of abstraction. Each level provides a motivation to formulate the next higher level of abstraction, and, the other way round, each higher level provides in some sense an explanation why the construction at the preceding level works.

First, in section 2, we consider some concrete examples obtained by slighty generalizing Inge Bethke's extensionalisation of Engeler's $\mathbf{D}_A$-model (Bethke(1986)).

In section 3 we then abstract upon the general pattern of these examples: given a graph model $\mathcal{P}(X)$ and a preorder $\preceq$ on X, we first define the *collaps* $M_\preceq$ of $\mathcal{P}(X)$ as a quotient of $\mathcal{P}(X)$ by $\preceq$. It turns out that $M_\preceq$ is a complete lattice, and we formulate sufficient and necessary conditions on $\preceq$ for $M_\preceq$ to be strict reflexive. It is then easy to see that the examples of section 2 satisfy these conditions.

In section 4 we provide an even more general *categorical* view on the process of extensionalising graph models by placing it in the framework of extensionalising categorical models of the *typed* lambda caculus. We define a non-extensional model of the *untyped* lambda calculus as a strict reflexive object 'living' in a non-extensional model of the *typed* lambda calculus (i.e. in a weak cartesian closed category (Hayashi(1985))). Analogously, an extensional model of the untyped lambda calculus is a strict reflexive object in an extensional model of the typed lambda calculus (i.e. in a cartesian closed category). Because the *canonical* way of extensionalising a weak cartesian closed category $\mathcal{C}$ is to take the *Karoubi envelope* K of $\mathcal{C}$, an obvious method for extensionalising a model $C \in \mathcal{C}$ of the untyped lambda calculus is transferring it to $\mathrm{K}(\mathcal{C})$. As it turns out this involves finding an *idempotent* arrow $f : C \to C$ (i.e. $f \circ f = f$) with certain properties.

In section 5 we show that the construction of section 3 is an instance of the categorical extensionalisation method. We define the wCCC **POW** of powersets and continuous functions and note that graph models are strict reflexive objects in **POW**. Then we show that preorders on graph models give rise to idempotents in **POW**, and that the requirements of section 3 on a preorder coincide with the requirements of section 4 on the corresponding idempotent. This result explains why, by means of preorders, we are able to extensionalise graph models and, moreover, indicates that this method of extensionalising graph models by preorders is, in some sense, canonical. Because the extensionalising **K(POW)** of **POW** is equivalent to the category of continuous complete lattices and continuous functions (Hoofman(1990/1)), extensionalising graph models by preorders is a method of transferring graph models from **POW** to this category of complete lattices.

In section 6 much of the technical work is done using the category **RL** which is isomorphic to **POW**. It turns out that preorders on a graph model are a special case of the more general notion of an *entailment relation* on a graph model (Scott(1982), Hoofman(1990/1)).

Finally, in section 7 we describe the extensionalising of *partial* graph models as an interesting modification of the extensionalising of ordinary graph models.

2 Extensionalising graph models: examples

We will start by briefly describing some examples, after which we will exhibit the general pattern that is behind these.

Let us first review the definition of Engeler's graph model $\mathbf{D}_A$.

2.1. DEFINITION. Let A be any non-empty set and put:

$$\begin{aligned} G_0(A) &:= A \\ G_{n+1}(A) &:= G_n(A) \cup (G_n(A)^{<\omega} \times G_n(A)) \\ G(A) &:= \bigcup_{n \in \mathbb{N}} G_n(A). \end{aligned}$$

(So $G(A)$ is the smallest set $X \supset A$ such that for all finite $\beta \subseteq X$ and $b \in X$ we have that $(\beta, b) \in X$. The application-defining embedding is just the identity.)

$\mathbf{D}_A$ will denote the graph model $(\mathcal{P}(G(A)), \bullet)$. □

Inge Bethke in Bethke(1986) showed how one may obtain an extensional combinatory algebra by slightly redefining the notion of application on $\mathcal{P}(G(A))$ and taking a quotient with respect to a certain equivalence relation. For this redefined notion of application as well as for the equivalence relation used, the definition of a preorder that extends the set inclusion relation on $\mathcal{P}(G(A))$ is crucial. The following generalizes this definition and gives us a whole collection of such preorders.

2.2. DEFINITION. Let A be any non-empty set, $G(A)$ as in the definition of $\mathbf{D}_A$. Suppose $f : A \to A$ is a 1-1 mapping, $\epsilon : A \to A^{<\omega}$ an arbitrary mapping. Define, for $x, y \in \mathcal{P}(G(A))$,

$$x \sqsubseteq_{f\epsilon} y \quad \textit{iff} \quad \forall a \in x \exists b \in y . a \preceq_{f\epsilon} b,$$

where $a \preceq_{f\epsilon} b$ holds iff either

(1) $a = b$, or
(2) $\exists\beta\exists c \quad (a = (\beta, c) \wedge b \in A \wedge c \preceq_{f\epsilon} f(b) \wedge \epsilon_b \sqsubseteq_{f\epsilon} \beta)$, or
(3) $\exists\alpha\exists c \quad (a \in A \wedge b = (\alpha, c) \wedge f(a) \preceq_{f\epsilon} c \wedge \alpha \sqsubseteq_{f\epsilon} \epsilon_a)$, or
(4) $\exists\alpha\exists\beta\exists c\exists d \quad (a = (\alpha, c) \wedge b = (\beta, d) \wedge \beta \sqsubseteq_{f\epsilon} \alpha \wedge c \preceq_{f\epsilon} d)$. □

2.3. REMARK. Given A, f, ϵ as above, $\preceq_{f\epsilon}$ is well-defined, in the sense that for all n and $a, b \in G_n(A)$, $a \preceq_{f\epsilon} b$ is defined in terms of the restriction of $\preceq_{f\epsilon}$ to $\bigcup_{m<n} G_m(A)$. For this we need that the range of f is contained in A and that the range of ϵ is contained in $A^{<\omega}$. □

2.4. Lemma. *Given A, f, ϵ again as above*

- *(i)* $\forall a \in A \quad \forall b. \quad a \preceq_{f\epsilon} b \Longleftrightarrow (\epsilon_a, f(a)) \preceq_{f\epsilon} b;$
- *(ii)* $\forall a \in A \quad \forall b. \quad b \preceq_{f\epsilon} a \Longleftrightarrow b \preceq_{f\epsilon} (\epsilon_a, f(a));$
- *(iii)* $\preceq_{f\epsilon}$ *is transitive.*

Proof: For this we need the injectivity of f. The proof is similar to that of proposition 1.3 in Bethke(1986). □

Now using the procedure described in Bethke(1986), via these preorders we obtain extensional combinatory algebras (namely strict reflexive lattices $\left(\mathcal{P}(X)/\equiv_{f\epsilon}\right)$, where $\equiv_{f\epsilon}$ is the canonical equivalence relation on $\mathcal{P}(G(A))$ obtained from $\sqsubseteq_{f\epsilon}$), which we will denote by $\mathbf{M}(A, f, \epsilon)$. The models $M(A)$ defined by Bethke are precisely the structures $\mathbf{M}(A, id, \emptyset)$ we get by taking f to be the identity mapping on A and ϵ the constant mapping $\epsilon_a = \emptyset$, for all $a \in A$.

It is not too difficult to show that, generally, different preorders will give rise to different (i.e. *non-isomorphic*) extensional combinatory algebras. For more details on this we refer to our technical report (Hoofman & Schellinx(1991)).

In fact the procedure turns out to be not at all limited to graph models of the $\mathbf{D}_A$-type. It works just as well for e.g. the Plotkin-Scott model $\mathcal{P}\omega$, where an embedding $\langle\cdot,\cdot\rangle : \mathbb{N}^{<\omega} \times \mathbb{N} \hookrightarrow \mathbb{N}$ is given by means of a bijective coding $p : \mathbb{N} \times \mathbb{N} \hookrightarrow \mathbb{N}$ of pairs of natural numbers as natural numbers and a bijective coding $e : \mathbb{N} \hookrightarrow \mathbb{N}^{<\omega}$ of finite sets of natural numbers by natural numbers, as follows:

$$\langle \alpha, m \rangle = p(e^{-1}(\alpha), m).$$

The codings p and e are given by:

for all $n, m \in \mathbb{N} : p(n, m) = \frac{1}{2}(n+m)(n+m+1) + m;$

for all $n \in \mathbb{N} : e(n) = \{k_0, k_1, \ldots, k_{m-1}\}$ iff $n = \sum_{i<m} 2^{k_i} \quad (k_i \neq k_j \text{ if } i \neq j);$

$e(0) = \emptyset.$

We obtain a suitable preorder by defining, for $x, y \in \mathcal{P}(\mathbb{N})$,

$$x \sqsubseteq y \quad \text{iff} \quad \forall a \in x \exists b \in y . a \preceq b,$$

where $a \preceq b$ iff either

- (i) $a = b$, or
- (ii) $\exists n_1 \exists n_2 \exists m_1 \exists m_2 \quad (a = p(n_1, m_1) \wedge b = p(n_2, m_2) \wedge e(n_2) \sqsubseteq e(n_1) \wedge m_1 \preceq m_2)$.

By easy induction arguments one may show that $\preceq$ on $\mathbb{N}$ is well-defined (in the sense that for all n, all $a, b \leq n$, $a \preceq b$ is defined in terms of $\preceq$ restricted to $\{k \mid k < n\}$) and transitive. (For details see Hoofman & Schellinx (1991).) Then, again as in Bethke(1986), via this preorder we obtain an extensional combinatory algebra $\mathbf{P}\omega$.

Clearly these examples indicate that there is something of a general pattern here, and indeed there is.

3 Extensionalising graph models: general construction and some remarks

Take any non-empty set X and a mapping $\langle\cdot,\cdot\rangle : X^{<\omega} \times X \longrightarrow X$. Next fix some reflexive, transitive $\preceq$ on X, making $(X, \preceq)$ a preorder.

Then

- (i) $\preceq$ induces an extension ε of the membership relation $\in$ between elements of X and subsets of X:
$$a \,\varepsilon\, x \quad \text{iff} \quad \exists a' \in x . a \preceq a';$$
- (ii) ε induces an extension $\sqsubseteq$ of the set inclusion relation $\subseteq$ on $\mathcal{P}(X)$:
$$x \sqsubseteq y \quad \text{iff} \quad \forall a \,\varepsilon\, x . a \,\varepsilon\, y \qquad (\text{iff} \quad \forall a \in x . a \,\varepsilon\, y);$$
- (iii) $\sqsubseteq$ induces an equivalence relation $\equiv$ on $\mathcal{P}(X)$:
$$x \equiv y \quad \text{iff} \quad x \sqsubseteq y \sqsubseteq x.$$

Now let $[x]$ be the $\equiv$-equivalence class of x; let $[x] \leq [y]$ iff $x \sqsubseteq y$; then $[x] = [y]$ iff $x \equiv y$. (Sometimes we will write $a \,\varepsilon\, [x]$, meaning $a \,\varepsilon\, y$ for some $y \equiv x$; as $y \equiv x$ iff $a \,\varepsilon\, y \leftrightarrow a \,\varepsilon\, x$, this is harmless and often facilitates notation.)

Define the collaps $M_\preceq$ of $\mathcal{P}(X)$ by $\preceq$ as $\left(\mathcal{P}(X) \,/\, \equiv , \leq \right)$.

3.1. Proposition. *$M_\preceq$ is a complete lattice with bottom $[\emptyset]$ and supremum* $\sup A = [\bigcup\{y \mid [y] \in A\}]$ *for all $A \subseteq M_\preceq$.*

Proof: Obvious from the definition of $\leq$ in terms of $\sqsubseteq$ and the fact that $\sqsubseteq$ extends $\subseteq$ and has the property that for any $D \subseteq \mathcal{P}(X)$, if $\forall y \in D, y \sqsubseteq z$, then $\bigcup D \sqsubseteq z$. □

Now, following the construction of the graph model $(\mathcal{P}(X), F, G)$, we define a mapping $G' : [M_\preceq \to M_\preceq] \longrightarrow M_\preceq$ by
$$G'(f) := [\{\langle \gamma, c\rangle \mid c \,\varepsilon\, f([\gamma])\}],$$
and $F' : M_\preceq \to M_\preceq^{M_\preceq}$ (the set of *all* mappings $M_\preceq \to M_\preceq$) by
$$F'([x])([y]) := [\{b \mid \exists \beta \sqsubseteq y . \langle \beta, b\rangle \,\varepsilon\, x\}].$$

3.2. Proposition. *(i)* $F' \in [M_\preceq \longrightarrow [M_\preceq \to M_\preceq]]$;
(ii) $G' \in [[M_\preceq \to M_\preceq] \longrightarrow M_\preceq]$.

Proof: Easy. Details may be found in Hoofman & Schellinx(1991). □

Now the following holds:

3.3. Theorem. *$\mathcal{M}_{\preceq} := (M_{\preceq}, F', G')$ is strict reflexive if and only if the preorder $(X, \preceq)$ satisfies*

- *(i) $\langle \beta, b\rangle \preceq \langle \alpha, a\rangle$ iff $\alpha \sqsubseteq \beta$ and $b \preceq a$;*
- *(ii) for all d : $\exists \beta \in X^{<\omega} \exists b \in X . d \preceq \langle \beta, b\rangle \preceq d$.*

Proof: ($\Leftarrow$) *reflexivity:* $F' \circ G' = id_{[M_{\preceq} \to M_{\preceq}]}$.

For, $(F'(G'(f)))([x]) =$

$$=\Big[\{b \mid \exists \beta \sqsubseteq x . \langle \beta, b\rangle \mathrel{\scriptstyle\mathrm{B}} \{\langle \gamma, c\rangle \mid c \mathrel{\scriptstyle\mathrm{B}} f([\gamma])\}\}\Big]$$

$$=\Big[\{b \mid \exists \gamma \sqsubseteq x . b \mathrel{\scriptstyle\mathrm{B}} f([\gamma])\}\Big] \qquad \text{(by (i)}\rightarrow\text{)}$$

$$=\Big[\bigcup\{\{b \mid b \mathrel{\scriptstyle\mathrm{B}} f([\gamma])\} \mid \gamma \sqsubseteq x\}\Big]$$

$$= \sup_{[\gamma] \leq [x]} f([\gamma]) = f([x]) \qquad \text{(by continuity of } f\text{)}.$$

strictness: $G' \circ F' = id_{M_{\preceq}}$.

For, $(G'(F'([x])) =$

$$=\Big[\{\langle \gamma, c\rangle \mid c \mathrel{\scriptstyle\mathrm{B}} \{b \mid \exists \beta \sqsubseteq \gamma . \langle \beta, b\rangle \mathrel{\scriptstyle\mathrm{B}} x\}\}\Big]$$

$$=\Big[\{\langle \gamma, c\rangle \mid \langle \gamma, c\rangle \mathrel{\scriptstyle\mathrm{B}} x\}\Big] \qquad \text{(by (i)}\leftarrow\text{)}$$

$$= [x] \qquad \text{(by (ii))}.$$

($\Rightarrow$) Let $\mathcal{M}_{\preceq}$ be strict reflexive. By extensionality of the induced applicative structure we have, for all x, $\{\langle \gamma, c\rangle \mid \langle \gamma, c\rangle \mathrel{\scriptstyle\mathrm{B}} x\} \equiv x$. Now take any d and put $x = \{d\}$. Then clearly we find $\exists \gamma \exists c . d \preceq \langle \gamma, c\rangle$, and $\langle \gamma, c\rangle \mathrel{\scriptstyle\mathrm{B}} x$, so $\langle \gamma, c\rangle \preceq d$. This proofs (ii).

By reflexivity the equality $F' \circ G' = id_{[M_{\preceq} \to M_{\preceq}]}$ is valid. So, from the transition above marked "(by (i)$\rightarrow$)",

$$\forall x \forall f \forall b \Big(\exists \beta \sqsubseteq x \exists \gamma \exists c (\langle \beta, b\rangle \preceq \langle \gamma, c\rangle \wedge c \mathrel{\scriptstyle\mathrm{B}} f([\gamma]))$$

$$\rightarrow \exists \delta \sqsubseteq x \exists d (d \mathrel{\scriptstyle\mathrm{B}} f([\delta]) \wedge b \preceq d)\Big). \qquad (\star)$$

In order to prove (i)$\rightarrow$, suppose $\langle \mu, m\rangle \preceq \langle \nu, n\rangle$. We define a mapping f_n by putting $f_n([x]) = [\{n\}]$, for all x. Obviously f_n is continuous. By taking $x = X$ and $f = f_n$ in $(\star)$ we have

$$\forall b \Big(\exists \beta \exists \gamma \exists c (\langle \beta, b\rangle \preceq \langle \gamma, c\rangle \wedge c \preceq n) \rightarrow \exists d (d \preceq n \wedge b \preceq d)\Big).$$

As $\langle \mu, m\rangle \preceq \langle \nu, n\rangle$ and $n \preceq n$, putting $b = m$, we conclude: $\exists d . d \preceq n \wedge m \preceq d$. So $m \preceq n$ by transitivity of $\preceq$.

To prove that also $\nu \sqsubseteq \mu$, we define a mapping $f_{n\mu}$ as follows:

$$f_{n\mu}([x]) = \begin{cases} [\emptyset], & \text{if } x \sqsubseteq \mu; \\ [\{\mathrm{n}\}], & \text{otherwise.} \end{cases}$$

It is easy to check continuity of $f_{n\mu}$. Now take $x = \mu$ and $f = f_{n\mu}$ in ($\star$). Then

$$\forall b\big(\exists\beta \sqsubseteq \mu\exists\gamma\exists c(\langle\beta, b\rangle \preceq \langle\gamma, c\rangle \wedge c \,\mathrm{E}\, f_{n\mu}([\gamma]))$$

$$\rightarrow \exists\delta \sqsubseteq \mu\exists d(d \,\mathrm{E}\, f_{n\mu}([\delta]) \wedge b \preceq d)\big).$$

Suppose $\nu \not\sqsubseteq \mu$. In that case $f_{n\mu}([\nu]) = [\{n\}]$, so $n \,\mathrm{E}\, f_{n\mu}([\nu])$. As $\langle\mu, m\rangle \preceq \langle\nu, n\rangle$ we conclude

$$\exists\delta \sqsubseteq \mu\exists d(d \,\mathrm{E}\, f_{n\mu}([\delta]) \wedge m \preceq d).$$

But $\delta \sqsubseteq \mu$ implies $f_{n\mu}([\delta]) = [\emptyset]$ by definition of $f_{n\mu}$. Contradiction. Therefore $\nu \sqsubseteq \mu$, finishing the proof (i)$\rightarrow$.

Finally, as by strictness the equality $G' \circ F' = id_{M_\preceq}$ is valid, we find (from the transition above marked "(by (i)$\leftarrow$)") that

$$\forall x\forall\gamma\forall c\big(\exists\beta\exists b(\beta \sqsubseteq \gamma \wedge \langle\beta, b\rangle \,\mathrm{E}\, x \wedge c \preceq b) \rightarrow \langle\gamma, c\rangle \,\mathrm{E}\, x\big).$$

Taking $x = \{\langle\alpha, a\rangle\}$ we find

$$\forall\gamma\forall c.\alpha \sqsubseteq \gamma \wedge c \preceq a \rightarrow \langle\gamma, c\rangle \preceq \langle\alpha, a\rangle.$$

This proves (i)$\leftarrow$, and ends the proof of our theorem. □

As the above shows us, it is not necessary to start from an *embedding* $\langle\cdot,\cdot\rangle : X^{<\omega} \times X \hookrightarrow X$. Any mapping will do. In fact, from the proof of theorem 3.3 it follows that we also have the following

3.4. PROPOSITION. *$\mathcal{M}_\preceq := (M_\preceq, F', G')$ is reflexive iff the preorder $(X, \preceq)$ satisfies*

$$\langle\beta, b\rangle \preceq \langle\alpha, a\rangle \quad \Rightarrow \quad \alpha \sqsubseteq \beta \text{ and } b \preceq a. \qquad \square$$

Taking equality ($=$) for $\preceq$, this shows us that we recover our 'plain' graph models precisely in case the mapping $\langle\cdot,\cdot\rangle : X^{<\omega} \times X \rightarrow X$ we start from is an embedding.

We should note that, independent from our work, the same result (theorem 3.3) has been obtained by Jean-Louis Krivine (see Krivine(1990)). Krivine moreover observes that for surjective embeddings there always exists a preorder satisfying the conditions of the theorem. In fact it is not difficult to extend his argument to non-surjective embeddings. We therefore can state the following

3.5. Proposition. *Every graph model can be collapsed by a preorder.*
Proof: Krivine(1990), page 106/107. □

The procedure outlined shows us in very general terms *why* Inge Bethke's extensionalisation of $\mathbf{D}_A$ works: clearly the preorder she defined, as well as the preorders in the examples of section 2, satisfy the conditions of theorem 3.3. Also we see that the method is not limited to these examples: to *any* graph model $\mathcal{M} := (\mathcal{P}(X), \bullet)$ (which is fully determined by the embedding $\langle \cdot, \cdot \rangle : X^{<\omega} \times X \hookrightarrow X$ defining the application-function $\bullet$) we may associate a (non-empty) collection $\mathcal{E}(\mathcal{M})$ of extensional combinatory algebras (fully determined by the collection of preorders $\preceq$ on X that satisfy the necessary conditions).

Observe that any element of $\mathcal{E}(\mathcal{M})$ can be isomorphically embedded in $\mathcal{M}$: let $\left(\mathcal{P}(X)/_{\equiv}, \star\right) \in \mathcal{M}_e$ and define $\varphi : \left(\mathcal{P}(X)/_{\equiv}, \star\right) \hookrightarrow (\mathcal{P}(X), \bullet)$ by $\varphi([a]) = \bigcup\{b \mid [b] = [a]\}$. One easily checks that φ is 1 - 1 and moreover $\varphi([a]) \bullet \varphi([a']) = \varphi([a] \star [a'])$.

The cardinality of the collection $\mathcal{E}(\mathcal{M})$ depends on the graphmodel $\mathcal{M}$ at hand: with a graph model $\mathbf{D}_A$ we can associate many non-isomorphic extensionalisations; on the other hand there are also graph models $\mathcal{M}$ for which $\mathcal{E}(\mathcal{M})$ is a singleton (e.g. for the Plotkin/Scott-model $\mathcal{P}\omega$ one may show that the preorder defined in section 2 is the *unique* preorder satisfying the necessary and sufficient conditions, so $\mathcal{E}(\mathcal{P}\omega) = \{\mathbf{P}\omega\}$.)

Let us finally note that, obviously, isomorphic graph models $\mathcal{G}_1 \cong \mathcal{G}_2$ will give rise to isomorphic extensionalisations $\mathcal{E}(\mathcal{G}_1) \cong \mathcal{E}(\mathcal{G}_2)$ (in the sense that for all $A \in \mathcal{E}(\mathcal{G}_i)$ there exists a $B \in \mathcal{E}(\mathcal{G}_j)$ such that $A \cong B$). Conversely though, we may have $A \in \mathcal{E}(\mathcal{G}), B \in \mathcal{E}(\mathcal{H})$ such that $A \cong B$, but $\mathcal{G} \ncong \mathcal{H}$. As $\mathbf{D}_{\{a\}} \ncong \mathcal{P}\omega$ (see e.g. Schellinx(1991)), the following gives an example.

3.6. Proposition. $\mathbf{P}\omega \cong \mathbf{M}(\{a\}, id, \emptyset)$.
Proof: Hoofman & Schellinx(1991), proposition 3.14 □

4 A categorical view on extensionalisation

As is well known, cartesian closed categories (CCC) are models of the extensional typed lambda calculus (Scott(1982)). In Hayashi(1985) categorical models of the *non-extensional* typed lambda calculus, called *weak* cartesian closed categories (wCCC), are defined. We recall his algebraic definition of wCCC (see also Martini(1987)).

4.1. Definition. A *weak cartesian closed category* is a category $\mathcal{C}$ with a terminal object $\mathbf{1}$ and binary products $A \times B$, and with the following data:

- For each pair of objects $A, B \in \mathcal{C}$ an object $A \Rightarrow B \in \mathcal{C}$, and an arrow $\mathbf{ev}_{A,B} \in \mathcal{C}((A \Rightarrow B) \times A, B)$. Furthermore, for each arrow $f \in \mathcal{C}(D \times A, B)$ an arrow $\mathbf{cur}(f) \in \mathcal{C}(D, A \Rightarrow B)$.

satisfying the following equations (omitting subscripts):

1. $\mathbf{ev} \circ (\mathbf{cur}(f) \times id) = f$

2. $\mathbf{cur}(f \circ (g \times id)) = \mathbf{cur}(f) \circ g$. □

Hayashi also gives an alternative, but equivalent, definition of wCCC, based on the notion of a *semi-functor*, i.e. a mapping between categories having the same properties as a functor, except that it need not preserve identities.

Intuitively, a wCCC may be conceived of as a CCC in which the η-rule $\mathbf{cur}(\mathbf{ev} \circ (f \times id)) = f$ does not necessarily hold. As a consequence, in wCCC's there is in general only a *retraction* instead of an *isomorphism* between $\mathcal{C}(A, B)$ and the set of 'points' of $A \Rightarrow B$. Given an arrow $f : A \to B$ there is a point $\phi(f) : \mathbf{1} \to (A \Rightarrow B)$ of $A \Rightarrow B$ defined by $\phi(f) := \mathbf{cur}(f \circ \pi'_{\mathbf{1},A})$, where π' denotes projection on the second coordinate.

The other way round, for each point $x : \mathbf{1} \to (A \Rightarrow B)$ there is an arrow $\psi(x) : A \to B$ defined by $\psi(x) := \mathbf{ev} \circ (x \times id_A)\langle t_A, id_A \rangle$, where t_A denotes the unique arrow $A \to \mathbf{1}$. By the first requirement in the definition of wCCC it follows that $\psi\phi(f) = f$ for an arrow $f : A \to B$, but because the η-rule does not hold, we generally can not infer that $\phi\psi(x) = x$ for a point x of $A \Rightarrow B$. Hence for each arrow $f \in \mathcal{C}(A, B)$ there may be many *representing* points x in $A \Rightarrow B$, i.e. points x such that $\phi(x) = f$. Intuitively, the points of $A \Rightarrow B$ may be conceived of as *programs* and arrows $A \to B$ as *functions*. It is then quite natural that for each function f there are several representing programs, i.e. programs that compute the function f.

Note that by the second requirement in the definition of wCCC we have $\mathbf{cur}(\mathbf{ev} \circ (f \times id)) = \mathbf{cur}(\mathbf{ev}) \circ f$. Hence the η-rule holds in a wCCC iff $\mathbf{cur}(\mathbf{ev}) = id$. It follows that a wCCC is in fact a CCC iff the above identity holds.

A very simple example of a wCCC to be defined in the next section is the category **POW** of powersets and continuous functions. In fact, although **POW** is not a CCC (exercise!), several weak exponents may be defined on **POW**. This is a general property: a category may admit several different (i.e. non-isomorphic) weak exponents, whereas exponents on the same category are always isomorphic. Note that the category **POW** contains all graph-models.

As shown in Hayashi(1985), it is possible to *extensionalise* wCCC's, i.e. to transform wCCC's in CCC's. This is done by means of the *Karoubi envelope* construction, of which we recall the definition.

4.2. Definition. Given a category $\mathcal{A}$, the *Karoubi envelope* of $\mathcal{A}$ is the category $\mathrm{K}(\mathcal{A})$ having as objects pairs (A, f), where A is an object from $\mathcal{A}$ and $f \in \mathcal{A}(A, A)$ an idempotent $\mathcal{A}$-arrow, i.e. $f \circ f = f$. Morphisms $(A, f) \to (B, g)$ in $\mathrm{K}(\mathcal{A})$ are arrows $h : A \to B$ in $\mathcal{A}$ such that $g \circ h \circ f = h$ (or equivalently $g \circ h = h$ and $h \circ f = h$). Composition of arrows is composition in $\mathcal{A}$; $id_{(A,f)} = f$.

As any idempotent arrow f uniquely determines its target and source A, we will often identify an object (A, f) in $\mathrm{K}(\mathcal{A})$ with its 'arrow part' f. □

4.3. Theorem. *If $\mathcal{C}$ is a wCCC, then $\mathrm{K}(\mathcal{C})$ is a CCC.*

Proof: See Hayashi(1985). In particular, for objects $f, g \in \mathrm{K}(\mathcal{C})$, $f \times g$ and $f \Rightarrow g$ in $\mathrm{K}(\mathcal{C})$ are equal to the application of the arrow-part of the functors $\times, \Rightarrow$ on $\mathcal{C}$ to f, g. □

In Hoofman(1991,1992), it is shown that the Karoubi envelope is a *canonical* way of extensionalising wCCC's (the Karoubi envelope is shown to be the canonical way to transform *semi-notions*, i.e. categorical notions based on *semi*-functors rather than on functors, to the corresponding ordinary categorical notions).

As it turns out, taking the Karoubi envelope of **POW** (or rather of the equivalent category **RL**, see section 6) boils down to equipping all objects with some sort of *entailment relation*. Hence the objects of K(**POW**) look like the information systems of Scott(1982), and are in fact *continuous information systems* as defined in Hoofman(1990/1). By the results of this last paper it follows that K(**POW**) is equivalent to the CCC of *continuous complete lattices* and continuous functions.

Let us now turn our attention to models of the *untyped* lambda calculus.

4.4. DEFINITION. Let $\mathcal{C}$ be a wCCC. An object $C \in \mathcal{C}$ is called *strict reflexive* iff $C \cong (C \Rightarrow C)$. □.

A strict reflexive object in a CCC is a model of the *extensional* untyped lambda calculus. More general, a strict reflexive object in a wCCC is a model of the *non-extensional* untyped lambda calculus [1]. Note that this is a fully *uniform* definition of models of the untyped lambda calculus in the sense that in the extensional as well as in the non-extensional case models are given by an *isomorphism* between an object and its (weak) exponent. The usual approach is to require an *isomorphism* in a CCC between an object and its function-space for a model of the extensional untyped lambda calculus, and only a *retraction* for a model of the non-extensional untyped lambda calculus. Hence in the ordinary approach the requirements on the object underlying a model differs in the two cases, while the environment (CCC) is kept fixed. In the wCCC-approach on the other hand, the environment of the model varies. A model of the extensional untyped lambda calculus is a strict reflexive object living in an *extensional environment* (i.e. a CCC), whereas a model of the non-extensional untyped lambda calculus is a strict reflexive object living in a *non-extensional environment* (i.e. a wCCC). Of course, if $j : C \cong (C \Rightarrow C)$ is a strict reflexive object in a CCC then there is an isomorphism ϕ' between $\mathcal{C}(C, C)$ and points of C defined by $\phi'(f) = j^{-1} \circ \phi(f)$ with inverse $\psi'(x) = \psi(j \circ x)$. For a strict reflexive object in a wCCC this only defines a retraction.

An example of a strict reflexive object in **POW** is the Plotkin-Scott model $\mathcal{P}\omega$. In **POW** we may define $\mathcal{P}\omega$ by means of an *isomorphism* $\mathcal{P}\omega \cong (\mathcal{P}\omega \Rightarrow \mathcal{P}\omega)$ instead of via a *retraction* in the category of lattices and continuous functions as in the usual approach. And in fact *every* graph model is easily seen to be strict reflexive in **POW**.

An obvious strategy for extensionalising a model $C \in \mathcal{C}$ of the non-extensional untyped lambda calculus is transferring the model from the non-extensional environment $\mathcal{C}$ to the extensional environment K($\mathcal{C}$). First, because the objects of K($\mathcal{C}$) are idempotents in $\mathcal{C}$, we should find an idempotent arrow $f : C \to C$ to represent C in K($\mathcal{C}$). Second, if

[1] In the precise technical sense this holds only for categories with '*enough points*' (see Barendregt(1982), 5.5). The category **POW** satisfies this demand. Also it is easy to see that if a category $\mathcal{C}$ has enough points, then so does K($\mathcal{C}$).

$j : C \cong (C \Rightarrow C)$ in C, then it is clear that

$$(f \Rightarrow f) \circ j \circ f : f \to (f \Rightarrow f)$$

and

$$f \circ j^{-1} \circ (f \Rightarrow f) : (f \Rightarrow f) \to f$$

are arrows in $\mathbf{K}(C)$. However, in general these arrows need not determine an isomorphism $f \cong (f \Rightarrow f)$ in $\mathbf{K}(C)$.

4.5. PROPOSITION. *If* $j : C \to (C \Rightarrow C)$ *and* $i : (C \Rightarrow C) \to C$ *are arrows in* C *and* $f : C \to C$ *is an idempotent in* C, *then the arrows* $(f \Rightarrow f)jf : f \to (f \Rightarrow f)$ *and* $fi(f \Rightarrow f) : (f \Rightarrow f) \to f$ *are inverses in* $\mathbf{K}(C)$ *iff*

1. $fi(f \Rightarrow f)jf = f$
2. $(f \Rightarrow f)jfi(f \Rightarrow f) = (f \Rightarrow f)$

PROOF: Easy. □

Hence a strict reflexive object C (or more general an object C with arrows $j : C \to (C \Rightarrow C)$ and $i : (C \Rightarrow C) \to C$) in a wCCC, can be transformed into a strict reflexive object in a CCC if we can find an idempotent satisfying the requirements of the theorem. Note that such an idempotent need not always exist.

In the next section we will see that preorders on graph models may be conceived of as idempotents in the wCCC **POW**. Furthermore, we will show that in this special case the requirements of the previous theorem on an idempotent f coincide exactly with the requirements of theorem 3.3 on a preorder. Hence the choice of a preorder on a graph model as in theorem 3.3 is a special case of the choice of an idempotent on a strict reflexive object in a wCCC as in the previous theorem, and the concrete construction of the previous section is a special case of the more general approach described in this section.

We think that this observation is interesting for two reasons. First, it explains *why* it is possible to extensionalise graph models by preorders. It places the extensionalising of models of the *untyped* lambda calculus, in particular the concrete constructions of the previous sections, in the more general and *well-known* frame work of the extensionalising of models of the *typed* lambda calculus. Second, it shows that the concrete constructions based on preorders, such as the constructions of the first section, are, in a certain sense, *canonical* ways of extensionalising graph models, just like the Karoubi envelope is a canonical way to extensionalise wCCC's.

5 The category POW

POW will denote the category having as objects all *powersets*, i.e. all sets of the form $\mathcal{P}(X)$, where X is a set; morphisms are all functions continuous with respect to the Scott-topology on the lattices $(\mathcal{P}(X), \subseteq)$. It is easy to see that **POW** is a wCCC:

- $\mathcal{P}(\emptyset) = \{\emptyset\}$ is terminal object in **POW**.
- Writing $X \uplus Y$ for the disjoint union of the sets X and Y, one easily checks that **POW** has binary products $\mathcal{P}(X) \times \mathcal{P}(Y) := \mathcal{P}(X \uplus Y)$.
- For the semi-exponents, we put $(\mathcal{P}(X) \Rightarrow \mathcal{P}(Y)) := \mathcal{P}(X^{<\omega} \times Y)$.

Let $F \in \mathcal{P}(X^{<\omega} \times Y)$. We define an evaluation function $\mathbf{ev} : (\mathcal{P}(X) \Rightarrow \mathcal{P}(Y)) \times \mathcal{P}(X) \longrightarrow \mathcal{P}(Y)$ by

$$\mathbf{ev}(F, a) := \{c \mid \exists \gamma \subseteq a.(\gamma, c) \in F\}.$$

For $f : \mathcal{P}(X) \times \mathcal{P}(Y) \rightarrow \mathcal{P}(Z)$ define $\mathbf{cur}(f) : \mathcal{P}(X) \rightarrow \mathcal{P}(Y^{<\omega} \times Z)$ by

$$\mathbf{cur}(f)(b) := \{(\gamma, c) \mid c \in f(b, \gamma)\}.$$

Clearly **ev** and $\mathbf{cur}(f)$ are continuous. We leave it to the reader to check that conditions 1 and 2 of definition 4.1 are satisfied. (Note that there are other possible choices for the semi-exponent, e.g. $\mathcal{P}(X^{<\omega} \times Y^{<\omega})$.)

In terms of the *program/function* intuition given in the previous section, a representing point in $\mathcal{P}(X) \Rightarrow \mathcal{P}(Y)$ for an arrow $\mathcal{P}(X) \rightarrow \mathcal{P}(Y)$ now is a 'program' consisting of a set of pairs (β, b) which can be seen as elementary instructions of the form '*on input* β, *output* b'. Of course there are many different 'programs' that represent the same arrow. Note that the above provides an alternative description of the concept of 'graph model'.

Any concrete embedding $\langle \cdot, \cdot \rangle : X^{<\omega} \times X \hookrightarrow X$ induces, through the mappings **cur** and **ev** a retraction (F, G) between $\mathcal{P}(X)$ and the lattice $\mathbf{POW}(\mathcal{P}(X), \mathcal{P}(X)) = [\mathcal{P}(X) \rightarrow \mathcal{P}(X)]$ of continuous functions from $\mathcal{P}(X)$ to $\mathcal{P}(X)$ known from the usual definition. In particular, also strict reflexive objects in **POW** (given by *surjective* embeddings) induce these retractions, which never are isomorphisms. The result always is a *non-extensional* lambda model.

As witnessed by the embedding $\langle \cdot, \cdot \rangle : X^{\omega} \times X \hookrightarrow X$ defining it, any graph model is a strict reflexive object in **POW** (for the existence of this injection implies the existence of (an) isomorphism(s)). Note however that the *properties* of the model do not depend on the fact that such an isomorphism exists. Also, the extensional collaps of a graph-model is defined with respect to the *given* embedding, rather than with respect to some isomorphism.

We will show that the construction of *extensional* combinatory algebras as described in section 3 in fact boils down to the *construction of strict reflexive objects* in the Karoubi envelope **K(POW)** of the category of graph models. As **K(POW)** is cartesian closed, these strict reflexive objects induce an *isomorphism* between the object and its hom-set, and therefore give rise to extensional lambda models.

Let f be some continuous idempotent mapping $\mathcal{P}(X) \rightarrow \mathcal{P}(X)$. Then f is an object of **K(POW)**, with exponent $(f \Rightarrow f) := \mathbf{cur}(f \circ \mathbf{ev} \circ (id \times f))$.

Suppose $\langle \cdot, \cdot \rangle$ is some mapping $X^{<\omega} \times X \rightarrow X$. Then the mappings $j : \mathcal{P}(X^{<\omega} \times X) \rightarrow \mathcal{P}(X)$ given by $x \mapsto \{a \mid a = \langle \beta, b \rangle \& (\beta, b) \in x\}$, and $i : \mathcal{P}(X) \rightarrow \mathcal{P}(X^{<\omega} \times X)$ given by $y \mapsto \{(\beta, b) \mid \langle \beta, b \rangle = a \in y\}$ are continuous, so they are arrows in **POW** and proposition 4.5 tells us under what conditions f is a *strict reflexive object* in **K(POW)**.

Returning to the construction described in section 3, we observe the following: take $\preceq$ to be a preorder on some set X, then, for $x \in \mathcal{P}(X)$, define $[x] := \{b \mid \exists b' \in x.b \preceq b'\}$. Now it's easy to see that $p := \lambda x.\{b \mid \exists \beta \subseteq x.b \in [\beta]\}$ is an idempotent arrow in **POW**. Note that $p(x) \equiv [x]$. Furthermore $(p \Rightarrow p)(x) = \mathbf{cur}(p \circ \mathbf{ev} \circ (id \times p))(x) = \{(\gamma, c) \mid c \in [\{c' \mid \exists \gamma' \subseteq [\gamma].(\gamma', c') \in x\}]\}$.

Using proposition 4.5 we then can prove the following

5.1. THEOREM. *The arrows* $(p \Rightarrow p)ip$ *and* $pj(p \Rightarrow p)$ *determine an isomorphism* $p \cong (p \Rightarrow p)$ *in* $\mathbf{K}(\mathbf{POW})$ *if and only if the preorder* $(X, \preceq)$ *satisfies*

- *(i)* $\langle \beta, b \rangle \preceq \langle \alpha, a \rangle$ *iff* $\alpha \sqsubseteq \beta$ *and* $b \preceq a$;
- *(ii) for all* $d : \exists \beta \in X^{<\omega} \exists b \in X.d \preceq \langle \beta, b \rangle \preceq d$. □

In fact, we will not give the proof here, but leave the calculations to the zealous reader. We shall however provide the (quite similar) details in the next section, while working in a more basic category, equivalent to **POW**.

6 A category of relations equivalent to POW

Define the category **RL** as follows: objects are sets, arrows $R : A \to B$ are relations $R \subseteq A^{<\omega} \times B$ satisfying the following monotonicity condition:

$$\beta \subseteq \beta' \;\&\; \beta R b \Rightarrow \beta' R b.$$

For the identity on an object A we take the arrow id_A defined by $\alpha(id_A)a \Leftrightarrow a \in \alpha$. Composition $S \star R$ of arrows $R : A \to B, S : B \to C$ is defined by

$$\alpha(S \star R)c \Leftrightarrow \exists \beta(\alpha R \beta \wedge \beta S c),$$

where $\alpha R \beta$ is an abbreviation for $\forall b \in \beta.\alpha R b$. Similarly we will write aRb for $\{a\}Rb$.

The category thus defined is in fact known as the semi-Kleisli category of the category of relations (see Hoofman, 1990/2).

6.1. PROPOSITION. **RL** *is equivalent to the category of graph models,* $\mathbf{RL} \equiv \mathbf{POW}$.

PROOF: We will define appropriate functors, and leave the details of verification to the reader.

- The functor $F : \mathbf{RL} \to \mathbf{POW}$ is defined on objects by $X \mapsto \mathcal{P}(X)$, on arrows by $R \mapsto \lambda x.\{b \mid \exists \beta \subseteq x.\beta R b\}$;
- The functor $G : \mathbf{POW} \to \mathbf{RL}$ is defined on objects by $\mathcal{P}(X) \mapsto X$, on arrows by $f \mapsto \{(\beta, b) \mid b \in f(\beta)\}$.

(In fact these functors even define an *isomorphism*). □

6.2. DEFINITION. Let $R : A \to B$ be an arrow in **RL**. We say that R is *linear* iff it satisfies $\alpha R b \Leftrightarrow \exists a \in \alpha . a R b$. □

So a linear arrow is fully determined by its values on singletons.
As can be readily verified (e.g. using proposition 6.1) **RL** is a wCCC and hence we now have a (bi-)functor $\cdot \Rightarrow \cdot : \mathbf{RL}^{op} \times \mathbf{RL} \to \mathbf{RL}$ acting on objects A, B by $(A \Rightarrow B) := A^{<\omega} \times B$ and on arrows $R : A \to A', S : B' \to B$ by $\Xi(R \Rightarrow S)(\beta, b) \Leftrightarrow \{a \mid \exists(\alpha, a) \in \Xi . \beta R \alpha\} S b$.

6.3. LEMMA. *If $S : A \to B$ and $T : B \to C$ are linear, then so is $T \star S : A \to C$. If $R : A \to A$ is linear, then so is $(R \Rightarrow R)$; furthermore we have $(\beta, b)(R \Rightarrow R)(\alpha, a)$ iff $\alpha R \beta$ and bRa.*
PROOF: Easy. □

Now take an idempotent arrow $R : A \to A$ in **RL**, so $R \star R = R$. Let $I : (A \Rightarrow A) \to A, J : A \to (A \Rightarrow A)$ also be arrows in **RL**. Again, using proposition 4.5, R is a strict reflexive object in **K(RL)** if and only if
(i) $(R \Rightarrow R) = (R \Rightarrow R) J R I (R \Rightarrow R)$;
(ii) $R = R I (R \Rightarrow R) J R$.

In order to obtain in **RL** the analogue of theorem 5.1 we have to impose some further restrictions on the arrows R, I, J: if we take R, I and J to be *linear* arrows we can, using the definition of composition and lemma 6.3, rewrite these conditions as:
(i) $\alpha R \beta \wedge bRa$ *iff*

$$\exists \gamma, \gamma', c, c', d, d' . \gamma R \beta \wedge bRc \wedge (\gamma, c) I d \wedge dRd' \wedge d' J(\gamma', c') \wedge \alpha R \gamma' \wedge c' R \alpha .$$

(ii) bRa *iff*

$$\exists \gamma, \gamma', c, c', d, D' . bRd' \wedge d' J(\gamma', c') \wedge \gamma R \gamma' \wedge c' Rc \wedge (\gamma, c) I d \wedge dRa .$$

In order to increase legibility, let us write $<$ for R. Hence $\alpha < \beta$ stands for $\forall b \in \beta(\alpha < b)$, and by linearity of $<$ this becomes $\forall b \in \beta \exists a \in \alpha(a < b)$. Furthermore, we write $(\alpha, a) < (\beta, b)$ for $\beta < \alpha \wedge a < b$. Now the two conditions can be written as:

1. $(\beta, b) < (\alpha, a)$ *iff* $\exists(\gamma, c), (\gamma', c'), d, d' . (\beta, b) < (\gamma, c) I d < d' J(\gamma', c') < (\alpha, a)$;
2. $b < a$ *iff* $\exists(\gamma, c), (\gamma', c'), d, d' . b < d' J(\gamma', c') < (\gamma, c) I d < a$.

Note that by idempotency of the relation $<$ we have for all α, β, a, b
(i) $a < b \Leftrightarrow \exists c . a < c < b$;
(ii) $\alpha < \beta \Leftrightarrow \exists \gamma . \alpha < \gamma < \beta$;
(iii) $(\beta, b) < (\alpha, a) \Leftrightarrow \exists(\gamma, c) . (\beta, b) < (\gamma, c) < (\alpha, a)$.

By restricting the relations I, J we obtain various specific instances of the conditions for $<$ to be a reflexive object in **K(RL)**. For example, let I be a function $\langle \cdot, \cdot \rangle : A^{<\omega} \times A \to A$, and define J by $aJ(\beta, b) \Leftrightarrow a = \langle \beta, b \rangle$. Our two conditions may then be written as:

1' $(\beta, b) < (\alpha, a)$ *iff* $\exists \gamma, \gamma', c, c' . (\beta, b) < (\gamma, c) \wedge \langle \gamma, c \rangle < \langle \gamma', c' \rangle \wedge (\gamma', c') < (\alpha, a)$

2' $b < a$ *iff* $\exists \gamma, \gamma', c, c' . b < \langle \gamma', c' \rangle \wedge (\gamma', c') < (\gamma, c) \wedge \langle \gamma, c \rangle < a$.

Adding one more restriction, namely reflexivity of the relation, now once more leads us to the conditions we encountered in the extensionalisation procedure of section 3:

6.4. THEOREM. *Let $<, I, J$ be as defined above, and $<$ reflexive. Then $(< \Rightarrow <)$ and $<$ are isomorphic in the Karoubi envelope if and only if*

1. $(\beta, b) < (\alpha, a)$ iff $\langle\beta, b\rangle < \langle\alpha, a\rangle$
2. $\forall a \exists \beta, b.\ a < \langle\beta, b\rangle < a.$

PROOF: We show that 1, 2 are equivalent to 1', 2' above. First suppose 1, 2 hold, then 1' holds:

- If $(\beta, b) < (\alpha, a)$, then $(\beta, b) < (\beta, b) \wedge \langle\beta, b\rangle < \langle\beta, b\rangle \wedge (\beta, b) < (\alpha, a)$ by reflexivity of $<$.
- If $\exists \gamma, \gamma', c, c'.(\beta, b) < (\gamma, c) \wedge \langle\gamma, c\rangle < \langle\gamma', c'\rangle \wedge (\gamma', c') < (\alpha, a)$, then $\exists \gamma, \gamma', c, c'.(\beta, b) < (\gamma, c) < (\gamma', c') < (\alpha, a))$ by 1. Hence $(\beta, b) < (\alpha, a)$ by transitivity of $<$.

and 2' holds:

- If $b < a$, then by 2 there exists (γ, c) such that $b < \langle\gamma, c\rangle < b < a$. Hence $b < \langle\gamma, c\rangle \wedge (\gamma, c) < (\gamma, c) \wedge \langle\gamma, c\rangle < a$.
- If $\exists \gamma, \gamma', c, c'.b < \langle\gamma', c'\rangle \wedge (\gamma', c') < (\gamma, c) \wedge \langle\gamma, c\rangle < a$ then $\exists \gamma, \gamma', c, c'.b < \langle\gamma', c'\rangle < \langle\gamma, c\rangle < a)$ by 1. Hence $b < a$ by transitivity of $<$.

The other way round, suppose 1', 2' hold, then 1 holds:

- If $(\beta, b) < (\alpha, a)$, then $\langle\beta, b\rangle < \langle\beta, b\rangle \wedge (\beta, b) < (\alpha, a) \wedge \langle\alpha, a\rangle < \langle\alpha, a\rangle$ by reflexivity of $<$. Hence $\langle\beta, b\rangle < \langle\alpha, a\rangle$ by 2'.
- If $\langle\beta, b\rangle < \langle\alpha, a\rangle$, then $(\beta, b) < (\beta, b) \wedge \langle\beta, b\rangle < \langle\alpha, a\rangle \wedge (\alpha, a) < (\alpha, a)$ by reflexivity of $<$. Hence $(\beta, b) < (\alpha, a)$ by 1'.

and 2 holds:

- We have $a < a$ by reflexivity of $<$, and hence $\exists \gamma, \gamma', c, c'.a < \langle\gamma', c'\rangle \wedge (\gamma', c') < (\gamma, c) \wedge \langle\gamma, c\rangle < a$ by 2'. From $(\gamma', c') < (\gamma, c)$ it follows that $\langle\gamma', c'\rangle < \langle\gamma, c\rangle$ by 1 (which we have already established). Hence $a < \langle\gamma, c\rangle < a$ by transitivity of $<$. □

Also interesting is the case in which I is defined as before, but J is defined independently of I by $aJ(\beta, b) \Leftrightarrow a = [\beta, b]$, where $[-,-]$ is a function $A^{<\omega} \times A \rightarrow A$. The conditions for isomorphism in the Karoubi envelope now reduce to

1" $(\beta, b) < (\alpha, a)$ *iff* $\exists \gamma, \gamma', c, c'.(\beta, b) < (\gamma, c) \wedge \langle\gamma, c\rangle < [\gamma', c'] \wedge (\gamma', c') < (\alpha, a)$;

2" $b < a$ *iff* $\exists \gamma, \gamma', c, c'.b < [\gamma', c'] \wedge (\gamma', c') < (\gamma, c) \wedge \langle\gamma, c\rangle < a$.

Again, if $<$ is reflexive we can give more simple sufficient, but this time not necessary, requirements.

6.5. Theorem. *Let $<, I, J$ be as defined above, and $<$ reflexive. If*

1. $(\beta, b) < (\alpha, a)$ iff $\langle\beta, b\rangle < \langle\alpha, a\rangle$,
2. $\forall a \exists \beta, b.\ a < \langle\beta, b\rangle < a$,
3. $\forall \alpha, a.\ [\alpha, a] < \langle\alpha, a\rangle < [\alpha, a]$,

then $(< \Rightarrow <)$ and $<$ are isomorphic in the Karoubi envelope.

Proof: It is an easy exercise to prove that the requirements of the theorem imply the requirements 1",2" above. □

To complete the picture, let us study the way in which theorems 3.3, 5.1 and 6.4 are related. For this the following lemma will be of use.

6.6. Lemma. *Let $\mathcal{A}, \mathcal{B}$ be categories. If $\mathcal{A} \equiv \mathcal{B}$, then $\mathbf{K}(\mathcal{A}) \equiv \mathbf{K}(\mathcal{B})$.* □

Now using the functor G from proposition 6.1, from the idempotent arrow p in **POW** (as in 5.1), we obtain a reflexive idempotent linear relation $G(p)$ in **RL**.

Conversely, given a reflexive idempotent linear relation $<$ in **RL**, we obtain an idempotent arrow $F(<)$ in **POW** defined by $F(<) := \lambda x.\{b \mid \exists \beta \subseteq x.\beta < b\}$. Then

$$a \preceq b \quad \textit{iff} \quad \{b\} < a$$

defines a preorder on X and $F(<) = \lambda x.\{b \mid \exists \beta \subseteq x. b \in [\beta]\}$.

Therefore, as $\mathbf{POW} \equiv \mathbf{RL}$, in fact theorems 5.1 and 6.4 are the same statements, expressed within different, but equivalent, categories.

Finally, to establish the relation with theorem 3.3, we note the following

6.7. Proposition. $\mathbf{K(RL)} \equiv \mathbf{CCLat}$, *where* **CCLat** *denotes the category of continuous complete lattices.*

Proof: For (X, R) in $\mathbf{K(RL)}$ we define $\mathtt{Pt}(X, R) := (\{[A] \mid A \in \mathcal{P}(X)\}, \subseteq)$, with $[A] := \{a \mid \exists \beta \subseteq A.\beta R a\}$. Then $\mathtt{Pt}(X, R)$ is a continuous complete lattice. For arrow $T : (A, R) \to (B, S)$ we define $\mathtt{Pt}(T) := \lambda x.\{b \mid \exists \beta \subseteq x.\beta T b\}$. This defines a functor $\mathtt{Pt} : \mathbf{K(RL)} \to \mathbf{CCLat}$.

Conversely, let $(D, \leq)$ be a continuous complete lattice with basis B_D. Define $R : \mathsf{B}_D^{\leq\omega} \to \mathsf{B}_D$ by

$$\beta R b \quad \textit{iff} \quad b \ll \bigvee \beta,$$

(where $x \ll y$ (" *x way below y*") iff for each directed subset S of B_D we have that $y \leq \bigvee S$ implies that there is an $y' \in S$ such that $x \leq y'$.)

Now one can check that $\mathtt{Rep}(D, \leq) := (\mathsf{B}_D, R)$ is in $\mathbf{K(RL)}$.

For continuous $f : (D, \leq) \to (D', \leq')$, define $\mathtt{Rep}(f) : \mathtt{Rep}(D, \leq) \to \mathtt{Rep}(D', \leq')$ by

$$\beta \mathtt{Rep}(f) b \quad \textit{iff} \quad b \ll f(\bigvee \beta)$$

This defines an arrow in $\mathbf{K(RL)}$, and $\mathtt{Rep}$ is a functor $\mathbf{CCLat} \to \mathbf{K(RL)}$. In fact the functors $\mathtt{Pt}$, $\mathtt{Rep}$ establish an equivalence of categories. For details we refer to Hoofman (1990/1). □

Now, by applying the equivalence, we find that the isomorphism in **K(RL)** under the conditions of theorem 6.4 induces precisely the isomorphism between continuous lattices given by the construction of section 3.

The following diagram summarizes our observations:

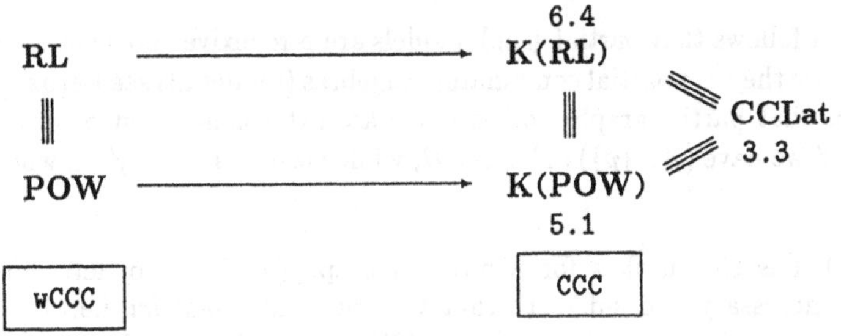

7 Getting partial

In this final section we will briefly describe an interesting modification of the construction given in section 3. Recall that the concept of combinatory algebra can be extended to applicative structures on which the application is not everywhere defined (see Bethke(1987)). We will refer to such structures as *partial* combinatory algebras (*pca*). Concrete examples of pca's can be constructed as follows: take some infinite set X, and $\langle\cdot,\cdot\rangle$ a *non-surjective* embedding $(X^{<\omega}\backslash\emptyset) \times X \hookrightarrow X$. By non-surjectivity we may fix some p not in the range of $\langle\cdot,\cdot\rangle$.

Also recall that given some cpo M with bottom $\perp$, a continuous function f from M to M is called *strict* if $f(\perp) = \perp$. We write $[M \xrightarrow{s} M]$ for the cpo of *strict continuous* functions from M to M. Then, for $A, B \in \mathcal{P}(X)$, define $\mathtt{F}(A)(B) := \{b \mid \exists \beta \subseteq B.\langle\beta, b\rangle \in A\}$. For $f \in [\mathcal{P}(X) \xrightarrow{s} \mathcal{P}(X)]$ define $\mathtt{G}(f) := \{\langle\gamma, c\rangle \mid c \in f(\gamma)\} \cup \{p\}$.

Next a partial application $*$ on $\mathcal{P}(X)\backslash\emptyset$ is given by

$$A * B := \begin{cases} \mathtt{F}(A)(B), & \text{if } \mathtt{F}(A)(B) \neq \emptyset; \\ \textit{undefined}, & \text{otherwise.} \end{cases}$$

Let $\mathcal{G} := (\mathcal{P}(X)\backslash\emptyset, *)$.

We will call the partial applicative structures thus obtained *partial graph models*.

7.1. LEMMA. *(i)* $\mathtt{F} \in \left[\mathcal{P}(X) \longrightarrow [\mathcal{P}(X) \xrightarrow{s} \mathcal{P}(X)]\right]$;

(ii) $\mathtt{G} \in \left[[\mathcal{P}(X) \xrightarrow{s} \mathcal{P}(X)] \longrightarrow \mathcal{P}(X)\right]$ *and Range*$(\mathtt{G}) \subseteq \mathcal{P}(X)\backslash\emptyset$;

(iii) $\mathtt{F} \circ \mathtt{G} = id_{[\mathcal{P}(X) \xrightarrow{s} \mathcal{P}(X)]}$.

PROOF: Straightforward. □

7.2. DEFINITION. A cpo $\mathcal{M}$ for which there exists mappings $\mathtt{F} : \mathcal{M} \longrightarrow [\mathcal{M} \xrightarrow{s} \mathcal{M}]$ and $\mathtt{G} : [\mathcal{M} \xrightarrow{s} \mathcal{M}] \longrightarrow \mathcal{M}$ satisfying the conditions of lemma 7.1 is called *p-reflexive*. □

7.3. DEFINITION. As usual, we write $a \simeq b$ for "*if either a or b is defined, then both are defined and equal*". We say that a partial applicative structure $(M, \bullet)$ is *extensional* iff for all $a, b \in M$ we have that $(\forall c \in M. a \bullet c \simeq b \bullet c)$ implies that $a = b$. □

7.4. PROPOSITION. *Partial graph models are non-extensional partial combinatory algebras.*

PROOF: From 7.1 it follows that partial graph models are p-reflexive cpo's via the mappings F and G. Therefore they are partial combinatory algebras (for details see Bethke(1987), theorem 2.8). To see that partial graph models never are extensional, just observe that for all $A, B \in \mathcal{P}(X)\backslash\emptyset$ we have $(A \cup \{p\}) * B \simeq A * B$, while clearly $A \cup \{p\} \neq A$, whenever $p \notin A$. □

In Bethke(1987) it is shown that for a p-reflexive cpo $(\mathcal{A}, F, G)$ to determine an extensional pca it is necessary and sufficient that the cpo is also *p-strict*, i.e. $G \circ F = id_{\mathcal{A}\backslash\{\bot\}}$. So by 7.4 we have that a p-reflexive cpo $(\mathcal{P}(X), \mathtt{F}, \mathtt{G})$ as defined above can never be p-strict.

Constructing extensional partial combinatory algebras

Take a non-empty set X and let us fix some mapping $\langle \cdot, \cdot \rangle : (X^{<\omega}\backslash\emptyset) \times X \to X$ as well as a preorder $\preceq$ on X. Define $\mathcal{M} := \left(\mathcal{P}(X)/_{\equiv}, \leq\right)$ as in section 3. By Proposition 3.1 $\mathcal{M}$ is a complete lattice with bottom $[\emptyset]$ and supremum $\sup A = [\bigcup\{y \mid [y] \in A\}]$, for all $A \subseteq \mathcal{M}$. We now fix some element $p \in X$ and following the construction of the partial graph model $\mathcal{G}$, define mappings $\mathtt{G}'$ by

$$\mathtt{G}'(f) := \left[\{\langle \gamma, c\rangle \mid c \mathrel{\mathsf{E}} f([\gamma])\} \cup \{p\}\right]$$

and $\mathtt{F}'$ by

$$\mathtt{F}'([x])([y]) := \left[\{b \mid \exists \beta \sqsubseteq y. \langle \beta, b\rangle \mathrel{\mathsf{E}} x\}\right].$$

7.5. LEMMA. *(i)* $\mathtt{F}' \in \left[\mathcal{M} \longrightarrow [\mathcal{M} \xrightarrow{s} \mathcal{M}]\right]$;

(ii) $\mathtt{G}' \in \left[[\mathcal{M} \xrightarrow{s} \mathcal{M}] \longrightarrow \mathcal{M}\right]$;

(iii) $Range(\mathtt{G}') \subseteq \mathcal{M}\backslash[\emptyset]$.

PROOF: Left to the reader. □

Again we can define a partial application $\star$ on $(\mathcal{M}\backslash[\emptyset])$ by

$$[a] \star [b] := \begin{cases} \mathtt{F}'([a])([b]), & \text{if } \mathtt{F}'([a])([b]) \neq [\emptyset]; \\ undefined, & \text{otherwise.} \end{cases}$$

Thus we obtain a partial applicative structure $\mathcal{G}_{\preceq} := (\mathcal{M}\backslash[\emptyset], \star)$.

The next theorem tells us which conditions on the preorder $(X, \preceq)$ are sufficient and necessary for $\mathcal{G}_{\preceq}$ to be an *extensional* pca.

7.6. Theorem. *$\mathcal{G}_{\preceq} := (\mathcal{M}, \mathrm{F}', \mathrm{G}')$ is p-strict p-reflexive if and only if the preorder $(X, \preceq)$ satisfies*

- *(i) for all $\beta, b : \langle \beta, b\rangle \not\preceq p$;*
- *(ii) for all $d : p \preceq d$;*
- *(iii) for all $d : d \not\preceq p$ iff $\exists \beta, b. d \preceq \langle \beta, b\rangle \preceq d$;*
- *(iv) for all $\alpha, \beta \neq \emptyset : \langle \beta, b\rangle \preceq \langle \alpha, a\rangle$ iff $\alpha \sqsubseteq \beta$ and $b \preceq a$.*

Proof: ($\Leftarrow$) *p-reflexivity:* $\mathrm{F}' \circ \mathrm{G}' = id_{[\mathcal{M} \stackrel{\centerdot}{\to} \mathcal{M}]}$.

For, $(\mathrm{F}'(\mathrm{G}'(f)))([x]) =$

$$= \Big[\Big\{b \mid \exists \beta \sqsubseteq x. \langle \beta, b\rangle \mathrel{\mathsf{E}} \{\langle \gamma, c\rangle \mid c \mathrel{\mathsf{E}} f([\gamma])\} \cup \{p\}\Big\}\Big]$$

$$= \Big[\Big\{b \mid \exists \beta \sqsubseteq x. \langle \beta, b\rangle \mathrel{\mathsf{E}} \{\langle \gamma, c\rangle \mid c \mathrel{\mathsf{E}} f([\gamma])\}\Big\}\Big] \qquad \text{(by (i))}$$

$$= \Big[\{b \mid \exists \gamma \sqsubseteq x. \gamma \neq \emptyset \wedge b \mathrel{\mathsf{E}} f([\gamma])\}\Big] \qquad \text{(by (iv)}\rightarrow\text{)}$$

$$= f([x]) \qquad \text{(by strictness and continuity of } f\text{).}$$

p-strictness: $\mathrm{G}' \circ \mathrm{F}' = id_{\mathcal{M} \setminus [\emptyset]}$.

For, let $[x] \in \mathcal{M}, x \neq \emptyset$.

Then, $(\mathrm{G}'(\mathrm{F}'([x])) =$

$$= \Big[\Big\{\langle \gamma, c\rangle \mid c \mathrel{\mathsf{E}} \{b \mid \exists \beta \sqsubseteq \gamma. \langle \beta, b\rangle \mathrel{\mathsf{E}} x\}\Big\} \cup \{p\}\Big]$$

$$= \Big[\{\langle \gamma, c\rangle \mid \langle \gamma, c\rangle \mathrel{\mathsf{E}} x\} \cup \{p\}\Big] \qquad \text{(by (iv)}\leftarrow\text{)}$$

$$= [x] \qquad \text{(by (ii) and (iii)).}$$

($\Rightarrow$) Suppose $(\mathcal{M}, \mathrm{F}', \mathrm{G}')$ is a p-strict p-reflexive cpo. By p-reflexivity the equality $\mathrm{F}' \circ \mathrm{G}' = id_{[\mathcal{M} \stackrel{\centerdot}{\to} \mathcal{M}]}$ is valid. Now let $\langle \beta, b\rangle \preceq p$. Then define

$$f([x]) := \begin{cases} [\emptyset], & \text{if } x \sqsubseteq \beta; \\ [\beta], & \text{otherwise.} \end{cases}$$

Clearly f is strict and continuous, and $f([\beta]) = [\emptyset]$. But $b \mathrel{\mathsf{E}} \mathrm{F}'(\mathrm{G}'(f))([\beta])$, contradicting p-reflexivity. This proves (i).

By extensionality of the induced partial applicative structure and (i) we have for $x \neq \emptyset$, that $[\{\langle \gamma, c\rangle \mid \langle \gamma, c\rangle \mathrel{\mathsf{E}} x\} \cup \{p\}] \equiv [x]$. Then $p \preceq d$, for any d. So (ii) is clear. Also by extensionality, if $\langle \gamma, c\rangle \preceq d$, then $\{\langle \gamma, c\rangle \mid \langle \gamma, c\rangle \mathrel{\mathsf{E}} \{d\}\} \equiv \{d\}$, so $\exists \beta \exists b. d \preceq \langle \beta, b\rangle \preceq d$. Therefore, if for no $\langle \beta, b\rangle$ we have $d \preceq \langle \beta, b\rangle \preceq d$, we cannot have $\langle \gamma, c\rangle \preceq d$. But then, by extensionality, $\{d, p\} \equiv \{p\}$, so $d \preceq p$. Conversely, if $\langle \beta, b\rangle \preceq d$ and $d \preceq p$ we have $\langle \beta, b\rangle \preceq p$ by transitivity, contradicting (i). So $d \not\preceq p$, and we have proved also (iii).

The proof of (iv) is similar to the proof of (i) of theorem 3.3. We leave the details to the reader. □

As a corollary of the proof of the theorem we obtain:

7.7. PROPOSITION. *$\mathcal{G}_{\preceq} := (\mathcal{M}, \mathrm{F}', \mathrm{G}')$ is p-reflexive if and only if the preorder $(X, \preceq)$ satisfies*

- *(1)* for all $\beta, b : \langle\beta, b\rangle \not\preceq p$;
- *(2)* for all $\alpha, \beta \neq \emptyset : \langle\beta, b\rangle \preceq \langle\alpha, a\rangle \quad \Rightarrow \quad \alpha \sqsubseteq \beta$ and $b \preceq a$. □.

This shows (by taking equality (=) for $\preceq$) that we recover our 'plain' partial graph models precisely in case the mapping $\langle\cdot,\cdot\rangle : X^{<\omega}\backslash\emptyset \times X \to X$ we start from is an embedding, which moreover is non-surjective (as p may not be in its range).

For an example of a structure satisfying the conditions of theorem 7.6 we refer the reader to Bethke(1987) (definitions 2.9, 2.10).

Like for the construction in section 3, we may alternatively give a description within the category **POW** and its Karoubi-envelope **K**(**POW**): when f is the idempotent mapping in **POW** induced by a preorder $\preceq$ (see section 5), then $(f \Rightarrow f) := \mathbf{cur}(f \circ \mathbf{ev} \circ (id \times f))$ is an idempotent mapping $\mathcal{P}(X^{<\omega} \times X) \to \mathcal{P}(X^{<\omega} \times X)$, but *also* an idempotent mapping $\mathcal{P}(X^{<\omega}\backslash\emptyset \times X) \to \mathcal{P}(X^{<\omega}\backslash\emptyset \times X)$. Take a mapping $\langle\cdot,\cdot\rangle : X^{<\omega}\backslash\emptyset \times X \to X$ and fix $p \in X$. Then define $i : \mathcal{P}(X^{<\omega}\backslash\emptyset \times X) \to \mathcal{P}(X)$ by $x \mapsto \{a \mid a = \langle\beta, b\rangle \& (\beta, b) \in x\} \cup \{p\}$ and $j : \mathcal{P}(X) \to \mathcal{P}(X^{<\omega}\backslash\emptyset \times X)$ by $y \mapsto \{(\beta, b) \mid \langle\beta, b\rangle = a \in y\}$.

In **K**(**POW**) we need to satisfy the following:

- (1) $(f \Rightarrow f)jfi(f \Rightarrow f) = (f \Rightarrow f)$;
- (2) for $x \neq \emptyset : fi(f \Rightarrow f)jf = f$.

It is not difficult to check that (1) and (2) hold if and only if the mapping $\langle\cdot,\cdot\rangle$ and the preorder $\preceq$ fulfill the conditions (i) - (iv) of theorem 7.6.

Acknowledgement

The considerations in this paper provide a full answer to a question of prof. Anne Troelstra, namely whether, and if so, in which sense, the constructions described in Bethke(1986,1987) can be considered as being canonical ways of obtaining extensional (partial) combinatory algebras. We'd like to thank him and Inge Bethke for their patient attention and useful comments.

References

H.P. BARENDREGT[1984] *The Lambda Calculus. Its Syntax and Semantics.* Nort-Holland.

I. BETHKE[1986] How to construct extensional combinatory algebras. *Indagationes Mathematicae 89*: 243-257.

I. BETHKE[1987] On the existence of extensional partial combinatory algebras. *Journal of Symbolic Logic 52-3.*

S. HAYASHI[1985] Adjunction of semi-functors: categorical structures in non-extensional lambda calculus. *Theoretical Computer Science 45*: 95-104.

R. HOOFMAN[1990/1] Continuous information systems. *Department of Computer Science, Utrecht University, preprint RUU-CS-90-25*

R. HOOFMAN[1990/2] Linear logic, domain theory and semi-functors. *Department of Computer Science, Utrecht University, preprint RUU-CS-90-34.*

R. HOOFMAN[1991] A note on semi-adjunctions. *Department of Computer Science, Utrecht University, preprint RUU-CS-91-41.*

R. HOOFMAN[1992] Ph.D.-thesis. *Department of Computer Science, Utrecht University.*

R. HOOFMAN & H. SCHELLINX[1991] Collapsing graph models by preorders. *ITLI - Prepublication Series for Mathematical Logic and Foundations, University of Amsterdam, ML-91-04.*

J.L. KRIVINE[1990] *Lambda-calcul, Types et Modèles.* Masson.

S. MACLANE[1971] *Categories for the Working Mathematician.* Springer Verlag.

S. MARTINI[1987] An interval model for second order lambda calculus. In: D.H. Pitt, A. Poigné & D. Rydeheard (eds.), *Category Theory and Computer Science, Edinburgh 1987.* LNCS vol. 283, Springer Verlag.

H. SCHELLINX[1991] Isomorphisms and non-isomorphisms of graph models. *Journal of Symbolic Logic 56-1*: 227-249

D.SCOTT[1982] Domains for denotational semantics. In: M. Nielsen & E.M. Schmidt (eds.), *Automata, Languages and Programming. Ninth Colloquium, Aarhus 1982.* LNCS vol.140, Springer Verlag.

Linear Logic and Interference Control
(Preliminary Report)

P. W. O'Hearn

School of Computer and Information Science
Syracuse University
Syracuse, New York, USA 13244
e-mail: ohearn@top.cis.syr.edu

1 Introduction

Two imperative programming language phrases interfere when one writes to a storage variable (or location, or memory cell) that the other reads from or writes to. Reynolds' syntactic control of interference (SCI) is a linguistic approach to controlling interference in Algol-like languages ([Rey78,89]). Girard's Linear Logic (LL) is often described as a logical approach to exercising control over resources ([Gir87,89]). In this report we will show that constraints arising from LL – or more precisely, Intuitionistic Linear Logic (ILL) with Weakening – control access to the (updatable) state in a manner very similar to SCI. To do this, we consider a simple language with an ILL-based type system, which we illustrate by treating a number of typical examples of interference control.

The most interesting aspect of our work is that it suggests a substantive connection between SCI and linear functional programming, as set out in [Laf88; Hol88; Abr90; GH90; Wad90,91b]. These in fact might be considered as two heads of the same coin. One aims to make imperative programming more elegant, by limiting difficulties caused by aliasing and interference, while the other aims to make functional programming more efficient, by permitting destructive updating in a purely functional context and by limiting the need for garbage collection. That the two have similar *formal* structure is perhaps more than a coincidence. A full understanding of this point may eventually shed new light on the relation between imperative and functional programming.

We should emphasize at the outset that the fit between ILL and SCI is not a perfect one. [Rey89] utilizes a sophisticated subtype discipline based on conjunctive types ([CD78]), and to forge a direct connection with [Rey89] would likely necessitate a study of subtypes in linear languages; an interesting topic which is nevertheless beyond the scope of this preliminary report. As a result, from the perspective of interference control, the ILL-based system to be described is inadequate in some respects. In particular, it is perhaps *overly restrictive*. This is not too surprising. The situation is rather like that in linear functional programming, where researchers have argued on practical grounds for modifications and additions to pure linear languages ([Wad91; GH90]). Similar remarks apply in the context of interference control.

We now turn our attention to the principles of interference control that form the basis for Reynolds' approach. To illustrate these principles we consider a simple form of parallel composition. The intention is to allow $C_1 \parallel C_2$ to be well-formed only when commands C_1 and C_2 don't interfere, thus assuring that their concurrent execution is determinate (so C_1 and C_2 are "disjoint processes" in the terminology of [Hoa71b]).

The first, and most obvious, constraint is to require that any identifier that appears on the left of an assignment statement in one command of a parallel composition isn't used in the other. Thus, we would certainly disallow $x := 1 \parallel x := 2$, because the operands to $\parallel$ interfere. But we would hope to allow

$$x := 1 \parallel y := 2$$

on the grounds that the two commands operate on different storage variables.

$$
\begin{array}{cc}
x & y \\
\hline
 & \\
\hline
\end{array}
$$

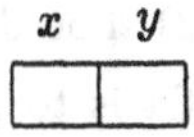

If x and y denote different variables, then executing $x := 1$ and $y := 2$ in parallel would result in a state where the contents of the variable associated with x is changed to 1, and that associated with y to 2.

$$
\begin{array}{cc}
x & y \\
\hline
1 & 2 \\
\hline
\end{array}
$$

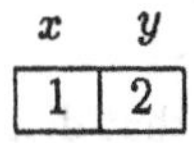

However, in most programming languages with procedures it is possible for different identifiers to be *aliases*; i.e. different names for the same variable. If x and y are aliases

$$
\begin{array}{c}
x\ y \\
\hline
 \\
\hline
\end{array}
$$

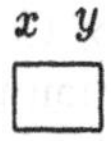

then the statements $x := 1$ and $y := 2$ interfere, because they both write to the same variable. We can only allow $x := 1 \parallel y := 2$ when x and y are not aliases.

Two procedures may interfere as well, if they share access to global variables. In a context in which procedures p and q interfere, the concurrent composition $p(1) \parallel q(3)$ should be illegal. Thus, we need a method that allows us to predict when procedures, as well as commands and variable identifiers, don't interfere.

The basic design principle proposed by Reynolds is to require that

distinct identifiers never interfere.

This gives us a particularly simple method of predicting non-interference: if the free identifiers in two phrases are disjoint, then they don't interfere. This justifies forming $x := 1 \parallel y := 2$ (or $p(1) \parallel q(2)$) when x and y (or p and q) are different identifiers. (We are implicitly assuming the principle that all "channels" of interference are named by identifiers, so that if no identifier free in a phrase P interferes with any identifier free in Q then P and Q don't interfere.)

The disjointness condition just mentioned can be relaxed by taking into account that certain types of phrases don't write to any variables, and therefore don't cause interference. For example, in a language where expressions (which occur on the right-hand side of assignment statements) cause no side effects, we can safely form

$x := z \parallel y := z + 1$

if z is an integer-expression identifier. That is, sharing is not harmful, in that it does not cause interference, when only *read access* to storage variables is shared. This is encapsulated in Reynolds' further principle;

> *passive* phrases, which don't write to any variables, don't interfere with one another.

Other kinds of phrases, besides expressions, may be passive as well. For example, if all free identifiers in a procedure definition are passive then the procedure itself is passive; e.g. $\lambda x \, . \, x := z + 1$. If we know that identifier p denotes a passive procedure then a composition of the form $p(1) \parallel p(3)$ should be legal.

To sum up, the basic approach of SCI is to require that different identifiers don't interfere, while recognizing that sharing between certain types of phrases is permissible because of their passivity. With this as background, we now outline the contents of the paper.

We begin in section 2 with a description of primitive types and sample Algol-like combinators for our illustrative language.

In section 3 we consider the principle that distinct identifiers don't interfere, which is an assumption about the nature of *environments* (the association of meanings to identifiers). We explain how, from a semantical perspective, this corresponds to a "tensor" product that has projections but not pairing. From the logical/lambda-calculus point of view, we are lead to abandon the rule of Contraction, which is the mechanism that permits sharing. The abolition of Contraction is the first point of convergence between SCI and ILL. However, we will retain Weakening.

The second point of convergence concerns passivity and Girard's "of course" modality "!". In section 4 we explain how, from a semantical perspective, passive types allow for a diagonal map which can model a limited form of Contraction, for passive types only. We further argue that passive types are certain co-monoids, corresponding to the explanation of "!" given in [Laf88]. On the lambda-calculus level, this leads us to consider passive types as being governed by the "!" modality.

Section 5 considers the problem of whether the typings of phrases are preserved by reduction.

In section 6 we describe a modified type system, based on a partitioning of types into passive and non-passive, that more closely resembles [Rey78,89] (though we still don't consider conjunctive types). Comparing it to the linear system reveals how Reynolds' term-forming rules for passive types are remarkably similar to Girard's rules for "!".

Section 7 contains conclusions and discussions on related work.

A final remark on the role of category theory in this work. Categorical aspects of ILL, as described in [Laf88], in conjunction with a model for SCI developed in [O'H90], are primarily responsible for our present understanding. But it must be admitted that the connection to be presented could now be understood, to some degree, without mentioning category theory at all. However, I believe that recounting some of the underlying categorical principles leads to a genuinely clearer exposition, especially with regard to

$$\frac{\Gamma \vdash v : \delta\mathbf{var} \quad \Gamma \vdash e : \delta}{\Gamma \vdash v := e : \mathbf{comm}} \text{ Assignment} \qquad \frac{\Gamma \vdash e_1 : \mathbf{int} \quad \Gamma \vdash e_2 : \mathbf{int}}{\Gamma \vdash e_1 + e_2 : \mathbf{int}} \text{ Plus}$$

$$\frac{\Gamma \vdash c_1 : \mathbf{comm} \quad \Gamma \vdash c_2 : \mathbf{comm}}{\Gamma \vdash c_1 \,;\, c_2 : \mathbf{comm}} \text{ Sequencing} \qquad \frac{\Gamma \vdash v : \delta\mathbf{var}}{\Gamma \vdash v : \delta} \text{ Dereferencing}$$

$$\frac{\Gamma \vdash c_1 : \mathbf{comm} \quad \Delta \vdash c_2 : \mathbf{comm}}{\Gamma, \Delta \vdash c_1 \parallel c_2 : \mathbf{comm}} \text{ Parallel Composition}$$

Table 1: Rules for Sample Primitives

similarities *and* differences between "!" and passivity. In any case, the categorical explanation of Reynolds' principles of interference control will be somewhat informal, and the relevant structure is mentioned without reference to the interference-specific aspects of [O'H90].

2 Primitive Types

An Algol-like language, in the sense of [Rey81], is a typed call-by-name λ-calculus, with imperative aspects of the language encompassed in primitive types. We will consider an illustrative language with the following primitive types:

$$P \ ::= \ \mathbf{comm} \mid \delta\mathbf{var} \mid \delta$$

where δ is a data type, e.g. **bool**(ean) or **int**(eger). **comm** is the type of commands, $\delta\mathbf{var}$ is the type of δ-variables, and δ (as a member of P) is the type of δ-expressions.

Syntax is described using typing judgements of the form $x_1, A_1, ..., x_n : A_n \vdash t : A$, where A, A_i are types, t is a term, and $x_1, ..., x_n$ is a list of identifiers without repetitions. Some illustrative phrases of primitive type are given in Table 1. Any variable can be considered as an expression (using Dereferencing), but not vice versa. $c_1 \,;\, c_2$ is sequential composition, while $c_1 \parallel c_2$ is parallel composition. The reason for the different contexts Δ and Γ in the rule for parallel composition will become evident in the next section.

3 Contraction and Interference

CATEGORICAL ASPECTS. First we consider the principle that distinct identifiers don't interfere, from the semantical perspective. This principle would seem to preclude the use of categorical products for forming environments. For example, we would not expect to have a diagonal $b \mapsto (b, b)$ that makes two copies, denoted by different identifiers, of a semantic entity b; if b interferes with itself, then the resulting environment would violate Reynolds' principle.

This suggests to consider a restricted form of product whose components never interfere; in set-theoretic notation

$$A \otimes B \;=\; \{(a,b) \in A \times B \mid a \text{ and } b \text{ don't interfere}\}$$

The idea is that, to make it clear that different identifiers never interfere, $\otimes$ should interpret the comma on the left-hand side of typing judgements. That is, the meaning $[\![t]\!]$ associated with a typing judgement $x_1 : A_1, ..., x_n : A_n \vdash t : A$ should be a map of the form ($[\![A]\!]$ and $[\![A_i]\!]$ are objects associated with types)

$$[\![A_1]\!] \otimes \cdots \otimes [\![A_n]\!] \xrightarrow{\quad [\![t]\!] \quad} \mathrm{A}$$

instead of a map of the form

$$[\![A_1]\!] \times \cdots \times [\![A_n]\!] \xrightarrow{\quad [\![t]\!] \quad} \mathrm{A}$$

for a categorical product $\times$.

Clearly $\otimes$ should be symmetric and associative up to isomorphism. Furthermore, assuming that our category has finite products, it makes sense to consider the terminal object as the unit of $\otimes$; the unit interprets the "empty" environment, which trivially satisfies the requirement that distinct identifiers don't interfere, and so its interpretation should not depend on whether we are using $\times$ or $\otimes$ for environments. Thus, we require that $(\otimes, \mathbf{1})$ is symmetric monoidal structure ([Mac71]), in which $\mathbf{1}$ is the terminal object. Note that for this kind of monoidal structure (with a terminal unit) we get "projections"; e.g. if f is the unique map from A to $\mathbf{1}$ then we can form

$$A \otimes B \xrightarrow{\quad f \otimes \mathrm{id} \quad} \mathbf{1} \otimes B \xrightarrow{\quad \sigma \quad} B$$

where id is the identity and σ is the evident "unit" isomorphism.

In order for procedures to mesh properly with environments, it is natural to require that, for each object B, $(\cdot) \otimes B$ has a right adjoint $B \multimap (\cdot)$. Thus, we get the usual isomorphism

$$Hom(A \otimes B, C) \;\cong\; Hom(A, B \multimap C)$$

for interpreting λ-abstraction and a map for application.

$$(A \multimap B) \otimes A \xrightarrow{\quad \mathbf{app} \quad} B$$

It is worthwhile to note how the structure of the application map corresponds to syntactic restrictions used in [Rey78,89]. Reynolds' basic constraint was to require that a procedure and its argument never interfere, and this is just what the $\otimes$ between the procedure and argument types in the domain of the application map suggests.

TYPE-THEORETIC ASPECTS. We now turn to our attention to the logical/λ-calculus side. Type assignments are manipulated by structural rules, which are interpreted by operations on environments. With terms included, the structural rules are

$$\frac{\Gamma, y : B, x : A, \Delta \vdash t : C}{\Gamma, x : A, y : B, \Delta \vdash t : C}\ \text{Exchange} \qquad \frac{\Gamma \vdash t : C}{\Gamma, x : A \vdash t : C}\ \text{Weakening}$$

$$\frac{\Gamma, x : A, y : A \vdash t : C}{\Gamma, z : A \vdash t[z/x, z/y] : C}\ \text{Contraction}$$

($t[z/x, z/y]$ denotes the substitution of z for x and y in t.) Exchange can be modeled using the exchange maps that result from the symmetric monoidal structure, and Weakening is interpreted by the "projections" for $\otimes$. Contraction cannot be modeled, however, because we have not assumed that $\otimes$ has a "diagonal" map that can be used to duplicate z (because z might interfere with itself). Put another way, Exchange and Weakening respect the property that distinct identifiers don't interfere, while Contraction does not. Thus, we abandon Contraction, but keep other typical term-formation rules.

Our types are given by the following grammar.

$$T ::= P \,|\, T \multimap T \,|\, T \times T$$

The typing rules are given in Table 2. We discuss the possibility of adding a type constructor for $\otimes$ in section 5.

The typing rules are presented in a sequent-calculus style. Notice that the only way to create a redex – $\mathsf{fst}\langle a, b\rangle, \mathsf{snd}\langle a, b\rangle, (\lambda x.t)u$ – is by using Cut, and Cut-free proofs give rise to terms in normal form. See [Abr90; GLT89; Asp90] for nice discussions on the relation between Cut-elimination and normalization.

In rules such as Cut and the one for parallel composition, there is the implicit assumption that identifiers appearing in different type assignments are disjoint, so that when we write Γ, Δ for concatenation the resulting sequence satisfies the requirement that an identifier appears at most once. This disjointness requirement is essential for the elimination of sharing.

EXAMPLES. First we consider the parallel composition $x := 1 \,||\, x := 2$. To derive this we might try

$$\frac{x : \mathbf{intvar} \vdash x := 1 : \mathbf{comm} \qquad x : \mathbf{intvar} \vdash x := 2 : \mathbf{comm}}{x : \mathbf{intvar} \vdash x := 1 \,||\, x := 2 : \mathbf{comm}}$$

But this does not fit the form of the rule for $||$, which requires the type assignments in the premises to be disjoint. If we had Contraction then we could proceed as follows, using phrases with the occurrences of x renamed apart.

$$\cfrac{\cfrac{y : \mathbf{intvar} \vdash y := 1 : \mathbf{comm} \qquad z : \mathbf{intvar} \vdash z := 2 : \mathbf{comm}}{y : \mathbf{intvar}, z : \mathbf{intvar} \vdash y := 1 \,||\, z := 2 : \mathbf{comm}}\ \text{Parallel Composition}}{x : \mathbf{intvar} \vdash x := 1 \,||\, x := 2 : \mathbf{comm}}\ \text{Contraction}$$

But without Contraction we cannot get this conclusion.

A similar example illustrates the abolition of aliasing. x and y would be aliases in the body of the λ-expression in the following procedure call:

Exchange, Weakening

$$\frac{\Gamma, x : A \vdash t : B \qquad \Delta \vdash u : A}{\Gamma, \Delta \vdash t[u/x] : B}\ \text{Cut}$$

$$\frac{}{x : A \vdash x : A}\ \text{Identifier} \qquad\qquad \frac{\Gamma \vdash a : A \quad \Gamma \vdash b : B}{\Gamma \vdash \langle a, b\rangle : A{\times}B}\ R{\times}$$

$$\frac{\Gamma, x : A \vdash t : C}{\Gamma, y : A{\times}B \vdash t[\mathrm{fst}(y)/x] : C}\ L_1{\times} \qquad\qquad \frac{\Gamma, x : B \vdash t : C}{\Gamma, y : A{\times}B \vdash t[\mathrm{snd}(y)/x] : C}\ L_2{\times}$$

$$\frac{\Gamma \vdash t : A \quad \Delta, x : B \vdash u : C}{\Gamma, \Delta, f : A{\multimap}B \vdash u[f(t)/x] : C}\ L{\multimap} \qquad\qquad \frac{\Gamma, x : A \vdash t : B}{\Gamma \vdash \lambda x\,.\,t : A{\multimap}B}\ R{\multimap}$$

Table 2: The Basic Contraction-Free Language

$$(\lambda\, x.\, x := 1 \;\|\; y := 2)\, y.$$

Cut is typically used to create a β-redex, e.g. as follows.

$$\frac{\Gamma, f : A{\multimap}B \vdash f(u) : C \qquad \Delta \vdash \lambda z.t : A{\multimap}B}{\Gamma, \Delta \vdash (\lambda z.t)u : B}\ \text{Cut}$$

But notice that Γ and Δ must be disjoint, and so the free identifiers occurring in the procedure and its argument must be disjoint. As in the previous example, we would need Contraction to get an identifier x occurring in both the procedure and the argument; without Contraction, aliasing does not arise.

DISCUSSION. Though our presentation has been from a different perspective, focusing on the rule of Contraction, we have now arrived at one of the type systems described early in [Rey78], before passivity was taken into account. At this point there is no sharing of storage variables whatsoever, even sharing of only read access. (The role of Contraction in sharing is of course clear from previous work on linear functional programming [Laf88; Hol88; Abr90; Wad90].)

4 Passivity and !

CATEGORICAL ASPECTS. Reynolds' principle of passivity says that passive phrases don't interfere with one another. Furthermore, phrases of certain types, such as **int** (integer expressions), are guaranteed to be passive. In the semantical category, then, there should be a notion of a *passive object*; in set-theoretic terms again, A is a passive object if all of its "elements" are themselves passive ($\forall a \in A\,.\ a$ is passive).

Suppose that a and b are passive entities, of types A and B respectively. Since passive entities don't interfere with one another, the pair (a, b) should be of type $A \otimes B$, because the basic requirement on $\otimes$ discussed (informally) in the previous section is that the components of a pair don't interfere. Thus, if A and B are passive objects then we should have an equality (or isomorphism) $A \otimes B = A \times B$ and, in particular, passive objects should admit a "diagonal" map $\epsilon : A \rightarrow (A \otimes A)$.

To move closer to LL, we require that that there is an endofunctor on our category that takes any object A to a passive object $!A$. One can think of $!A$ as the passive "subset" or "subobject" of A. By this reading, it also makes sense to require that $!A = A$ when A is passive and, in particular, $!!A = !A$. As an example, $\lambda x.x := z + 1$ would be a passive procedure, i.e. it would be in !(**intvar**⊸**comm**), while $\lambda x.x := z\,; y := 1$ would not (for y : **intvar**) because it assigns to the global variable denoted by y.

We can now restate the equality noted above for passive objects as $!A \otimes !B = !A \times !B$. This is not a general principle of "!" in LL; while LL does require a map $\epsilon : !A \rightarrow (!A \otimes !A)$ to model Contraction, this map is not required to be a "true" diagonal, when $!A \otimes !B = !A \times !B$ fails. Lafont ([Laf88]) instead requires a weaker property; that $!A$ should have a co-commutative co-monoid structure (σ, ϵ), where $\sigma : A \rightarrow \mathbf{1}$. That is, the following diagrams should commute, where α, γ, and ϱ are the evident isomorphisms associated with symmetric monoidal structure $\otimes, \mathbf{1}$ (γ is the exchange map taking (a, b) to (b, a)).

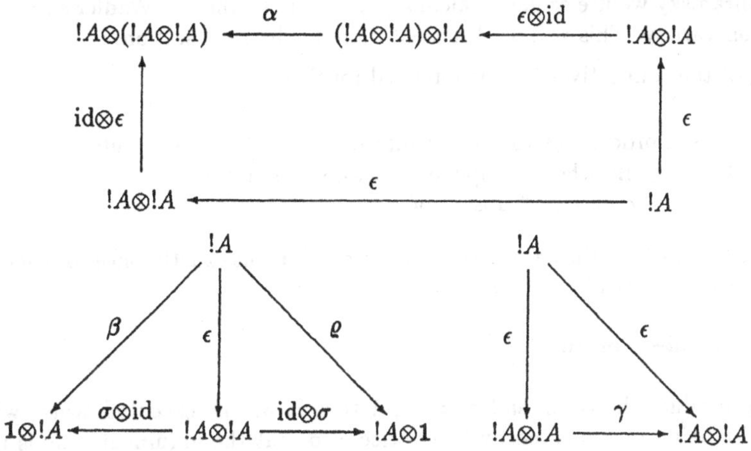

These diagrams (co-associativity, co-unit, co-commutativity) allow the interpretation of contraction (ϵ) to commute appropriately with the monoidal structure that models environments. While we have argued, from the SCI perspective, for a property ($!A \otimes !B = !A \times !B$) not implied by Lafont's conditions, our (informal) explanation of ! as passivity should certainly possess this co-monoid structure.

Finally, according to [Laf88], there should be a co-monad structure ($\eta : ! \rightarrow !^2, \mu : ! \rightarrow I$) such that $!A$ is the co-free co-commutative co-monoid over A. In terms of passivity, $\mu_A : !A \rightarrow A$ converts a passive entity to a non-passive one (here, non-passive means "not necessarily passive"), and $\eta_A : !A \rightarrow !!A$ should act as the identity. We will not try to motivate the co-freeness requirement (for discussion see [GL87; See89; Bar90]), but it is met in [O'H90].

TYPE-THEORETIC ASPECTS. To incorporate passivity we now consider "!" as a new type constructor, to be governed by Girard's rules

$$\frac{\Gamma, !A, !A \vdash C}{\Gamma, !A \vdash C}\ \text{Contraction} \qquad \frac{\Gamma \vdash C}{\Gamma, !A \vdash C}\ \text{Weakening}$$

$$\frac{\Gamma, A \vdash B}{\Gamma, !A \vdash B}\ \text{Dereliction} \qquad \frac{!\Gamma \vdash B}{!\Gamma \vdash !B}\ R!\ \ (!\Gamma \text{ is } !A_1, \ldots, !A_n)$$

As we have included Weakening in general, it won't be necessary to consider the weakening rule for "!". Contraction is the basic mechanism that allows sharing of passive entities. Dereliction allows any passive identifier to be considered also as a non-passive one, and is interpreted by the map μ associated with the co-monad structure of "!". $R!$ says that if all free identifiers of a phrase are passive then the phrase itself is passive.

The type system will now be one for the ($\times\multimap$!)-fragment of ILL with Weakening ($\times$ is often written &). We use the λ-calculus versions of Girard's rules for "!" from Table 3, in addition to the rules from tables 1 and 2. Notice that there is no destructor for the left rule for ! (Dereliction), and neither is there a constructor for the right rule. This "implicit" interpretation of the "!" types is as in the λ-calculus for ILL considered recently by Wadler ([Wad91]); in contrast, Abramsky and Holmström ([Abr90; Hol88]) use explicit binding forms for Contraction and Dereliction, and a constructor for $R!$. However, like Abramsky we use sequent calculus typing rules, whereas Wadler's are of a natural deduction variety. This will be discussed more in the next section.

The reading of the connectives that is intended for SCI is as follows.

$A \multimap B$	–	procedures that don't interfere with their arguments
$A \times B$	–	pairs whose components possibly interfere
$!A$	–	the passive "subset" of A

There is no requirement here that a procedure use its argument exactly once, or even at most once. This is illustrated by

$$\vdash \lambda c.\, c\,;c : \mathbf{comm} \multimap \mathbf{comm}.$$

If we feed this procedure the command $x := x + 1$ then it will be executed twice, with the net effect of x being incremented by 2. (Notice also that the argument type is not prefixed by "!".)

We also include a rule that expresses the passive nature of expressions.

$$\frac{}{x : \delta \vdash x : !\delta}\ \text{Expression Passivity}$$

In lieu of subtypes, one could also consider a rule that saying that δ and $!\delta$ are equivalent wherever they appear in a type expression, e.g. $\delta \multimap A$ vs. $!\delta \multimap A$.

$$\frac{\Gamma, x :!A, y :!A \vdash t : C}{\Gamma, z :!A \vdash t[z/x, z/y] : C}\ \text{Contraction} \qquad \frac{\Gamma, x : A \vdash t : B}{\Gamma, x :!A \vdash t : B}\ \text{Dereliction}$$

$$\frac{!\Gamma \vdash t : B}{!\Gamma \vdash t :!B}\ R! \quad (!\Gamma \text{ is } x_1 :!A_1, \ldots, x_n :!A_n)$$

Table 3: Rules for Passive Types

EXAMPLES. We illustrate these rules by considering examples from the Introduction. First, if z is an integer-expression identifier, then we want to form the parallel composition $x := z \parallel y := z + 1$. This is done as follows, letting V(ariable) stand for **intvar** and E(xpression) for **int**.

$$\cfrac{\cfrac{\cfrac{\cfrac{x : V, a : E \vdash x := a : \mathbf{comm} \qquad y : V, b : E \vdash y := b + 1 : \mathbf{comm}}{x : V, y : V, a : E, b : E \vdash x := a \parallel y := b + 1 : \mathbf{comm}}\ \text{PC+Ex}}{x : V, y : V, a :!E, b :!E \vdash x := a \parallel y := b + 1 : \mathbf{comm}}\ \text{Dereliction (twice)}}{x : V, y : V, z :!E \vdash x := z \parallel y := z + 1 : \mathbf{comm}}\ \text{Contraction}}{x : V, y : V, z : E \vdash x := z \parallel y := z + 1 : \mathbf{comm}}\ \text{Expression Passivity}$$

The top inference consists of the rule for Parallel Composition, and several applications of Exchange (which we often apply without mention). The key step of the derivation is the use of Contraction, which allows the sharing of the passive identifier z. Expression Passivity discharges the ! on the type of z. (Actually, the last step combines Expression Passivity and Cut.)

A passive procedure type is a type of the form $!(A \multimap B)$. We can show that the procedure $(\lambda\, x\, .\, x := z + 1)$ is passive as follows.

$$\cfrac{\cfrac{\cfrac{\cfrac{z : E, x : V \vdash x := z + 1 : \mathbf{comm}}{z : E \vdash \lambda x.x := z + 1 : V \multimap \mathbf{comm}}\ R\multimap}{z :!E \vdash \lambda x.x := z + 1 : V \multimap \mathbf{comm}}\ \text{Dereliction}}{z :!E \vdash \lambda x.x := z + 1 :!(V \multimap \mathbf{comm})}\ \text{R!}}{z : E \vdash \lambda x.x := z + 1 :!(V \multimap \mathbf{comm})}\ \text{Expression Passivity}$$

Finally, if p is a passive procedure identifier then we can derive $p(1) \parallel p(3)$ (letting C stand for **comm**).

$$\cfrac{\cfrac{\cfrac{r : E \multimap C \vdash r(1) : C \qquad q : E \multimap C \vdash q(1) : C}{r : E \multimap C, q : E \multimap C \vdash r(1) \parallel q(3) : C}\ \text{PC}}{r :!(E \multimap C), q :!(E \multimap C) \vdash r(1) \parallel q(3) : C}\ \text{Dereliction (twice)}}{p :!(E \multimap C) \vdash p(1) \parallel p(3) : C}\ \text{Contraction}$$

Dereliction essentially allows r to be considered as a non-passive procedure, so that the application $r(1)$ can be formed using L–∘ and a suitable rule for the constant 1 (similarly for $q(3)$).

$$\dfrac{\dfrac{}{\vdash 1 : E}\ \text{Constant} \qquad \dfrac{}{x : C \vdash x : C}\ \text{Identifier}}{r : E \multimap C \vdash r(1) : C}\ L\multimap$$

ON VARIABLE DECLARATIONS. The presence of types such as $!\delta\mathbf{var}$ and $!\mathbf{comm}$ may cause some confusion. For instance, if x is of type $!\mathbf{intvar}$ then we can form the composition $x := 1 \parallel x := 2$, apparently violating the requirement that the same variable not be assigned to by concurrent processes. But this really doesn't cause a problem, because the way that variables get declared in a program guarantees that they are not of passive type. (Semantically, $!\delta\mathbf{var}$ might be an empty type, or might contain an entity $\perp$ which, when appearing on the left of an assignment statement, causes divergence.)

Declarations are typically performed using a combinator of the form ([Rey81])

$$\mathbf{new}_\delta : (\delta\mathbf{var} \multimap \mathbf{comm}) \multimap \mathbf{comm}.$$

$\mathbf{new}_\delta\,(P)$ is executed by allocating a new storage variable, supplying it to the procedure P, and de-allocating the variable after P finishes executing. Only non-passive variables get declared via **new**, by virtue of its type, and so we cannot use Contraction to obtain anomalous commands like $x := 1 \parallel x := 2$ when x is a declared variable.

DISCUSSION. The above examples illustrate how Girard's rules for "!" allow us to treat a number of examples concerning passivity and interference control. However, the pure linear type system does not take into account some extra principles of passivity, such as the equivalence $!A\times!B = !A\otimes!B$ $(=!(A\times B))$ discussed earlier. Contraction for passive types reflects the idea that a passive entity doesn't interfere with *itself*, but not that two *different* passive entities don't interfere with one another. Thus, we cannot derive a typing judgement such as

$$x : \mathbf{intvar}, y : \mathbf{boolvar}, z : \mathbf{int}\times\mathbf{bool} \vdash x := \mathrm{fst}(z) \parallel y := \mathrm{snd}(z) : \mathbf{comm}.$$

But this parallel composition would be reasonable, from the SCI perspective, because the phrases $\mathrm{fst}(z)$ and $\mathrm{snd}(z)$ are of passive type (by Expression Passivity), and therefore don't interfere. This suggests, e.g., to add an axiom asserting $z :!A\times!B \vdash z :!(!A\times!B)$, allowing us to consider z in this example as passive. We will do so in section 6.

5 The Subject Reduction Problem

In this section we consider *subject reduction*; i.e. the property that reduction preserves typings. The failure of this property was the main problem with the treatment of passivity in [Rey78]; this was solved in [Rey89] using conjunctive types. The main point of this section is to explain subtleties surrounding "!" and reduction which have not yet been fully resolved. (This is one area where ideas from [Rey89] could conceivably influence other work on applying ILL.)

Define the relation $\rhd$ as follows:

$$(\lambda x.t)u \rhd t[u/x] \quad \mathrm{fst}\langle a,b\rangle \rhd a \quad \mathrm{snd}\langle a,b\rangle \rhd b$$

Proposition 1 *If* $\Gamma \vdash t : A$ *and* $t \rhd t'$ *then* $\Gamma \vdash t' : A$.

This result holds because of the similarity between reduction and Cut-elimination (e.g. [GLT89; Abr90; Asp90]). The key cases are when Cut is used to create a redex, and a left rule meets a right rule. For example,

$$\dfrac{\dfrac{\Delta \vdash a : A \quad y : B \vdash y : B}{\Delta, f : A \multimap B \vdash f(a) : B}\,L\text{-}\!\multimap \qquad \dfrac{\Gamma, x : A \vdash t : B}{\Gamma \vdash \lambda x.t : A \multimap B}\,R\text{-}\!\multimap}{\Gamma, \Delta \vdash (\lambda x.t)a : B}\,\text{Cut}$$

can be replaced by

$$\dfrac{\Gamma, x : A \vdash t : B \quad \Delta \vdash a : A}{\Gamma, \Delta \vdash t[a/x] : B}\,\text{Cut}$$

A redex can also be created by substituting in a passive procedure of type $!(A \multimap B)$, and another key case is when Dereliction meets $R!$.

$$\dfrac{\dfrac{\Delta, f : A \multimap B \vdash f(a) : B}{\Delta, f :!(A \multimap B) \vdash f(a) : B}\,\text{Dereliction} \qquad \dfrac{!\Gamma \vdash \lambda x.t : A \multimap B}{!\Gamma \vdash \lambda x.t :!(A \multimap B)}\,R!}{\Delta, !\Gamma \vdash (\lambda x.t)a : B}\,\text{Cut}$$

We reduce this problem to one not involving "!" on the procedure type.

$$\dfrac{\Delta, f : A \multimap B \vdash f(a) : B \qquad !\Gamma \vdash \lambda x.t : A \multimap B}{\Delta, !\Gamma \vdash (\lambda x.t)a : B}\,\text{Cut}$$

This is not the whole story; in particular, the form of the $R!$ rule, with $!\Gamma$ on the left of $\vdash$, is important for treating cases involving Contraction. But we will not give the proof of Proposition 1 here; the details would be quite lengthy, and shortly we will encounter some difficulties involving reduction that have not yet been resolved.

Before continuing we would like to bring up an interesting point. If we had formulated the type system in a natural-deduction (ND) style, removing Cut and using elimination rules in place of left rules

$$\dfrac{\Delta \vdash t : A \times B}{\Delta \vdash \mathrm{fst}(t) : A}\,\times E_1 \qquad \dfrac{\Delta \vdash t : A \times B}{\Delta \vdash \mathrm{snd}(t) : B}\,\times E_2$$

$$\dfrac{\Gamma \vdash p : A \multimap B \quad \Delta \vdash q : A}{\Gamma, \Delta \vdash p(q) : B}\,\multimap E \qquad \dfrac{\Gamma \vdash t :!A}{\Gamma \vdash t : A}\,!E$$

then Proposition 1 would fail. To see this, one only has to note that

$$g : C \multimap !A, y : C, f :!A \multimap !A \multimap B | - (\lambda x.f(x)(x))(g(y)) : B$$

is easy to derive, while β-reducing leads to

$$g : C \multimap !A, y : C, f : !A \multimap !A \multimap B | - f(g(y))(g(y)) : B.$$

Without the Cut rule, there would be no way to derive the contractum because Contraction can't be used to get the multiple occurrences of the non-passive identifiers g and y. With Cut, we can consider the term $f(x)(x)$, where $x : !A$, and then substitute $g(y)$ in for x. Of course, if we erase λ-terms from derived typing judgements then the ND and sequent calculus systems lead to the same sets of derivable sequents, but some terms typecheck in the sequent calculus version that don't under ND.

One might argue that this example goes against the spirit of Linear Logic, because of the multiple occurrences of the identifiers g and y, whose types are not prefixed by "!". It is then notable that this example is effectively disallowed by restrictions on procedure types proposed in [Rey89] (see section 7).

SUBJECT REDUCTION. By a "context" $C[\cdot]$ we mean a term with a "hole" that, when filled with a term t, yields another term $C[t]$ (possibly allowing capture by bound identifiers). The following is "subject reduction".

$$\text{If } \Gamma \vdash C[t] : A \text{ and } t \rhd t' \text{ then } \Gamma \vdash C[t'] : A.$$

This property fails in our system (Proposition 1 does not apply because the definition of $\rhd$ does not take into account reduction on subterms). For example, using Cut to substitute $(\lambda x.g(x))y$ for a in $f(a)(a)$, where $a : !A$, one can derive

$$g : C \multimap !A, y : C, f : !A \multimap !A \multimap B | - f((\lambda x.gx)y)(\lambda x.gx)y) : B.$$

But if we reduce one redex, and not the other, then the result is not a valid typing. Applying Cut-elimination results in *both* redexes being reduced. Thus, $f(g(y))(g(y))$ would be well-typed (of type B in the indicated type assignment), while $f((\lambda x.gx)y)(g(y))$ would not.

One obvious way to deal with this example is to consider a rule that allows a number of passive entities, of the same type, to replace different identifiers. For example, (this is similar in some respects to the Mix rule; see [Asp90] for a λ-calculus version)

$$\frac{\Gamma, x_1 : !A, x_2 : !A \vdash u : B \qquad \Delta \vdash t_1 : !A \qquad \Delta \vdash t_2 : !A}{\Gamma, \Delta \vdash u[t_1/x_1, t_2/x_2] : B}$$

This allows us to easily deal with the counterexample to subject reduction just given, by substituting $(\lambda x.g(x))y$ for a and $g(y)$ for b in a phrase of the form $f(a)(b)$. It further appears that an extension along these lines – generalizing this rule to arbitrary n, instead of only 2 – does offer a "solution" to the subject reduction problem for our language. However, we will not pursue this point, as the solution is not altogether satisfactory.

The problem is that we have not yet found a treatment of the tensor product which (in the presence of Weakening) does not compromise subject reduction (neither are we sure of how to add tensors in the framework of [Rey89]). One example that illustrates the possible use of tensor products in SCI is record types, as in a language like Pascal. A record may be thought of as a collection of variables that (in the absence of pointers) don't interfere with one another. Thus, an accurate representation of records, for the purposes of interference control, would be as tensor types, and not as product types. The fact that different uses (selecting different fields) of an identifier of record type don't interfere would then be made manifest in the type structure

To see the problem, we might interpret the left rule for $\otimes$ by projections (since we have Weakening) and proceed as follows.

$$\frac{\Gamma, a : A, b : B \vdash t : C}{\Gamma, x : A \otimes B \vdash t[\mathrm{fst}(x)/a, \mathrm{snd}(x)/b] : C}\ L\otimes \qquad \frac{\Gamma \vdash t_1 : A \quad \Delta \vdash t_2 : B}{\Gamma, \Delta \vdash \langle t_1, t_2 \rangle : A \otimes B}\ R\otimes$$

But then

$$a : A, b : B \vdash \langle \mathrm{fst}\langle a, b\rangle, \mathrm{snd}\langle a, b\rangle\rangle : A \otimes B$$

is derivable while

$$a : A, b : B \vdash \langle \mathrm{fst}\langle a, b\rangle, b\rangle : A \otimes B$$

is not. (This example wouldn't be relevant to a language without Weakening.)

REMARK. It is true that these "problems" with subject reduction do not necessarily arise if one restricts attention to a *fixed reduction strategy that applies to closed terms only*. Thus, the remarks of this section don't apply to the *operational semantics* for ILL-calculi given in [Hol88;Abr90].

6 A Partitioning of Types

The type systems in [Rey78,89] do not have a "!" constructor, and neither do they include rules such as Dereliction and $R!$. However, restricted forms of these rules do show up implicitly. To illustrate this point, we consider a restricted type system which more closely resembles those from [Rey78,89], and we relate it to the linear one.

Instead of a type constructor "!", there is a partitioning of types into passive and non-passive (or active). The partitioning is as follows.

Metavariables

ϕ : passive phrase types
α : active phrase types
θ : arbitrary phrase types

Productions

$$\begin{aligned} \phi &::= \delta \mid \theta \xrightarrow[p]{} \theta \mid \phi \times \phi \\ \alpha &::= \delta\,\mathbf{var} \mid \mathbf{comm} \mid \theta \to \alpha \mid \alpha \times \theta \mid \theta \times \alpha \\ \theta &::= \phi \mid \alpha \end{aligned}$$

Notice that types such as !**intvar** and !**comm** do not appear.

Identifier, Cut, Exchange, Weakening, $L\times$, $R\times$, from Table 2
Rules from Table 1, and

$$\frac{\Gamma \vdash t:\theta \quad \Delta, x:\alpha \vdash u:\theta'}{\Gamma, \Delta, f:\theta \rightarrow \alpha \vdash u[f(t)/x]:\theta'}\ L\rightarrow \qquad \frac{\Gamma, x:\theta \vdash t:\alpha}{\Gamma \vdash \lambda x\,.\,t:\theta \rightarrow \alpha}\ R\rightarrow$$

$$\frac{\Gamma \vdash t:\theta \quad \Delta, x:\theta' \vdash u:\theta''}{\Gamma, \Delta, f:\theta \xrightarrow[p]{} \theta' \vdash u[f(t)/x]:\theta''}\ L\xrightarrow[p]{} \qquad \frac{\Phi, x:\theta \vdash t:\theta'}{\Phi \vdash \lambda x\,.\,t:\theta \xrightarrow[p]{} \theta'}\ R\xrightarrow[p]{}$$

$$\frac{\Gamma, z:\phi, y:\phi \vdash t:\theta}{\Gamma, x:\phi \vdash t[x/z, x/y]:\theta}\ \text{Contraction} \qquad \frac{\Gamma \vdash p:\theta \xrightarrow[p]{} \alpha}{\Gamma \vdash p:\theta \rightarrow \alpha}\ \text{Coercion}$$

Table 4: Rules for Partitioned Types

Typing rules for the partitioned types are in Table 4. Φ ranges over passive type assignments (each type in Φ is passive), and plays the role of contexts of the form $!\Gamma$ in the linear system. $A \xrightarrow[p]{} B$ is the type of passive procedures, and corresponds to $!(A \multimap B)$. The right rule for $\xrightarrow[p]{}$ is essentially as in [Rey78,89]: a procedure is passive if all of its free identifiers are. Coercion allows a passive procedure to be used where a non-passive one is expected. Restrictions on passive procedure types can be considered as well, e.g. by requiring that θ is passive whenever θ' is in $\theta \xrightarrow[p]{} \theta'$ ([Rey89]).

One important point is that $\phi \times \phi'$ is considered to be a passive type, while $!A \times !B$ is not in the linear system. This is related to the equivalence $!A \times !B = !A \otimes !B$ discussed in section 4, and allows us, e.g., to derive the example discussed at the end of section 4.

$$x:\mathbf{intvar}, y:\mathbf{boolvar}, z:\mathbf{int}\times\mathbf{bool} \vdash x := \mathrm{fst}(z) \parallel y := \mathrm{snd}(z) : \mathbf{comm}.$$

TRANSLATION TO LINEAR TYPES. The relation between partitioned types and the linear system can be seen via the following translation $(\cdot)^\star$ into linear types.

$$\mathbf{comm}^\star = \mathbf{comm} \qquad \delta\mathbf{var}^\star = \delta\mathbf{var} \qquad \delta^\star = !\delta$$

$$(\phi \times \phi')^\star = !(\phi^\star \times \phi'^\star) \qquad (\alpha \times \theta)^\star = \alpha^\star \times \theta^\star \qquad (\theta \times \alpha)^\star = \theta^\star \times \alpha^\star$$

$$(\theta \xrightarrow[p]{} \theta')^\star = !(\theta^\star \multimap \theta'^\star) \qquad (\theta \rightarrow \alpha)^\star = \theta^\star \multimap \alpha^\star$$

Notice that $A \rightarrow B$ corresponds directly to $A \multimap B$, without a decomposition such as $(!A) \multimap B$.

We write $\Gamma \vdash_p t : A$ to refer to derivability in the system of partitioned types, and $\Gamma \vdash_l t : A$ for derivability in the linear system from the previous sections. We will need to extend the linear system with the following rule

$$\frac{}{x : !A \times !B \vdash_l x : !(!A \times !B)}\ \times!$$

If we erase the occurrences of x then $(\times!)$ is not derivable in ILL with Weakening.

$(\cdot)^*$ extends from types to type assignments in the evident way. The relation between the systems can now be stated as follows.

Proposition 2 $\Gamma \vdash_p t : A \implies \Gamma^* \vdash_l t : A^*$

To prove the proposition, one only needs to check that each of the rules in the partitioned system is derivable under this translation. We illustrate with the following cases.

Coercion:

$$\dfrac{\dfrac{\dfrac{}{f : A \multimap B \vdash f : A \multimap B}\text{ Identifier}}{f :!(A \multimap B) \vdash f : A \multimap B}\text{ Dereliction} \qquad \Gamma \vdash t :!(A \multimap B)}{\Gamma \vdash t : A \multimap B}\text{ Cut}$$

$R \underset{P}{\rightarrow}$:

$$\dfrac{\dfrac{!\Gamma, x : A \vdash t : B}{!\Gamma \vdash \lambda x\,.\,t : A \multimap B}\,R\multimap}{!\Gamma \vdash \lambda x\,.\,t :!(A \multimap B)}\,R!$$

$R\times$, in the case of a passive product:

$$\dfrac{\dfrac{}{x :!A\times!B \vdash x :!(!A\times!B)}\times! \qquad \dfrac{\Gamma \vdash a :!A \quad \Gamma \vdash b :!B}{\Gamma \vdash \langle a, b\rangle :!A\times!B}\,R\times}{\Gamma \vdash \langle a, b\rangle :!(!A\times!B)}\text{ Cut}$$

This last case shows the need for $\times!$.

DISCUSSION. The first two cases in the proof of Proposition 2 illustrate how Reynolds' coercion from passive to non-passive procedures plays (part of) the role of Dereliction (Dereliction also shows up in $L \underset{P}{\rightarrow}$), and how his treatment of passive λ-abstraction ($R \underset{P}{\rightarrow}$) is closely related to $R!$. Also, Reynolds' rules for application implicitly incorporate Contraction for passive types; without coercions [Rey89], or a notion of "passive occurrence" of an identifier [Rey78], application for non-passive procedures would be formulated as

$$\dfrac{\Gamma, !\Delta \vdash p : A \multimap B \quad \Gamma', !\Delta \vdash q : A}{\Gamma, \Gamma', !\Delta \vdash p(q) : B}$$

This is derivable using $L\multimap$, Contraction, and Cut.

7 Conclusions and Related Work

As we mentioned in the Introduction, the fit between SCI and ILL is not perfect. In sections 4 and 6 we considered principles of passivity that go beyond basic properties of "!" in ILL. Furthermore, as will be discussed below, subtypes play an important role in [Rey89], and they have not been considered in the pure linear system (tables 2 and 3). However, despite their differences the most essential ingredients of the two remain very similar;

- general Contraction is abandoned, but
- a controlled form of Contraction is allowed, governed by "!" or by passivity.

This is suggestive of a relationship between SCI and linear functional programming; but further work is needed to determine whether the relation will be profitable for either. (Some possibilities along these lines will be touched on below.)

Controlling interference is an old problem in programming languages, and there are a number of existing programming languages that use syntactic restrictions which eliminate, or at least limit, aliasing. The first such language seems to have been Concurrent Pascal ([Bri73]), where interference constraints were used to ensure that all interaction between concurrent processes was exclusively mediated by monitors (see also [Hoa71a,b]). Others include Euclid ([Pop77]), Turing ([HMRC87]), FX ([GL86]) and Occam ([PM87]). The essential idea of controlling, but not eliminating, the use of Contraction appears (implicitly) to some extent in all of these languages.

We now discuss differences between this work and [Rey89]. To deal with the subject reduction problem, Reynolds designed a type system possessing the property that all identifiers occurring freely within a phrase of passive type themselves have passive type. This property does not hold in the linear system from sections 3-5, e.g. b in

$$a :!A, b : B \vdash \mathrm{fst}\langle a, b\rangle :!A.$$

or g and y in

$$g : C \multimap !A, y : C \vdash g(y) :!A.$$

Recall that these are just the kinds of cases that caused problems wrt subject reduction. Reynolds avoids these difficulties by a subtle combination of subtypes and restrictions on procedure types, packaging everything together using conjunctive types.

- Procedure types such as $A \underset{P}{\rightarrow} B$, where A non-passive and B passive, and $C \rightarrow D$, where D passive, are disallowed.
- Instead of products, whose components are selected by projection, "objects" with named fields are used, together with a forgetting-fields coercion that allows components of an object that are not used to be forgotten.

Restated in terms of ! and $\multimap$, the first restriction disallows procedures like $g : C \multimap !A$ above. The second allows a field selection to be typed without regard for identifiers in fields not being selected. For example, if $(i = \cdot, j = \cdot)$ is a constructor for an object with i and j fields, then the selection $(i = a, j = b).i$ can be typed as

$$a : A \vdash (i = a, j = b).i : A$$

and the type of b is not relevant at all.

It is worthwhile to note that the forgetting-fields coercion has the added benefit that it allows syntactic restrictions to be relaxed, by taking into account contexts in which a component of an object is not used. For example, if c_1, c_2, c_3 : **comm**, then here we cannot type

$\mathrm{fst}\langle c_1, c_3\rangle \parallel \mathrm{fst}\langle c_2, c_3\rangle$

because the non-passive identifier c_3 appears on both sides of $\parallel$, even though it is never used. In [Rey89], this kind of example can be treated by using a coercion that forgets c_3. Thus, as mentioned in the Introduction, the linear type system is *more restrictive* than [Rey89]. This is suggestive of the interesting role that subtypes might play in linear languages, by allowing syntactic restrictions to be relaxed without compromising semantic properties of linear types.

Turning to linear functional programming, the type systems presented here are most similar to ones recently described by Wadler ([Wad91]). In particular, his "steadfast" types involve a partitioning of types into non-linear and linear that bears a striking resemblance to Reynolds' passive/active distinction (section 6). Furthermore, [Wad90] requires that "a non-linear data structure must not contain any linear components"; compare this to Reynolds' requirement that all free identifiers in a passive phrase be passive. Here, it would seem to be a good idea to adopt Reynolds' restrictions on procedure types, e.g. by disallowing types of the form $C \multimap !A$, thus removing phrases of non-linear type such as $g(y)$, where g and y are linear. Phrases of this form do not appear to play an important role in [Wad90,91] and, as discussed in section 5, seem to be the source of certain anomalies wrt reduction.

The similarity with Wadler's work suggests an application of LL to interference control of a different flavour than what has been described here. The idea is to consider a more conventional interpretation of "!", where $!A$ would not restrict to a passive "subset", but rather would simply indicate that sharing (even of write access) is permissible. This would allow the restriction that distinct identifiers never interfere to be relaxed, but in a controlled fashion: interference would *only* be allowed between identifiers with types prefixed by "!". In this scheme, passivity might reappear via *read only* types [GH90; Wad90,91].

The most unsatisfactory aspect of SCI at present is the inability to treat recursive non-passive procedures. The difficulty can be seen by looking at the fixed-point equation $\mathbf{Y}(F) = F(\mathbf{Y}(F))$. If F is non-passive, then the right-hand side of the equation violates the requirement that a procedure never interfere with its argument, which is the basic restriction in [Rey78,89]. In the scheme just mentioned (with the more conventional "!"), the recursion combinator could be typed as $!(A \multimap A) \multimap A$, and the right-hand side of the fixed-point equation would be permissible because F would have type $!(A \multimap A)$, indicating that sharing is allowed. This proposal would require restrictions on Dereliction and $R!$, as discussed in [Wad91]. (Reynolds has hinted at an approach to recursion whereby abstraction on qualified identifiers of the form $p.a$ is used to indicate that an argument is allowed to interfere with p. We further speculate that this idea might fit together with our proposal by considering a suitable variation on *use* types, as in [GH90; Wad91]; see also the related effect systems of [GL86; Luc87]).

ACKNOWLEDGEMENTS: I am grateful to Bob Tennent for his encouragement and for comments on this work. Thanks also to Steve Brookes, Jawahar Chirimar, Prakash Panangaden, John Reynolds, and Phil Scott for helpful discussions; and to Samson Abramsky, whose comments (with those of Reynolds and Panangaden) influenced me to explore the connection to LL.

References

[Abr90] S. Abramsky. Computational interpretations of Linear Logic. 1990.

[Asp89] A. Asperti. Categorical topics in computer science. TD-7/90, Dipartmendo di Informatica, University di Pisa, 1989.

[Bar90] M. Barr. *-Autonomous categories and linear logic. 1990.

[Bri72] P. Brinch Hansen. Structured multiprogramming. *Communications of the ACM* 15, 574–577, 1972.

[Bri73] P. Brinch Hansen. *Operating Systems Principles.* Prentice Hall, 1973.

[CD78] M. Coppo and M. Dezani. A new type-assignment for λ-terms. *Archiv. Math. Logik.*, 19:139–156, 1978.

[GH90] J. Guzmàn and P. Hudak. Single-threaded polymorphic lambda calculus. *5th LICS Symposium*, Philadelphia, 1990.

[Gir87] J.-Y. Girard. Linear logic. *Theoretical Computer Science*, 50, 1–102, 1987.

[Gir89] J.-Y Girard. Towards a geometry of interaction. *Contemporary Mathematics 92: Categories in Computer Science and Logic*, 69–108, 1989.

[GL86] D.K. Gifford and J.M. Lucassen. Integrating functional and imperative programming. *Proc. ACM Conference on Lisp and Functional Programming*, 28–38, 1986.

[GL87] J.-Y. Girard and Y. Lafont. Linear logic and lazy computation. *TAPSOFT '87, vol.2, LNCS 250*, 52–66, 1987.

[GLT89] J.-Y. Girard, Y. Lafont, and P. Taylor. *Proofs and Types.* Cambridge University Press, 1989.

[HMRC87] R. C. Holt, P. A. Matthews, J. A. Rosselet, and J. R. Cordy. *The Turing Programming Language. Design and Definition.* Prentice Hall, 1987.

[Hoa71a] C. A. R. Hoare. Procedures and parameters: an axiomatic approach. In *Symposium on Semantics of Algorithmic Languages*, ed. E. Engeler, Lecture Notes in Mathematics 188, Springer Verlag, 102–116, 1971.

[Hoa71b] C. A. R. Hoare. Towards a theory of parallel programming. *International Seminar on Operating System Techniques*, Belfast, Northern Ireland, 1971.

[Hol88] S. Holmström. Linear functional programming. *Proceedings of the Workshop on Implementation of Lazy Functional Languages*, Chalmers University, 1988.

[Laf88] Y. Lafont. The linear abstract machine. *Theoretical Computer Science*, 59:157–180, 1988.

[Luc87] J. M. Lucassen. *Types and Effects: Towards the Integration of Functional and Imperative Programming.* PhD thesis, MIT, 1987.

[Mac71] S. Mac Lane. *Categories for the Working Mathematician.* Springer-Verlag, 1971.

[O'H90] P.W. O'Hearn. *Semantics of Noninterference: a Natural Approach.* PhD thesis, Queen's University, Kingston, Ontario, Canada, 1990.

[PM87] D. Pountain and D. May. *A tutorial introduction to occam programming.* McGraw-Hill, 1987.

[Pop77] G.J. Popek et. al. Notes on the design of EUCLID. *SIGPLAN Notices* 12(3), 11–18, 1977.

[Rey78] J. C. Reynolds. Syntactic control of interference. *Conference Record of the 5th ACM Symposium on Principles of Programming Languages*, pp. 39-46, 1978.

[Rey81] J. C. Reynolds. The essence of Algol. In J. W. de Bakker and J. C. van Vliet, editors, *Algorithmic Languages*, pages 345–372. North-Holland, Amsterdam, 1981.

[Rey89] J. C. Reynolds. Syntactic control of interference, part II. *International Colloquium on Automata, Languages, and Programming,* 1989.

[See89] R.A.G. Seeley. Linear logic, ∗-autonomous categories and cofree coalgebras. *Contemporary Mathematics 92: Categories in Computer Science and Logic,* 371–382, 1989.

[Wad90] P. Wadler. Linear types can change the world! In M. Broy and C. Jones, editors, *Programming Concepts and Methods*, North Holland, 1990.

[Wad91] P. Wadler. Is there a use for Linear Logic? *ACM/IFIP Symposium on Partial Evaluation and Semantics Based Program Manipulation*, 1991.

HIGHER DIMENSIONAL WORD PROBLEM

Albert BURRONI
Université PARIS 7, UFR de maths,
2 place Jussieu, 75005, Paris

The word problem on a monoid admits two natural generalizations :
- the first one is the extension from monoids to categories. In this case the words become the "paths" and the equality problems take the form of diagram commutation problems.
- the second is the extension from monoids to universal algebras. In this case, the words become the "terms", and the rewriting systems set the rules for their equality.

Is it possible to unify these two extensions ? In this paper we answer as follows : the rewriting problem for terms is nothing but a 2-dimensional path problem in a 2-category. This observation leads to the general problem for n-paths in an n-category, or even in an ∞-category. A lot of computations made by categoricians are 1-, 2- or 3-dimensional computations, and in fact n-dimensional computations take place in an (n+1)-category (see §1.2). Furthermore, beyond the unity so given to various word problems, the link with combinatorial topology appears, word problems being in this setting refinement of homotopy theory (see §1.3).

1. General setting

We are going to define the n-dimensional word problem ($n \in \mathbb{N}$), and more generally, a word problem of variable dimension which means computing in an ∞-category (such a computation being often a reduction to a canonical form, but more generally being the construction of a "homotopy" between two expressions).

1.1. ∞-Categories.

One will find in [Bu2] more synthetical definitions of ∞-categories, but the definition presented below fits pretty well our purpose. Finally the whole construction amounts to the juxtaposition of an infinity of 2-categories.

An ∞-*graph* **G** is the datum of a diagram of sets

$$(*) \qquad G_0 \underset{b_0}{\overset{a_0}{\leftleftarrows}} G_1 \underset{b_1}{\overset{a_1}{\leftleftarrows}} G_2 \underset{b_2}{\overset{a_2}{\leftleftarrows}} G_3 \ \dots\ G_n \underset{b_n}{\overset{a_n}{\leftleftarrows}} G_{n+1} \underset{b_{n+1}}{\overset{a_{n+1}}{\leftleftarrows}} G_{n+2} \dots$$

such that, for every $n \in \mathbb{N}$, the following equations hold :

$$a_n a_{n+1} = a_n b_{n+1}, \quad b_n a_{n+1} = b_n b_{n+1}.$$

The elements of G_n are named n-*cells*, and the following representations of 0-, 1- and 2-cells are well-known :

$$X \qquad\qquad X\xrightarrow{f}Y \qquad\qquad X \underset{g}{\overset{f}{\rightrightarrows}} Y \ (\Downarrow\lambda)$$

$$(n=0) \qquad\qquad (n=1) \qquad\qquad (n=2)$$

where $X,Y\in G_0$, $f,g\in G_1$, $\lambda\in G_2$.

Symbols are sometimes omitted, as for instance here : . $\rightrightarrows$ ($\Downarrow\lambda$) .

We also need higher dimensional cells, and representing them is possible (although difficult), as for instance, for a 3-cell : . $\Downarrow \to \Downarrow$.

In fact, as we shall see, , it seems that the main part of the calculus always works in 2-graphs.

A n-*graph* is an ∞-graph G such that $G_p = \emptyset$ for every p>n, and such an n-graph will be identified with the diagram (*) truncated at the level n. In particular a 0-graph is just a set, and a 1-graph is an (oriented) graph in the usual sense.

Starting from an ∞-graph G, we define, for every $0\le i<j$, a graph G_{ij} :

$$G_i \underset{b_{ij}}{\overset{a_{ij}}{\leftleftarrows}} G_j$$

with : $a_{ij} = a_i\, a_{i+1} \dots a_j$, $b_{ij} = b_i\, b_{i+1} \dots b_j$,

and we define, for every $0\le i<j<k$, a 2-graph G_{ijk} :

$$G_i \underset{b_{ij}}{\overset{a_{ij}}{\leftleftarrows}} G_j \underset{b_{jk}}{\overset{a_{jk}}{\leftleftarrows}} G_k.$$

In order to obtain an ∞-category on G, we need only a category structure on each G_{ij} in such a way that the 2-graphs G_{ijk} become 2-categories.

It remains to define a 2-category on a 2-graph $G_0 \underset{b_0}{\overset{a_0}{\leftleftarrows}} G_1 \underset{b_1}{\overset{a_1}{\leftleftarrows}} G_2$;

essentially such a category needs the additional data of two "unity" maps :

$$X \longmapsto X\xrightarrow{id(X)}X, \text{ and } X\xrightarrow{f}Y \longmapsto X \underset{f}{\overset{f}{\rightrightarrows}} Y \ (\Downarrow id(f)),$$

and of three "composition laws" :

$$X\xrightarrow{f}Y\xrightarrow{g}Z \longmapsto X\xrightarrow{g\circ f}Z,$$

$$X \underset{g}{\overset{f}{\rightrightarrows}} Y \underset{g'}{\overset{f'}{\rightrightarrows}} Y \ (\Downarrow\lambda,\ \Downarrow\lambda') \longmapsto X \underset{g'\circ g}{\overset{f'\circ f}{\rightrightarrows}} Z \ (\Downarrow\lambda\circ\lambda'),$$

$$X \to Y \ (\Downarrow\lambda \text{ over } \Downarrow\mu) \longmapsto X \rightrightarrows Y \ (\Downarrow\mu\perp\lambda).$$

These data must satisfy the neutrality and associativity axioms, and the "Godement rule".

We have the usual notions of homomorphisms between ∞-graphs or ∞-categories which we define as natural transformations through diagrams and projective limits. So we get two categories, namely **∞-Cat** and **∞-graph**, and a forgetfull functor, U_∞ : **∞-Cat** $\longrightarrow$ **∞-Graph**, and from general category theory, this functor U_∞ admits a left adjoint []. The ∞-category [G] is named the free ∞-category generated by the ∞-graph G.
We know also through general arguments that **∞-Graph** and **∞-Cat** are complete and co-complete.
In the same way we get a functor U_n : **n-Cat** $\longrightarrow$ **n-Graph**, with a left adjoint []. Let I_n : **n-Cat** $\longrightarrow$ **∞-Cat** be the the functor which just adds units in higher dimensions. Then we have a double commutative square

$$\begin{array}{ccc} \textbf{n-Cat} & \xrightarrow{I_n} & \infty\textbf{-Cat} \\ U_n \downarrow \uparrow [\,] & & U_\infty \downarrow \uparrow [\,] \\ \textbf{n-Graph} & \xrightarrow[\subseteq]{} & \infty\textbf{-Graph} \end{array}$$

1.2. CW-presentations.

The ∞-categories can be constructed by adjoining successive cells as for the CW-complexes, these adjonctions being completed by symetric constructions of collapses. This pair adjonction/collapse is the analogous of the pair generator/relation in the description of algebraic structures.

The *formal* n-*cell*, for every n$\in$N, is the free ∞-category $[e_n]$ generated by the ∞-graph e_n and determined through the conditions

$$(e_n)_p = \begin{cases} 2 & \text{if } 0\le p\le n-1 \\ 1 & \text{if } p=n \\ 0 & \text{if } n+1\le p \end{cases},$$

with $a_p, b_p : 2 \longrightarrow 2$ for $0\le p\le n-1$, and $a_n, b_n : 1 \longrightarrow 2$, given as the constant map on 0 and 1 respectively (we adopt the Von Neuman convention, namely that $k=\{0,1,\dots,k-1\}$ for every $k\in\mathbb{N}$).
We define the ∞-graph ∂e_n, edge of e_n, as being e_n truncated at dimension n-1, and two evident ∞-functors

$$[e_{n-1}] \xleftarrow{\ col_n\ } [\partial e_n] \xrightarrow{\ adj_n\ } [e_n].$$

The ∞-functor col_n is the "elementary collapse", and the ∞-functor adj_n is the "elementary adjonction".

An (n-1)-dimensional *attaching map* towards an ∞-category C is an ∞-functor of the form $\varphi : [\partial e_n] \longrightarrow C$. To such a data we associate two ∞-categories C/φ and $C[\varphi]$, through the two following pushouts :

$$\begin{array}{ccccc} [e_{n-1}] & \xleftarrow{\ col_n\ } & [\partial e_n] & \xrightarrow{\ adj_n\ } & [e_n] \\ \downarrow & & \downarrow \varphi & & \downarrow \\ C/\varphi & \longleftarrow & C & \longrightarrow & C[\varphi] \end{array}$$

C$/\varphi$ is named the *collapse* (of dimension n-1) of **C** by φ, and **C**$[\varphi]$ is named the *adjonction* (of dimension n) of φ to **C**.

Every "finitely presented"∞-category **D** can be described starting from **C** $= \emptyset$ through a finite sequence of operations of these types, namely a sequenceof p adjonctions of $\varphi_1,\dots,\varphi_p$, followed by q collapses by $\psi_1,\dots,\psi_q$. Such an analysis of **D** will be denoted through :

$$\mathbf{D} = \emptyset[\varphi_1,\dots,\varphi_p]/(\psi_1,\dots,\psi_q),$$

and we name it a *CW-presentation* of **D**.

In fact, technically, the collapse and the adjonction are essentially of the same nature (think of these operations in the case of CW-complexes, where they yield to the same homotopy type), and we have a canonical epimorphism

$$\emptyset[\varphi_1,\dots,\varphi_p,\ \psi_1,\dots,\psi_q] \twoheadrightarrow \emptyset[\varphi_1,\dots,\varphi_p]/(\psi_1,\dots,\psi_q).$$

But, *in practise*, we proceed in a slightly different manner, adjoining by "blocks", "dimensionwise", and collapsing at the end, all the collapses being of dimension n+1, with $n = \dim(\mathbf{D}) = \sup_{1\le i\le p}\dim(\varphi_i)$. In this way, **D** appears directly as an n-category. Our process is summarized as follows :

(**)

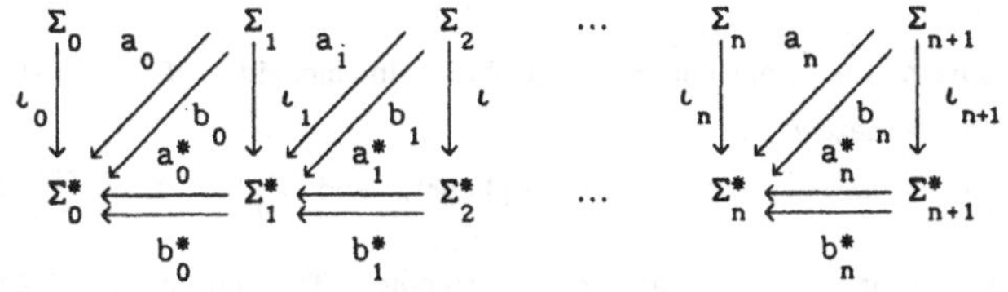

where, for each $0\le i\le n$, the partial diagram

$$\Sigma^*_0 \underset{b^*_0}{\overset{a^*_0}{\leftleftarrows}} \Sigma^*_1 \underset{b^*_1}{\overset{a^*_1}{\leftleftarrows}} \Sigma^*_2 \ \cdots\ \Sigma^*_{i-1} \underset{b^*_{i-1}}{\overset{a^*_{i-1}}{\leftleftarrows}} \Sigma^*_i$$

represents the underlying i-graph Σ^*_i to the i-category previously obtained through the introduction of j-cells for $0\le j\le i$ (the recurrence begins with $\Sigma_0 = \Sigma^*_0$ and $\iota_0 = \mathrm{id}(\Sigma_0)$) via the graphs $\Sigma^*_j \underset{b^*_j}{\overset{a^*_j}{\leftleftarrows}} \Sigma^*_{j+1}$ which represent the "blocks" made by the union of the j-dimensional attaching maps.

The examples given in 1.3 shall illustrate this construction.

Before concluding, let us informally explain the reasons why the collapses are n-dimensional. First, it is clear that they cannot be of higher dimension ; next, in the case n = 1 so familiar to categoricians, adding invertible morphisms in commutative diagrams is prefered to equalizing objects in a category. Of course, this is not indispensable.

1.3. Examples.

The complexity of the last construction increases with the dimension n.

Case n = 0.

The 0-dimensional word problem can be identified with the so called "maze problem". Let $\Sigma_0 \underset{b_0}{\overset{a_0}{\leftleftarrows}} \Sigma_1$ be a graph. Does it exist a path from X to Y for $X, Y \in \Sigma_0$? To this "semi-Thue"-problem is associated a Thue-problem (which is the actual maze problem) : $\bar{X} = \bar{Y}$? (with $X \longmapsto \bar{X}$ the canonical surjection $\Sigma_0 \longrightarrow \Sigma_0/(\Sigma_1)$ from Σ_0 to the quotient of Σ_0 by the equivalence relation generated by the relation defining the semi-Thue problem).
Then, in the diagram (**) for n = 0, Σ_0 represents a 0-graph and Σ_0^* the 0-category generated by Σ_0. Of course, as there are zero operations (in an n-category there are n operations on the n-cells), these two structures can be identified to the sets : $\Sigma_0 = \Sigma_0^*$ and $\iota_0 = id(\Sigma_0)$. One can also consider Σ_0 as a set of adjonctions of 0-cells to $\varnothing$. The graph $\Sigma_0 = \Sigma_0^* \underset{b_0}{\overset{a_0}{\leftleftarrows}} \Sigma_1$ represents the adjonction of 1-cells which are used to describe the collapse $\Sigma_0/(\Sigma_1)$.

Case n = 1.

We obtain the classical word problem in monoids (if $\Sigma_0 = 1$) and more generally in the categories.
Σ_1 is what is usually named the alphabet, and $a_0, b_0 : \Sigma_1 \rightrightarrows 1$ are too trivial to appear in the case of a monoid. The datum $\Sigma_0 = \Sigma_0^* \underset{b_0}{\overset{a_0}{\leftleftarrows}} \Sigma_1^*$ represents the underlying graph to the free category generated by the graph $\Sigma_0^* \underset{b_0}{\overset{a_0}{\leftleftarrows}} \Sigma_1$, and $\iota_1 : \Sigma_1 \longrightarrow \Sigma_1^*$ isthe level-1 canonical embedding of a graph Σ into the graph which is underlying the free category Σ^* that it generates.
$\Sigma_1^* \underset{b_1}{\overset{a_1}{\leftleftarrows}} \Sigma_2$ represents a set of diagrams of the form

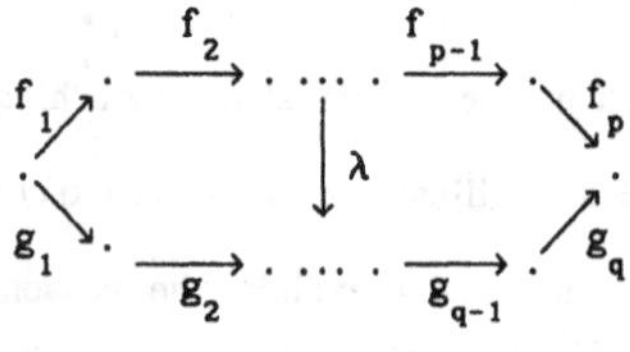

i.e. the adjonction of a set of 2-cells $\lambda \in \Sigma_2$.
As before, this problem entails a semi-Thue problem, and the associated Thue problem is the classical word problem.
The elements of Σ_2^* represent 2-cells admitting a unique decomposition, which is schematically given through

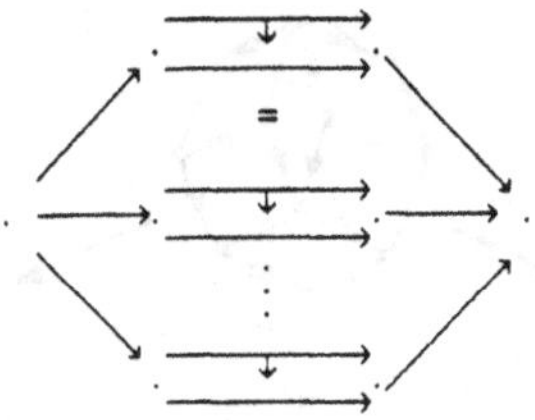

In this figure, every 1-cell represents in fact a path (i.e. an element of Σ_1^*), every 2-cell an element of Σ_2, and the equal sign ("=") within empty spaces means commutativity in the free category generated by the graph $\Sigma_0^* \leftleftarrows \Sigma_1$.

Case n = 2.

The construction goes on with the adjonction of 3-cells $\Sigma_2^* \leftleftarrows \Sigma_3$. It's easy to imagine, but tedious to describe, how the elements of Σ_3^* are built. Let us merely say that the corresponding Thue problem is nothing but a computation in 2-categories, or the 2-calculus. It has been recently an object of research, especially by P.L. Curien [Cur] and Y. Lafont [Laf], after the works of E.G. Rodeja and his school in Santiago de Compostella. Finally it's also the rewriting problem for terms as we shall show in part 2 below.

1.4. Relationship with combinatorial topology.

The ∞-graphs - as geometrical objects of combinatorial topology - are too poor to give interesting models. On the other hand the simplicial sets, the cubic sets, and eventually other objects of this nature provide the classical space models. But all (∞-graphs, simplicial sets, etc) can be interpreted as CW-presentations of ∞-categories. For the ∞-graphs it's evident, but, even if the intuition broadly suggests that this is true for the simplicial and cubic sets, the concrete realization of this fact seems to be a very complex problem. This is achieved for simplicial sets in the work of Street [Str].
To illustrate this fact, we shall restrain ourselves to extracting the hereafter picture from [Bu2] ; this picture shows how "reading" a 4-dimensional cubic cell may be interpreted in a 4-category. For the sake of simplicity we omit the 2-cells. In this figure they are all from left to right as shown hereafter and all fill the hexagons.

But there is an essential difference between the word problems and the combinatorial topological problems : here all the cells are oriented and these orientations are essential, except in the collapses where they have a purely **technical function** (Church-Rosser property).

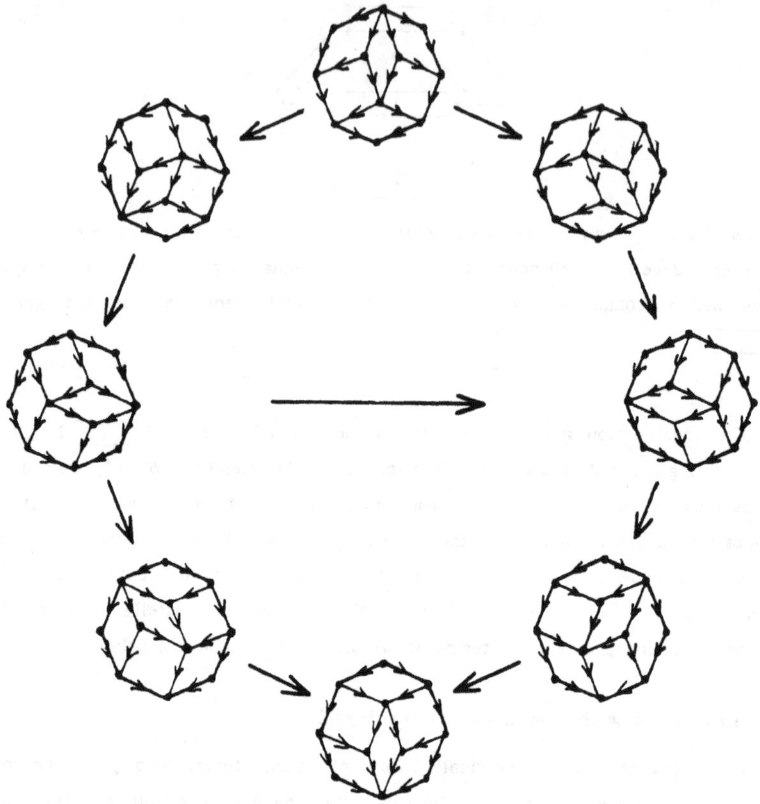

2. Rewriting systems.

The notion of algebraic theory was defined by Lawvere [Law] as equivalent to the datum of a (strict) cartesian category **T** in which every object is a cartesian power $S^n = S \times S \times \ldots \times S$ ($n \in \mathbb{N}$) of a unique object S of **T**. In this second part we show how such a structure is equivalent to the datum of a special 2-category (more precisely a special 2-monoid) and how the rewriting systems of terms in a universal algebra can be interpreted as CW-presentations of a 2-category. So we are in the context of 2-dimensional word problems.

2.1. 2-monoids.

Let n be an element of $\mathbb{N}$. Let us call *n-monoid* an n-category **C** such that $C_0 = \{*\}$ (i.e. a single 0-cell). A 1-monoid is of course a monoid.

A 2-monoid can also be interpreted as a *strict monoidal category*, that is a 3-uple $(\mathbf{C}, I, \otimes)$, where **C** is a category, I an object of **C**, and $\otimes : \mathbf{C} \times \mathbf{C} \longrightarrow \mathbf{C}$

a functor verifying the neutrality and associativity relations :

$$id(I) \otimes f = f = f \otimes id(I), \ (f \otimes g) \otimes h = f \otimes (g \otimes h)$$

for every morphisms f,g,h of **C** (the non necessarily strict monoidal categories are those where equalities are replaced with coherent natural equivalences).
The transformation which exhibits $(\mathbf{C}, I, \otimes)$ as a 2-monoid is similar to a "suspension" in topology : one "augments" the dimension of the elements of **C**. First one adds a 0-cell $*$, then each object X of **C** becomes a 1-cell $X : * \longrightarrow *$ and, finely, every morphism $f : X \longrightarrow Y$ of **C** becomes a 2-cell of the 2-monoid

$$* \ \underset{Y}{\overset{X}{\genfrac{}{}{0pt}{}{\longrightarrow}{\longrightarrow}}} \ * \quad (\downarrow f)$$

An homomorphism $F : (\mathbf{C}, I, \otimes) \longrightarrow (\mathbf{C}', I', \otimes')$ between two strict monoidal categories (identified to 2-categories) is a functor $F : \mathbf{C} \longrightarrow \mathbf{C}'$ which commutes with the operations : $F(I) = I'$ and $F(f \otimes g) = F(f) \otimes' F(g)$ (so it is a strict monoidal functor). We write **2-Mon** the category of 2-monoids and homomorphisms between them. It is of course a full sub-category of **2-Cat**.

An example of monoidal category is given with the cartesian categories $(\mathbf{C}, 1, \times)$, where 1 is a final object of **C** and $\times : \mathbf{C} \times \mathbf{C} \longrightarrow \mathbf{C}$ a cartesian product functor. Here we shall behave with these categories as if they were strict : in fact, we do nothing but follow the customs of algebrists and logicians (to make no difference between $S \times (S \times S)$, $(S \times S) \times S$ and $S \times S \times S$). The only cartesian category which will really interest us (see $\mathbb{F}^{op}$ in 2.2) is cartesian strict.

A first step towards the "elimination" of the universal property of the catesian product - for the benefit of an equational system - is given with the following result :

Proposition : *The strict monoidal category* $(\mathbf{C}, 1, \times)$ *is cartesian if and only if there exists two natural transformations* ε, δ, *which, for every* $X \in \mathbf{C}_0$, *define two morphisms*

$$1 \xleftarrow{\varepsilon(X)} X \xrightarrow{\delta(X)} X \times X$$

satisfying the relations :

$$(id(X) \times \varepsilon(X)) \circ \delta(X) = id(X) = (\varepsilon(X) \times id(X)) \circ \delta(X)$$

$$(id(X) \times \varepsilon(Y)) \times (\varepsilon(X) \times id(Y)) \circ \delta(X \times Y) = id(X \times Y)$$

$$\varepsilon(1) = id(1)$$

where X,Y *are arbitrary objects in* **C**.
When they exist, these data are unique up to isomorphism.

2.2. A CW-presentation of $\mathbb{F}$.

Let **Ens** be the category of sets and $\mathbb{F}$ the full subcategory of **Ens** whose objects are the natural integers ($\mathbb{F}_0 = \mathbb{N}$), where each integer n is the set

$\{0,1,\dots,n-1\}$.

$\mathbb{F}$ is a cartesian category, but it's also a cocartesian one, so $\mathbb{F}^{op}$ is cartesian and it is this last structure which interests us : the products in $\mathbb{F}^{op}$ are given by the coproducts in $\mathbb{F}$, and coincides with the usual addition in $\mathbb{N}$. For $p,p',q,q'\in\mathbb{N}$, the canonical injections of the coproduct $p \xrightarrow{\lambda} p+p' \xleftarrow{\lambda'} p'$ are defined through $\lambda(i) = i$ for $0\leq i\leq p-1$, and through $\lambda'(i) = p+1$ for $0\leq i\leq p'-1$. For $u : p \longrightarrow q$ and $u' : p' \longrightarrow q'$ in $\mathbb{F}$, the coproduct $u+u'$ is given with :

$$(u+u')(j) = \begin{cases} u(j) & \text{if } 0\leq j\leq p-1 \\ u'(j-1) & \text{if } p\leq j\leq p+p'-1 \end{cases}$$

Dualizing, this structure is then transformed in a cartesian category $(\mathbb{F}^{op},0,+)$ and one easily proves that it is strict. So we get an example of a 2-monoid which will play an essential part in 2.3.

Theorem : *The 2-monoid* $(\mathbb{F}^{op},0,+)$ *is finitely presented (in the sense of CW-presentations).*

A CW-presentation of this 2-monoid can be explicited. But, as $\mathbb{F}$ is more familiar than $\mathbb{F}^{op}$, we will give a presentation for the "dual" 2-monoid $(\mathbb{F},0,+)$, where only composition ("$\circ$") is dualized.

- adjonctions :
  - one 0-cell written : $*$
  - one 1-cell written : $1 : * \longrightarrow *$
  - three 2-cells written : $* \underset{1}{\overset{0}{\Downarrow\eta}} *$, $* \underset{1}{\overset{2}{\Downarrow\mu}} *$, $* \underset{2}{\overset{2}{\Downarrow\tau}} *$

where the integer n represents the "path" :1+1+...+1 (n times).

- collapses :

these ones are written hereafter by means of equations, instead of arrows, which lets the choice for their orientations (for the possibility of a choice giving a canonical system, see [La]). We did not avoid redundance (and we wrote n instead of id(n) for $n\in\mathbb{N}$) :

(1) $\mu\circ(\eta+1) = 1$

(2) $\mu\circ(1+\eta) = 1$

(3) $\mu\circ(\mu+1) = \mu\circ(1+\mu)$

(4) $\tau\circ(1+\eta) = \eta+1$

(5) $\tau\circ(\eta+1) = 1+\eta$

(6) $\tau\circ(\mu+1) = (1+\mu)\circ(\tau+1)\circ(1+\tau)$

(7) $\tau\circ(1+\mu) = (\mu+1)\circ(1+\tau)\circ(\tau+1)$

(8) $\tau\circ\tau = 2$

(9) $(1+\tau)\circ(\tau+1)\circ(1+\tau) = (\tau+1)\circ(1+\tau)\circ(\tau+1)$

(10) $\mu\circ\tau = \mu$

The formulation given in [Bu 1] was incomplete : the relation (10), which plays a subtile role in this system, was missing. We are thankful to Y. Laffont for pointing this out to us.

A graphic schematization of these equalities with the alphabet

is given through

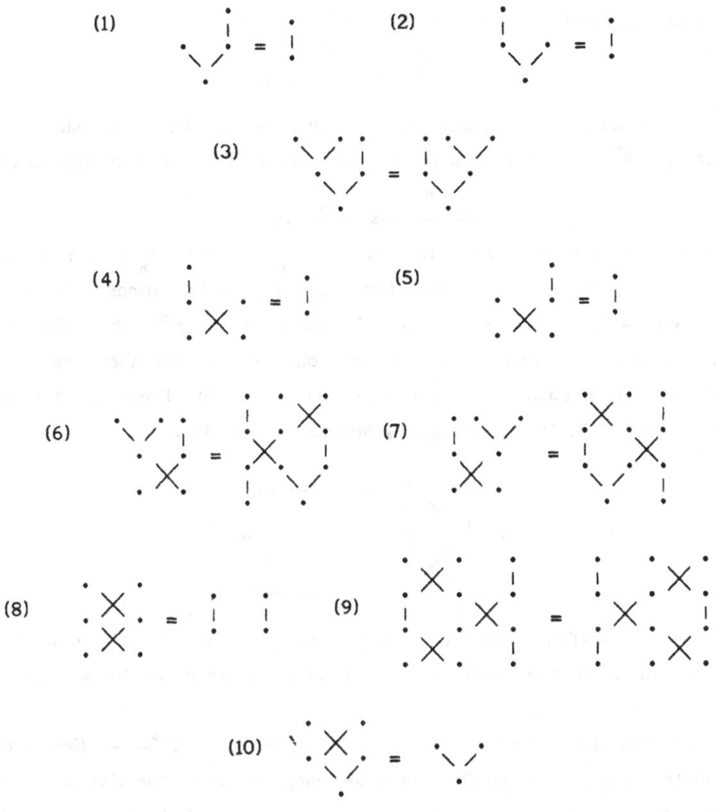

2.3. The main theorem.

An algebraic theory (with one sort) is presented by a system (Ω,E) of operations and equations (or rewriting system). This generates a strict cartesian category T = T(Ω,E) which has the following property : the datum of an (Ω,E)-algebra amounts to the datum of a strict cartesian functor F : T⟶Ens. In fact, if we take as canonical projections the products

$$X \xleftarrow{\mathrm{id}(X)\times\varepsilon(Y)} X\times Y \xrightarrow{\varepsilon(X)\times\mathrm{id}(Y)} Y ,$$

it suffices to consider F as a strict monoidal functor, because of the relations $F(1) = 1$, $F(f\times g) = F(f)\times F(g)$, and because of the unicity of $F(\varepsilon(X)) = \varepsilon(F(X))$.

Theorem : *The 2-monoid* **T** = T(Ω,E) *is finitely presented, in the sense* 1.2. *of CW-presentations, if and only if the algebraic theory* **T** *in the sense of Lawvere is finitely presented* (Ω *and* E *finite).*

In order to prove it, one first constructs, with the help of the ε,δ,τ as defined in 2.2, and for every $n\in\mathbb{N}$, two applications in $\mathbb{F}$:

$$0 \xrightarrow{\varepsilon_n} n \xleftarrow{\delta_n} n+n$$

Besides, the universal property of F determines a strict cartesian functor $i_T : \mathbb{F}^{op} \longrightarrow \mathbf{T}$ providing the images in **T** of the former applications :

$$1 \xleftarrow{\varepsilon_n} S^n \xrightarrow{\delta_n} S^{2n}.$$

(In order to simplify, we have written $\varepsilon_n = i_T(\varepsilon_n)$, $\delta_n = i_T(\delta_n)$). These morphisms provide candidates for two natural transformations ε,δ, through setting $\varepsilon(n) = \varepsilon_n$, $\delta(n) = \delta_n$. A priori (coming from $\mathbb{F}^{op}$) the relations in proposition 1.2 are satisfied, so we need only verify that they are actually transformations natural in **T** (and not only in $\mathbb{F}^{op}$). Precisely, for every $\alpha : S^p \longrightarrow S^q$ of **T**, the following diagrams must commute :

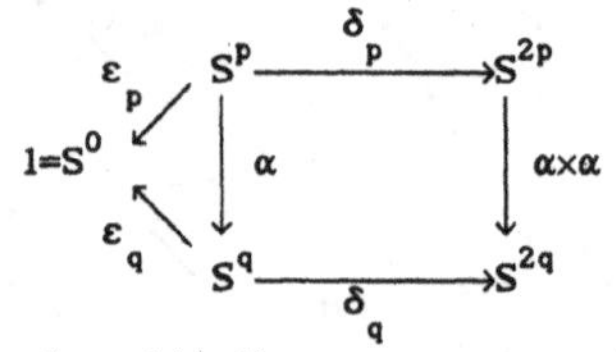

We show it is sufficient that this diagram commutes for the α's in Ω. Thus, if Ω is finite, we then have a finite number of equations to add to E.

Remark 1. In order to do computations in a **T**-algebra $F : \mathbf{T}^{op} \longrightarrow \mathbf{Ens}$ (and not only in the theory **T**) it suffices to make computations in the theory $\mathbf{T}_F$ which is obtained through the adjonction of the elements of F (i.e. of F(S)) as constants, and throughthe relations they satisfy. At last, if F is finitely presented, it is sufficient to add the finite data of this presentation : $\mathbf{T}_F$ will still be finitely presented (whenever **T** is).

Remark 2. This result extends to the many- sorted algebraic theories. The part played by $\mathbb{F}$ will be played by $\mathbb{F}^I$, for I the set of sorts.

This second part has been exposed in [Bu 1] and the proofs are in [Bu 2].

References.

[Bu 1] A. Burroni, Conférence (non-publiée) aux journées d'études E.L.I.T., 27 juin-2 juillet 1988.

[Bu 2] A. Burroni, *Le problème des mots dans les ∞-catégories*, Preprint (1991)

[Cur] P.L. Curien, *Substitution up to isomorphisms*, Actes des journées d'études E.L.I.T., 27 juin-2 juillet 1998, *Diagrammmes* vol. 23 (1990), 43-66.

[Laf] Y. Lafont, Private communication.

[Law] W. Lawvere, *Functorial semantics of algebraic theories*, Dissertation, Columbia University (1963).

[Str] R. Street, *The algebra of oriented simplexes*, J.P.A.A. 43 (1986) 235-242

BCK-FORMULAS HAVING UNIQUE PROOFS

Sachio Hirokawa
Department of Computer Science
College of General Education
Kyushu University, Fukuoka 810 JAPAN
e-mail : hirokawa@ec.kyushu-u.ac.jp

Abstract

The set of relevantly balanced formulas is introduced in implicational fragment of BCK-logic. It is shown that any relevantly balanced formula has unique normal form proof. Such formulas are defined by the 'relevance relation' between type variables in a formula. The set of balanced formulas (or equivalently one-two-formulas) is included in the relevantly balanced formulas. The uniqueness of normal form proofs is known for balanced formulas as the coherence theorem. Thus the result extends the theorem with respect to implicational formulas. The set of relevantly balanced formulas is characterized as the set of irrelevant substitution instances of principal type-schemes of BCK-λ-terms.

1 Introduction

What kind of formulas have unique normal form proof? The coherence theorem in cartesian closed categories gives a solution for the problem. (A historical comment is in Lambek and Scott [14].) Simple proofs in terms of typed λ-calculus are given by Mints [13] and by Bavaev and Solov'ev [1]. The theorem states that the balanced formulas have unique normal form proof. A formula is balanced iff no type variable occurs more than twice. We use the word 'types' and the word 'formulas' in the same meaning according to Curry-Howard isomorphism [10]. In the same line, a proof for a formula α is a λ-term which has α as its type-scheme. In this paper we show a new class of implicational formulas which have this uniqueness property. All the balanced formulas belong to this class. So the result is an extension of the coherence theorem with respect to implicational formulas. (Babaev and Solov'ev [1] and Mints [13] consider not only implicational formulas but also conjunctive formulas. The present paper treats only implicational fragment.)

Balanced formulas are identical to the formulas with one-two-property in [5, 11]. Jaskowski [11] proved that a one-two-formula is provable in intuitionistic logic iff it is provable in BCK-logic. The BCK-logic is a logic in which one can use an assumption only once. (In Gentzen style sequential formulation, the logic is equivalent to the intuitionistic logic LJ lacking weakening rule. (See [15].) Implicational part of linear logic [2] is contained in BCK-logic.) Moreover, it was proved in [9] that any proof for a one-two-formula is a proof in BCK-logic. So the result in the present paper is an extension of the coherence theorem even though we consider only BCK-formulas and proof figures in BCK-logic.

From a different motivation, Komori [12] introduced the notion of BCK-minimal formulas and conjectured that BCK-minimal formulas have unique BCK-proofs in normal form. A BCK-minimal formula is a BCK-provable formula which is not a non-trivial substitution instance of other BCK-provable formulas. Affirmative solutions were given independently by Wronski [17], by Hirokawa [6] and by Tatsuta [16]. In [6], the author proved the following theorem.

> If two BCK-λ-terms in β-normal form have the same principal type-scheme, then they are identical.

A principal type-scheme of a λ-term is a most general type-scheme of the term. Since the minimality implies the principality, Komi's conjecture can be proved by the above theorem.

The principality is essential in the above theorem. However the principality is not always necessary for the uniqueness. Consider a type $\alpha = (e \to f) \to f \to c \to c$ and a λ-term $\lambda xyz.z$ which can be type assigned to α by the following type-assignment.

$$\dfrac{\dfrac{\dfrac{z : c}{\lambda z.z : c \to c}}{\lambda yz.z : f \to c \to c}}{\lambda xyz.z : (e \to f) \to f \to c \to c}$$

Traverse this type-assignment figure upward. Then we notice that $\lambda xyz.z$ is the unique closed λ-term in β-normal form having this type. Moreover we see that a principal type-scheme of the term is $\beta = a \to b \to c \to c$ and that α is a substitution instance of β by a substitution $[a := e \to f, b := f]$. By principal type-scheme theorem (Theorem 15.26 in [3]), every type-scheme of a λ-term is a substitution instance of a principal type-scheme of the λ-term. Taking into account of the uniqueness of λ-terms for the same principal type-schemes, we can rephrase our problem as follows.

> What kind of substitution keeps the uniqueness of λ-term?

The substitution in the above example only changes the type variables which correspond to vacant variables x and y. If we call these type variables 'irrelevant', then θ does not identify a relevant type variable with distinct type variables. In section 2, we formally define the notion of these 'relevance' and 'irrelevance' of type variables in a type. Then we define the set of 'irrelevant' substitutions which do not cause any identification of a relevant type variables. Characterizations of relevant type variables in a principal type-scheme of a BCK-λ-terms are shown. In section 3, we clarify the set of types obtained by irrelevant substitutions from principal type-schemes of BCK-λ-terms. We prove that the set is identical to the set of 'relevantly balanced' types in which every relevant type variable is balanced. In section 4, we prove that relevantly balanced formulas have unique normal form proofs. Subset relation of these sets of formulas is shown in Figure 1.

Some possibilities are left that a relevantly balanced formula has other non-BCK-proofs. We conjecture that this is not the case. I.e., we conjecture that if a λ-term in β-normal form has a relevantly balanced type, then it is a BCK-λ-term.

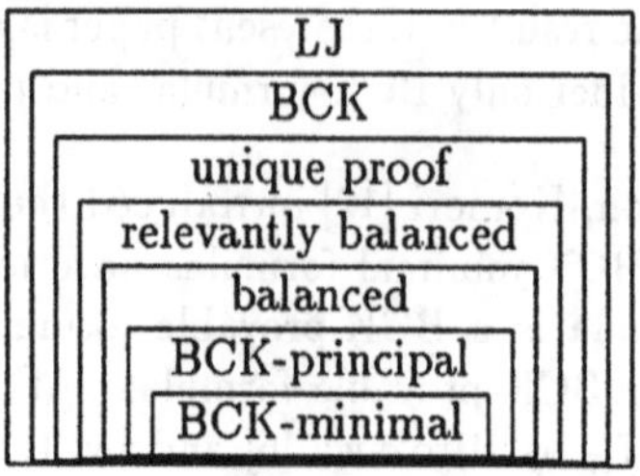

Figure 1: Hierarchy of BCK-formulas having unique proofs

We refer the basic notions to [3] for type assignment systems to λ-terms. We say TA-figures instead of type-assignment figures. Types are constructed from type variables and implicational symbol '$\rightarrow$'. We use the letters $a, b, c, \cdots$ for type variables, $\alpha, \beta, \gamma, \cdots$ for types, $x, y, z, \cdots$ for term variables, $L, M, N, \cdots$ for λ-terms and $\sigma, \tau, \theta, \cdots$ for simultaneous substitutions. The set of type variables in a type α is denoted by $var(\alpha)$. The set of free variables in a λ-term M is denoted by $FV(M)$. We write $B|\!-M : \alpha$ for a type assignment α to M using an assumption set $B = \{x_1 : \alpha_1, \cdots, x_n : \alpha_n\}$. Note that the set $\{x_1, \cdots, x_n\}$ of subjects in B is identical to $FV(M)$. When (B, α) is not a non-trivial substitution instance of any (B', α') such that $B'|\!-M : \alpha'$, it is said to be a principal type-assignment and the pair (B, α) is called a principal-pair. Then we write $B||\!-M : \alpha$. When M is a closed λ-term and $B = \emptyset$, α is said to be a principal type-scheme of M.

2 Relevant and irrelevant type variables

In this section, the relevance relation between type variables in a type is introduced. Then relevant type variables and irrelevant type variables in a type are defined. Intuitively, a type variable is relevant to a type when it is essentially used in the deduction of the type. Characterizations of relevant type variables is given for principal type-schemes of BCK-λ-terms.

A λ-term is a BCK-λ-term when no variable occurs twice in it. The set of BCK-λ-terms is inductively defined as follows.

Definition 1 (BCK-λ-terms) *Any variable x is a BCK-λ-term. If M and N are BCK-λ-terms such that $FV(M) \cap FV(N) = \emptyset$, then (MN) is a BCK-λ-term. If M is a BCK-λ-term and x is a variable, then $(\lambda x.M)$ is a BCK-λ-term.*

The notion of connection was introduced in [6] for the analysis of principal type assignment figures. A connection is a series of the occurrences of the same type in a TA-figure. Each occurrence of the type in the same connection has the same meaning through the deduction.

Definition 2 (connection [6]) *Let M ba a λ-term and P be a TA-figure for $B|\!-M : \beta$. Two occurrences of the same type γ in P are said to be directly connected to each other iff one of (1),(2) and (3) holds:*

(1) Both γ appear at the same position in α in distinct occurrences of the same assumption $x : \alpha$.

$$\cdots \quad x : \overset{\frown}{\alpha} \quad \cdots \quad x : \alpha \quad \cdots$$

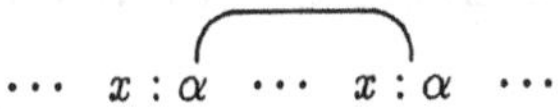

(2) (a) One γ appears at a position in α of the major premise $M : \alpha \to \beta$ of an $(\to E)$ rule and the other γ appears at the same position in α of the minor premise $N : \alpha$ of the $(\to E)$ rule, or

(b) one γ appears at a position in β of the major premise $M : \alpha \to \beta$ of an $(\to E)$ rule and the other γ appears at the same position in β of the consequence $MN : \beta$ of the $(\to E)$ rule.

$$\frac{M : \overset{\frown}{\alpha \to \beta} \quad N : \alpha}{MN : \beta} \ (\to E)$$

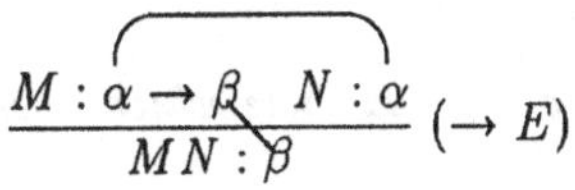

(3) (a) One γ appears at a position in α of the assumption $x : \alpha$ and the other γ appears at the same position in α of the consequence $\lambda x.M : \alpha \to \beta$, or

(b) one γ appears at a position in β of $M : \beta$ and the other γ appears at the same position in β of $\lambda x.M : \alpha \to \beta$.

$$\frac{\begin{array}{c}[x : \alpha] \\ \vdots \\ M : \beta\end{array}}{\lambda x.M : \alpha \to \beta} \ (\to I)$$

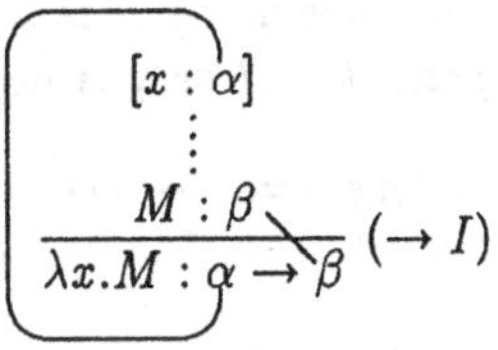

The connection relation is the reflexive, transitive and symmetric closure of the direct connection relation.

Remark 1 *Case (1) does not happen to BCK-λ-terms.*

Lemma 1 ([7]) *Let M be a closed λ-term and P be a TA-figure for $|-M : \alpha$. Then α is a principal type-scheme of M iff (1) and (2) hold for any type variable a in α.*

(1) All occurrences of a in P are connected.

(2) One of the following (a) and (b) holds.

(a) There is a subterm N of M such that P contains a TA-formula of the form $N : a$.

(b) There is a subterm $\lambda x.N$ of such that $x \notin FV(N)$ and P contains a TA-formula of the form $\lambda x.N : a \to \beta$. ∎

We call the terms N in (2-a) and $\lambda x.N$ in (2-b) the *adjoint terms* to the type variable a.

The following two lemmas determine the structure of subfigures of principal type assignment for BCK-λ-terms.

Lemma 2 ([7]) *Let $xM_1 \cdots M_n$ be a BCK-λ-term in β-normal form and $B||\!-xM_1 \cdots M_n : \alpha$. Then $\alpha = a$ for some type variable a, $x : \alpha_1 \to \cdots \to \alpha_n \to a \in B$ and the following (a),(b) and (c) hold.*

(a) *$a \notin var(B_i, \alpha_i)$ for $i = 1, \cdots, n$.*

(b) *$var(B_i, \alpha_i) \cap var(B_j, \alpha_j) = \emptyset$ for $i \neq j$.*

(c) *$B_i||\!-M_i : \alpha_i$ for $B_i = \{y : \gamma \in B \mid y \in FV(M_i)\}$.* ∎

Lemma 3 ([7]) *Let $\lambda x.M$ be a BCK-λ-term in β-normal form and $B||\!-\lambda x.M : \alpha \to \gamma$. Then (a) and (b) hold.*

(a) *If $x \notin FV(M)$ then α is a type variable which does not occur in B or in γ and $B||\!-M : \gamma$.*

(b) *If $x \in FV(M)$ then $B \cup \{x : \alpha\}||\!-M$.* ∎

The positive and negative occurrence of a subtype in a type is defined inductively as usual.

Definition 3 (positive/negative occurrence) *α is positive in α. γ is positive (negative) in $\alpha \to \beta$ iff γ is positive (negative) in β or γ is negative (positive) in α.*

Definition 4 (core) *The core of a type $\alpha = \alpha_1 \to \cdots \to \alpha_n \to a$ is the right-most type variable a in α. We denote it by $core(\alpha)$.*

Definition 5 (relevant type variable) *Let b,c be type variables in α. We write $c \succ_\alpha b$ iff there is a negative occurrence of subtype β in α of the form $\beta = (\gamma \to \cdots \to c) \to \cdots \to b$. We define $\succ_\alpha^*$ as the reflexive and transitive closure of $\succ_\alpha$. We say that c is relevant to b in α when $c \succ_\alpha^* b$. A type variable in α is relevant to α iff $a \succ_\alpha^* core(\alpha)$. a is irrelevant to α iff a is not relevant to α.*

Theorem 1 *Let M be a closed BCK-λ-term in β-normal form, $||\!-M : \alpha$ and $a \in var(\alpha)$.*

(1) *Following (a),(b),(c) and (d) are equivalent.*

  (a) *The adjoint term to a has the form $N : a$.*

  (b) *a occurs positive in α exactly once and a occurs negative in α exactly once.*

  (c) *a occurs exactly twice in α.*

  (d) *a is relevant to α.*

(2) *Following (a),(b),(c) and (d) are equivalent.*

  (a) *The adjoint term to a has the form $\lambda x.N : a \to \beta$, $x \notin FV(N)$.*

(b) a occurs exactly once in α and its occurrence is negative.

(c) a occurs exactly once in α.

(d) a is irrelevant to α.

To prove this theorem, we need to extend the notion of relevance for a pair (B, α) of assumption set B and a type α.

Definition 6 *Let $B = \{x_1 : \alpha_1, \cdots, x_n : \alpha_n\}$ be an assumption set, α be a type and a, b, c be type variables in B or in α. $c \succ_{(B,\alpha)} b$ iff $c \succ_\alpha b$ or $c \succ_{\alpha_i} b$ for some α_i. We define $\succ^*_{(B,\alpha)}$ as the reflexive and transitive closure of $\succ_{(B,\alpha)}$. a is relevant to (B, α) iff $a \succ^*_{(B,\alpha)} core(\alpha)$.*

When the assumption set B is empty, the above definition is identical to the previous definition of relevance. The following Lemma 4 is an extension of $(c) \Rightarrow (d)$.

Lemma 4 *Let M be a BCK-λ-term ,$B||\!-M : \alpha$ and a be a type variable in (B, α). If a occurs exactly twice in (B, α), then a is relevant to (B, α).*

Proof. By induction on M.

Case 1. $M = x$. Then α is a type variable, say $\alpha = a$. Thus $B = \{x : a\}$. Therefore a is relevant to $(\{x : a\}, a)$.

Case 2. $M = xM_1 \cdots M_n (n \geq 1)$. Then $\alpha = b$ for some type variable b and $x : \zeta_1 \to \cdots \to \zeta_n \to b \in B$. The TA-figure for $B||\!-M : b$ has the following form.

$$\frac{x : \zeta_1 \to \cdots \to \zeta_n \to b \quad M_1 : \zeta_1 \quad \cdots \quad Mn : \zeta_n}{xM_1 \cdots M_n : b}$$

Subcase 2.1. $a = b$. Then b is relevant to (B, b).

Subcase 2.2. $a \neq b$. Since a occurs exactly twice in (B, b), a occurs exactly twice in (B_i, ζ_i) for some unique i, by Lemma 2. By induction hypothesis for $B_i||\!-M_i : \zeta_i$, a is relevant to (B_i, ζ_i). So there is a sequence $a \succ_{(B_i,\zeta_i)} a_1, a_1 \succ_{(B_i,\zeta_i)} a_2, \cdots, a_n \succ_{(B_i,\zeta)} core(\zeta_i)$. These relevant relation hold in (B, b). Moreover we have $core(\zeta_i) \succ_{(B,b)} b$. Therefore $a \succ^*_{(B,b)} b$. Thus a is relevant to (B, b).

Subcase 3. $M = \lambda x.N$. Then $\alpha = \xi \to \zeta$. Then a TA-figure for $B : ||\!-\lambda x.N : \xi \to \zeta$ has the following form.

$$\frac{N : \xi}{\lambda x.N : \xi \to \zeta}$$

Subcase 3.1. $x \notin FV(N)$. Then by Lemma 3, $\xi = b$ for some type variable b, $b \notin var(B, \zeta)$ and $B||\!-N : \zeta$. Since b occurs only once, we have $a \neq b$. Therefore a occurs exactly twice in (B, ζ). By induction hypothesis for $B||\!-N : \zeta$, a is relevant to (B, ζ). Therefore a is relevant to $(B, \xi \to \zeta)$.

Subcase 3.2. $x \in FV(N)$. By Lemma 3 we have $B \cup \{x : \xi\}||\!-N : \zeta$. Since a occurs exactly twice in $(B, \xi \to \zeta)$, a occurs exactly twice in $(B \cup \{x : \xi\}, \zeta)$. By induction hypothesis a is relevant to $(B \cup \{x : \xi\}, \zeta)$. Therefore a is relevant to $(B, \xi \to \zeta)$. ∎

To prove $(d) \Rightarrow (a)$ we need the following lemma. The lemma says that a relevance relation between type variables in a sub-figure is determined by an occurrence of sub-type in the sub-figure.

Lemma 5 *Let $xM_1 \cdots M_n$ be a BCK-λ-term in β-normal form, $B||\!-xM_1 \cdots M_n : a$, $x : \alpha_1 \to \cdots \to \alpha_n \to a \in B$. If $b \succ_{(B,a)} c$ for type variable b and c in (B,a) and $c \neq a$, then there exists a unique (B_i, α_i) such that $b, c \in var(B_i, \alpha_i)$ and $b \succ_{(B_i,\alpha_i)} c$, where $B_i = \{y : \gamma \mid y \in FV(M_i)\}$.*

Proof. Let $B = \{x_1 : \zeta_1, \cdots, x_m : \zeta_m\}$. Since $x : \alpha_1 \to \cdots \to \alpha_n \to a \in B$, $x_q = x$ and $\zeta_q = \alpha_1 \to \cdots \to \alpha_n \to a$ for some q. By the definition of $b \succ_{(B,a)} c$, there is a negative occurrence of some subtype ξ in $\zeta_1 \to \cdots \to \overbrace{(\alpha_1 \to \cdots \to \alpha_n \to a^-)}^{\zeta_q} \to \cdots \to \zeta_m \to a^+$ of the form $\xi = (\nu_1 \to \cdots \to \nu_k \to b) \to \cdots \to c$. Since $c \neq a$, ξ is a subtype of $\zeta_p (p \neq q)$ or ξ is a subtype of α_i.

Case 1. ξ is a subtype of ζ_p. Since $x_p : \zeta_p \in B$, we have $x_p : \zeta_p \in B_i$ for some i. Therefore we have $b, c \in var(B_i, \alpha_i)$. Note that the sign of ξ in (B_i, α_i) is identical to the sign of ξ in (B, a). Thus we have $b \succ_{(B_i,\alpha_i)} c$.

Case 2. ξ is a subtype of α_i. Then we have $b, c \in var(B_i, \alpha_i)$. Moreover we have $b \succ_{(B_i,\alpha_i)} c$, since the sign of ξ in (B_i, α_i) is identical to the sign of ξ in (B, a). ∎

$(d) \Rightarrow (a)$ is proved as a special case of the following lemma.

Lemma 6 *Let M be a BCK-λ-term in β-normal form, P a TA-figure for $ptaM : \alpha$ and $a \in var(B, \alpha)$. If $a \succ^*_{(B,\alpha)} core(\alpha)$ then P contains a TA-formula of the form $N : a$ for some sub-term N of M.*

Proof. By induction on M and the length of $a \succ^*_{(B,\alpha)} core(\alpha)$.

Base step. $M = x$ is a variable. Then $\alpha = a$ is a type variable and $B = \{x : a\}$. Thus $N = x$ is the desired term.

Induction step. Case 1. $M = \lambda x.L$. Then $\alpha = \gamma \to \delta$.

Subcase 1.1. $x \notin FV(L)$. Then γ is a type variable which does not occur in δ or B. Therefore we have $a \succ^*_{(B,\delta)} core(\alpha)$. By induction hypothesis for $B||\!-L : \delta$, L contains a sub-term N such that $N : a$. Thus the lemma holds.

Subcase 1.2. $x \in FV(L)$. By lemma 3, $B \cup \{x : \gamma\}||\!-L : \delta$. On the other hand, we have $a \succ^*_{(B\cup\{x:\gamma\},\delta)} core(\delta)$. By induction hypothesis for $B \cup \{x : \gamma\}||\!-L : \delta$, there is a sub-term N of L such that $N : a$.

Case 2. $M = xM_1 \cdots M_n$. Then $\alpha = c$ is a type variable. The TA-figure for M has the following form.

$$\frac{x : \alpha_1 \to \cdots \to \alpha_n \to b \quad M_1 : \alpha_1 \quad \cdots \quad M_i : \alpha_i \quad \cdots \quad Mn : \alpha_n}{xM_1 \cdots M_i \cdots M_n : c}$$

From $a \succ^*_{(B,\alpha)} core(\alpha)$, there is a type variable $b \in var(B, c)$ such that $a \succ^*_{(B,\alpha)} b$ and $b \succ_{(B,\alpha)} c$.

Subcase 2.1. $b = c$. Then by induction hypothesis for $a \succ^*_{(B,c)} c$, lemma holds.

Subcase 2.2. $b \neq c$. Then by Lemma 5 we have $b, c \in var(B_i, \alpha_i)$ and $a \succ^*_{(B_i,\alpha_i)} b$ for some i, where $B_i = \{y : \gamma \in B \mid y \in FV(M_i)\}$. Since $b \succ_{(B,c)} c$ we have $b = core(\alpha_i)$ for some j. Note that $var(B_p, \alpha_p) \cap var(B_q, \alpha_q) = \emptyset$ for $p \neq q$. Therefore $i = j$. Therefore $a \succ^*_{(B_i,\alpha_i)} core(\alpha_i)$. By induction hypothesis for $B_i||\!-M_i : \alpha_i$, there is a sub-term N of M_i such that P contains a TA-formula $N : a$. Thus the lemma holds. ∎

Proof of Theorem 1. We prove (1) by showing $(a) \Rightarrow (b) \Rightarrow (c) \Rightarrow (d) \Rightarrow (a)$. (2) is similar.

To prove (a) $\Rightarrow$ (b), assume that $N : a$ appears in the TA-figure P for $||\!-M : \alpha$. Then N has the form $N = xN_1 \cdots N_n$ and P has the following form.

$$P_j \left\{ \begin{array}{c} P_1 \left\{ \dfrac{x : \alpha_1 \to \cdots \to \alpha_n \to a \quad \overset{\vdots}{N_1 : \alpha_1} \quad \cdots \quad \overset{\vdots}{N_n : \alpha_n}}{xN_1 \cdots N_n : a} \right. \\ \vdots \\ L_j : \gamma_j \end{array} \right.$$
$$\vdots$$
$$M : \alpha$$

Consider the sequence of TA-formulas $L_j : \gamma_j$ in P from $xN_1 \cdots N_n : a$ to $M : \alpha$ and sub-TA-figure P_j whose end-formula is $L_j : \gamma_j$. At the first step $j = 1$, a does not occur above $N_i : \alpha_i$. Therefore by Lemma 2, a occurs positive in $xN_1 \cdots N_n : a$ and in $x : \alpha_1 \to \cdots \to \alpha_n \to a$. We can see that at any step j, a occurs exactly twice in some assumption or in end-formula $L_j : \gamma_j$. Moreover the sign of such an occurrence is positive. At the final step, i.e., at $M : \alpha$, a occurs positive exactly once and occurs negative exactly once, since there is no assumption. Thus (a)$\Rightarrow$ (b) is proved. (b)$\Rightarrow$(c) is trivial. (c)$\Rightarrow$(d) is a special case of Lemma 4. (d)$\Rightarrow$(a) is proved as as special case of Lemma 6. ∎

3 Irrelevant substitutions and relevantly balanced types

An irrelevant substitution to a type is a substitution to the type which does not cause any identification of a relevant type variable and other type variable. It is shown in this section that the set of irrelevant substitution instances of principal type-schemes of closed BCK-λ-terms is identical to the set of relevantly balanced types. A type is relevantly balanced when each relevant type variable in the type occurs twice with opposite sign.

Definition 7 (irrelevant substitution/strictly irrelevant substitution) *Let α be a type and θ be a substitution. θ is irrelevant to α iff for any relevant type variable a in α, every type variable in $a\theta$ occurs exactly twice in $a\theta$. An irrelevant substitution θ is strictly irrelevant iff $a\theta = a$ for all relevant type variable a in α.*

We denote by BCK and by BCK-β the set of closed BCK-λ-terms and the set of closed BCK-λ-terms in β-normal form respectively. The set of principal type-schemes of closed BCK-λ-terms is denoted by pts(BCK). The set of principal type-schemes of closed BCK-λ-terms in β-normal form is denoted by pts(BCK). Since every BCK-λ-term has type-scheme (see [4]), we have $pts(BCK) = pts(BCK\text{-}\beta)$.

Lemma 7 *Let $\alpha \in pts(BCK\text{-}\beta)$ and θ be a strictly irrelevant substitution to α and b be a type variable in α. If b is irrelevant to α, then any type variable c in $b\theta$ is irrelevant to $\alpha\theta$.*

Proof. It suffices to show that if there is a type variable c in $var(b\theta)$ which is relevant to $\alpha\theta$, then b is relevant to α and $c = b\theta = b$. We prove this claim by induction on the length of $c \succ^*_{\alpha\theta} core(b\theta)$.

Base step. $c = core(\alpha\theta)$. Let a be the core of α and $\alpha = \alpha_1 \to \cdots \to \alpha_n \to a$. Then a is relevant to α. Therefore $a\theta = a$. Note that $core(\alpha\theta) = core(\alpha_1\theta \to \cdots \to \alpha_n\theta \to a\theta)$ $= core(a\theta) = core(a) = a$. Thus we have $c \in var(b\theta)$ and $c \in var(a\theta)$. Since θ is irrelevant and a is relevant to $\alpha\theta$, it follows that $a = b$. Therefore b is relevant to α.

Induction step. $c \succ_{\alpha\theta} d$ and $d \succ^*_{\alpha\theta} core(\alpha\theta)$ for some $d \in var(\alpha\theta)$. Since $d \in var(\alpha\theta)$, there is some type variable b' such that $d \in var(b'\theta)$. By induction hypothesis for $d \succ^*_{\alpha\theta} core(\alpha\theta)$, b' is relevant to α and $b'\theta = b'$. Therefore $d = b'$ and d is relevant to α. By the definition of $c \succ_{\alpha\theta} d$, there is an occurrence of subtype γ of $\alpha\theta$ of the form $\gamma = \gamma_1 \to \cdots \to \gamma_i \to \cdots \to \gamma_l \to d$ such that $c = core(\gamma_i)$ and the core d occurs negative in $\alpha\theta$. Since d is relevant to α, we have $d\theta = d$. Therefore α contains, at the same position of γ in $\alpha\theta$, a subtype $\xi = \xi_1 \to \cdots \to \xi_i \to \cdots \to \xi_l \to d$ such that $\xi_j\theta = \gamma_j$ $(j = 1, \cdots, l)$. Consider the core e of ξ_i. We have c=$core(\gamma_i)$=$core(\xi_i\theta)$=$core(e\theta)$. Since the core d of γ is negative in $\alpha\theta$, the core d of ξ is negative in α. Thus e is relevant to α. Thus we have $c \in var(e\theta)$ and $c \in var(b\theta)$. Since θ is irrelevant to α, it follows that $b = e$. Therefore b is relevant to α. ∎

Lemma 8 *Let $\alpha \in pts(BCK\text{-}\beta)$, θ be a substitution to α and b be a type variable in α. If b is relevant to α, then $core(b\theta)$ is relevant to $\alpha\theta$.*

Proof. Since b is relevant to α, $b \succ^*_\alpha core(\alpha)$. We prove the lemma by induction on the length of $\succ^*_\alpha$.

Base step. $b = core(\alpha)$. Then α has the form $\alpha = \alpha_1 \to \cdots \to \alpha_n \to b$. Therefore $\alpha\theta = \alpha_1\theta \to \cdots \to \alpha_n\theta \to b\theta$. So we have $core(\alpha\theta) = core(b\theta)$. Thus $core(b\theta)$ is relevant to $\alpha\theta$.

Induction step. $b \succ_\alpha b'$ and $b' \succ^*_\alpha core(\alpha)$ for some type variable b'. By the definition of $b \succ_\alpha b'$, there is an occurrence of subtype $\gamma = \gamma_1 \to \cdots \to \gamma_i \to \cdots \to \gamma_l \to b'$ of α such that the occurrence of b' is negative in α and $\gamma_i = \xi_1 \to \cdots \to \xi_m \to b$. Now consider the corresponding occurrence of $\gamma\theta$ in $\alpha\theta$. We have $\gamma\theta = \gamma_1\theta \to \cdots \to (\xi_1\theta \to \cdots \to \xi_m\theta \to b\theta) \to \cdots \to \gamma_l\theta \to b'\theta$. Since the occurrence of b' is negative in α, the occurrence of the $core(b'\theta)$ is negative in $\alpha\theta$. By induction hypothesis for $b' \succ^*_\alpha core(\alpha)$, $core(\alpha\theta)$ is relevant to $\alpha\theta$. Therefore $core(b\theta)$ is relevant to $\alpha\theta$. ∎

Definition 8 (structural/identification substitution) *Let $\theta = [a_1 := \alpha_1, \cdots, a_n := \alpha_n]$ be a substitution and α be a type. θ is structural to α iff for each i either (1) or (2) holds.*

(1) $\alpha_i = a_i$.

(2) α_i is not a type variable, $var(\alpha_i) \cap var(\alpha) = \emptyset$ and $var(\alpha_i) \cap var(\alpha_j) = \emptyset$ for $i \neq j$.

A structural substitution θ is said to be irrelevant iff $\alpha_i = a_i$. θ is said to be relevant iff θ is not irrelevant. θ is identification iff α_i is a type variable and θ does not contain trivial rewriting of type variables.

Lemma 9 (decomposition of substitution [8]) *Let M be a closed BCK-λ-term in β-normal form which has a principal type-scheme γ and α be an substitution instance of γ by θ, i.e, $\alpha = \gamma\theta$. Then there are closed λ-term L in β-normal form, a relevant structural substitution σ_1 to γ, an irrelevant structural substitution σ_2 to γ and an identification substitution to $\gamma(\sigma_1+\sigma_2)$ such that $||{-}L : \gamma\sigma_1$, L is η-reducible to M and $\gamma\theta = \gamma(\sigma_1+\sigma_2)\tau$.*

∎

$$
\begin{array}{rll}
||{-}M : & \gamma & \\
\eta\uparrow & \downarrow\sigma_1 & \begin{array}{l} \textit{relevant} \\ \textit{structural} \end{array} \\
||{-}L : & \gamma\sigma_1 & \\
 & \downarrow\sigma_2 & \begin{array}{l} \textit{irrelevant} \\ \textit{structural} \end{array} \\
|{-}L : & \gamma(\sigma_1+\sigma_2) & \\
 & \downarrow\tau & \textit{identification} \\
|{-}M : & \gamma(\sigma_1+\sigma_2)\tau &
\end{array}
$$

Definition 9 (relevantly balanced types) *Let a be a type variable in a type α. a is essentially balanced in α iff a occurs positive in α exactly once and a occurs negative in α exactly once. α is relevantly balanced iff every relevant type variable in α is essentially balanced in α.*

Lemma 10 *Let M be a closed BCK-λ-term in β-normal form and γ be a principal type-scheme of M. If θ is irrelevant to γ, then $\gamma\theta$ is relevantly balanced.*

Proof. By decomposition of θ we have $\theta = (\sigma_1+\sigma_2)\tau$, where σ_1 is a relevant structural substitution, σ_2 is an irrelevant substitution and τ is an identification substitution. Let a be an arbitrary relevant type variable in α. Since θ is irrelevant, τ identifies only irrelevant type variables. Therefore $\sigma_2\tau$ is strictly irrelevant to $\gamma\sigma_1$. By Lemma 7 we have $a \in var(\gamma\sigma_1)$, $a\sigma_2\tau = a$ and that a is relevant to $\gamma\sigma_1$. By Theorem 1, a is balanced in $\gamma\sigma_1$. Since τ does not identify a with other type variable, a is balanced in $\alpha = \gamma(\sigma_1+\sigma_2)\tau$. ∎

Lemma 11 *Let γ be a principal type-scheme of a closed BCK-λ-term M in β-normal form and θ be a substitution. If $\gamma\theta$ is relevantly balanced, then in the decomposition of θ, $\sigma_2\tau$ is strictly irrelevant to $\gamma\sigma_1$.*

Proof. To prove that $\sigma_2\tau$ is irrelevant to $\gamma\sigma_1$, let us assume the opposite and see what happens. If $\sigma_2\tau$ were not irrelevant to $\gamma\sigma_1$ then one of (1) and (2) holds.

(1) There are distinct type variables c and d in $\gamma\sigma_1$ such that c occurs twice in $\gamma\sigma_1$ and $var(c\sigma_2\tau) \cap var(d\sigma_2\tau) \neq \emptyset$.

(2) There is a type variable c in $\gamma\sigma_1$ such that c occurs twice in $\gamma\sigma_1$ and some type variable occurs twice in $c\sigma_2\tau$.

If c occurs twice in $\gamma\sigma_1$, then c is relevant to $\gamma\sigma_1$ by Theorem 1. Therefore $c\sigma_2 = c$ and $c\sigma_2\tau$ is a type variable. Therefore the second case can not happen. Put $f = c\sigma_2\tau$. In case (1), there is a type variable e in $d\sigma_2$ which is identified to f by τ. Then f occurs three times in $\gamma(\sigma_1+\sigma_2)\tau$. By Lemma 8, f is relevant to α. This contradicts that α is relevantly balanced. Therefore $\sigma_2\tau$ is irrelevant to $\beta\sigma_1$. ∎

$$\begin{array}{c c c}
||\!-M : \gamma & & \\
\eta \uparrow \sigma_1 \downarrow & & \\
||\!-L : \gamma\sigma_1 & c^+ & c^- \\
\sigma_2 \downarrow & \downarrow & \downarrow \\
|\!-L : \gamma(\sigma_1 + \sigma_2) & c & c \\
\tau \downarrow & \downarrow & \downarrow \\
|\!-L : \gamma(\sigma_1 + \sigma_2)\tau & f & f
\end{array}$$

Theorem 2 *Let α be a type-scheme of a closed BCK-λ-term. Then α is relevantly balanced iff α is an irrelevant substitution instance of some type in $pts(BCK\text{-}\beta)$.*

Proof. By Lemma 10 and Lemma 11. ∎

Remark 2 *The above theorem holds even if we replace irrelevant substitution by strictly irrelevant substitution.*

Remark 3 *Provability of α is essential in only-if-part of Theorem 2. A type $(((b \to a) \to c) \to b) \to c \to a$ is relevantly balanced, but it is not provable even in classical logic. So it is not a type-scheme of any λ-term.*

4 Uniqueness of normal form proofs for relevantly balanced formulas

In this section, we prove that if a relevantly balanced type is provable in BCK-logic, then it has unique normal form proof.

Lemma 12 *Let M be a closed BCK-λ-term in β-normal form, P be a TA-figure for $B||\!-M : \alpha$, $B = \{x_1 : \alpha_1, \cdots, x_n : \alpha_n\}$ and $a \in var(B, \alpha)$. If one of (a),(b) and (c) holds then P contains a TA-formula of the form $N : a$ for some subterm N of M.*

(a) $a = core(\alpha)$.

(b) α *contains a subtype of the form* $\gamma_1 \to \cdots \to \gamma_l \to a$ $(l \geq 1)$.

(c) Some α_j *contains a subtype of the form* $\gamma_1 \to \cdots \to \gamma_l \to a$ $(l \geq 1)$.

Proof. By induction on M.

Subcase 1. $M = x$. Then $\alpha = a$. Thus the lemma holds.

Subcase 2. $M = xM_1 \cdots M_m (m \geq 1)$. Then by Lemma 2, $\alpha = b$ for some type variable b and a TA-figure for $xM_1 \cdots M_m : b$ has the following form.

$$\frac{x : \xi_1 \to \cdots \to \xi_m \to b \quad M_1 : \xi_1 \quad \cdots \quad \left.\begin{array}{c}\vdots \\ M_j : \xi_j\end{array}\right\} P_j \quad \cdots \quad Mn : \xi_m}{xM_1 \cdots M_m : b}$$

Subcase 2.1. $a = b$. Then $xM_1 \cdots M_m$ itself is the desired N.

Subcase 2.2. $a \neq b$. Then (c) holds. By induction hypothesis for $B_j||\!-M_j : \xi_j$, P_j contains some TA-formula $N : a$ for some subterm N of M_j.

Case 3. $M = \lambda x.L$. Then we have $\alpha = \xi \rightarrow \zeta$ by Lemma 3.

Subcase 3.1. $x \notin FV(L)$. Then $\xi = b$ is a type variable, $b \notin var(B, \zeta)$ and $B||\!-L : \zeta$. Since b does not occur in $var(B, \zeta)$, we can replace α by ζ in the condition (b). Thus we can apply induction hypothesis for $B||\!-L : \zeta$. So there is a subterm N such that $N : a$ appears in P.

Subcase 3.2. $x \in FV(L)$. Then we have $B \cup \{x : \xi\}||\!-L : \zeta$. By induction hypothesis, there is a subterm N such that $N : a$ appears in P. ∎

We write $\beta_1 \approx \beta_2$ when β_2 is a trivial variant of β_1, i.e., when there are substitutions σ_1 and σ_2 such that $\beta_1 = \beta_2\sigma_2$ and $\beta_2 = \beta_1\sigma_1$. Remember that $pts(BCK - \beta)$ denotes the set of principal type-schemes of closed BCK-λ-terms in β-normal form.

Lemma 13 *Let $\beta_i \in pts(BCK\text{-}\beta)$ and θ_i be a strictly irrelevant substitution to β_i $(i = 1, 2)$. If $\beta_1\theta_1 = \beta_2\theta_2$ then $\beta_1 \approx \beta_2$.*

Proof. Let b_1 be an occurrence of a type variable in β_1. Consider the corresponding occurrence of type $b_1\theta_1$ in $\beta\theta_1$. Since $\beta_1\theta_1 = \beta_2\theta_2$, there is an occurrence of subtype γ in β_2 at the same position such that $b_1\theta_1 = \gamma\theta_2$.

$$\begin{array}{c} \beta_1 \\ \\ \downarrow \theta_1 \\ \beta_1\theta_1 = \beta_2\theta_2 \\ \uparrow \theta_2 \\ \\ \beta_2 \end{array}$$

b_1

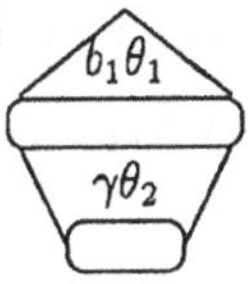

γ

Firstly, we claim that γ is a type variable.
When b_1 is relevant to β_1, then we have $b_1\theta_1 = b_1$ since θ_1 is strictly irrelevant to β_1. Therefore γ is a type variable. When b_1 is irrelevant to β_1, assume that γ is not a type variable. Then $\gamma = \gamma_1 \rightarrow \cdots \rightarrow \gamma_l \rightarrow b_2$ for $l \geq 1$. By Lemma 12, b_2 is relevant to β_2. Therefore $b_2\theta_2 = b_2$. Thus we have $b_1\theta_1 = \gamma\theta_2 = \gamma_1\theta_2 \rightarrow \cdots \rightarrow \gamma_l\theta_2 \rightarrow b_2$. Since b_1 is irrelevant to β_1, b_2 is irrelevant to $\beta_1\theta_1$ by Lemma 7. On the other hand, since b_2 is relevant to β_2, b_2 is relevant to $\beta_2\theta_2$ by Lemma 8. A contradiction. Thus γ is a type variable. This completes the proof for the first claim.

$$\begin{array}{ccc} \beta_1 & b_1 & b_1 \\ \downarrow \theta_1 & \downarrow & \downarrow \\ \beta_1\theta_1 & b_1\theta_1 & b_1\theta_1 \\ \| & \| & \| \\ \beta_2\theta_2 & c_1\theta_2 & c_2\theta_2 \\ \uparrow \theta_2 & \uparrow & \uparrow \\ \beta_2 & c_1 & c_2 \end{array}$$

Secondly we claim that this correspondence becomes a function from $var(\beta_1)$ to $var(\beta_2)$. Assume that β_1 contains two occurrences of the same type variable b_1 and that c_1 and c_2 are the corresponding occurrences of type variables in β_2 such that $b_1\theta_1 = c_1\theta_2$ and $b_1\theta_1 = c_2\theta_2$. Since b_1 occurs twice in β_1, b_1 is relevant to β_1 by Theorem 1. Therefore

$b_1\theta_1 = b_1$. By Lemma 8, b_1 is relevant to $\beta\theta_1$. By Lemma 7, c_1 and c_2 are relevant to β_2. Since θ_2 is strictly irrelevant to β_2, it follows that $c_1\theta_2 = c_2$ and $c_2\theta_2 = c_2$. Thus we have $c_1 = c_2$. This completes the proof of the second claim.

$$\begin{array}{ccc} \beta_1 & b_1 & b_1 \\ \downarrow\theta_1 & \downarrow & \downarrow \\ \beta_1\theta_1 & b_1\theta_1 & b_1\theta_1 \\ \| & \| & \| \\ \beta_2\theta_2 & c\theta_2 & c\theta_2 \\ \uparrow\theta_2 & \uparrow & \uparrow \\ \beta_2 & c & c \end{array}$$

Thirdly we claim that this function is one-to-one and surjective. The surjectivity is clear. To prove that this function is one-to-one, assume that there were two occurrences of the same type variable c in β_2 and occurrences of type variables b_1 and b_2 such that $b_1\theta_1 = c\theta_2$ and $b_2\theta_1 = c\theta_2$. Since c occurs twice in β_2, c_2 is is relevant to β_2 by Theorem 1. By Lemma 8, $c = c\theta_2 = b_1\theta_1 = b_2\theta_1$ is relevant to $\beta_2\theta_2$. By Lemma 7, b_1 and b_2 are relevant to β_1. Since θ_1 is strictly irrelevant to β, we have $b_1\theta_1 = b_1$ and $b_2\theta_1 = b_2$. Thus we have $b_1 = b_2$. Therefore this correspondence is one-to-one. This completes the proof of the third claim.

This function is the desired substitution σ_1. ∎

Theorem 3 ([6]) *If two close BCK-λ-terms in β-normal form have the same principal type-scheme, then they are identical.*

Theorem 4 *Let α be a type of a closed BCK-λ-term. If α is relevantly balanced, then the closed BCK-λ-term in βη-normal form which has α as its type-scheme is unique.*

Proof. Assume that $|\!-M : \alpha$ and $|\!-N : \alpha$ for closed BCK-λ-terms M and N in $\beta\eta$-normal form. By principal type-scheme theorem, M has a principal type-scheme γ_1 and a substitution θ_1 such that $||\!-M : \gamma_1$ and $\alpha = \gamma_1\theta_1$. The decomposition of θ_1 has the following form.

$$\begin{array}{rcl} ||\!-M: & \gamma_1 & \\ \sigma_2 & \downarrow & \textit{irrelevant structural} \\ |\!-M: & \gamma_1\sigma_2 & \\ \tau & \downarrow & \textit{identification} \\ |\!-M: & \gamma_1\sigma_2\tau & \end{array}$$

Since M is in η-normal form, the decomposition does not contain any relevant structural substitution. Since α is relevantly balanced, $\theta = \sigma_2\tau$ is irrelevant substitution by Lemma 11. On the other hand, we have $||\!-N : \gamma_2$ and $\alpha = \gamma_2\theta_2$ for some substitution θ_2, where γ_2 is a principal type-scheme of N. Similar to θ_1, θ_2 is an irrelevant substitution to γ_2. Now by Lemma 13 we have $\gamma_1 \approx \gamma_2$. By Theorem 3 we have $M = N$. ∎

Definition 10 (balanced types) *A type α is balanced iff every type variable in α occurs at most twice in α.*

Theorem 5 *Let α be a type-scheme of a closed λ-term. If α is balanced then α is relevantly balanced and the closed λ-term in βη-normal form which has α as its type-scheme is unique.*

Proof. Let M be a closed λ-term in β-normal form such that $|\!-M : \alpha$. Then M is a BCK-λ-term by [9]. By principal type-scheme theorem, M has a principal type-scheme β and a substitution θ such that $||\!-M : \beta$ and $\alpha = \beta\theta$. By Theorem 4, it suffices to show that θ is an irrelevant substitution to β. If θ is not irrelevant to β, there are distinct type variables $a \neq b$ in β such that $var(a\theta) \cap var(b\theta) \neq \emptyset$ and either a or b occurs more than twice. Therefore total number of the occurrences of a and b in β is more than 3. Since $a\theta$ and $b\theta$ contain a common type variable, say c, it follows that c occurs in $\beta\theta$ more than three times. This contradicts the balancedness of α. Therefore θ is irrelevant to β. Thus α is relevantly balanced. The uniqueness of normal form proof for α follows from Theorem 4. ∎

The converse of Theorem 4 does not hold in general. The relevantly balancedness is not a necessary condition for the uniqueness of proof. Consider a BCK-λ-term $M = \lambda xyz.xy(\lambda u.uz)$ and its principal type-scheme $\alpha = (a \to ((b \to c) \to c) \to d) \to a \to b \to d$. The principal TA-figure has the following form.

$$\dfrac{\dfrac{\dfrac{\dfrac{\dfrac{x : a \to ((b \to c) \to c) \to d \quad y : a}{xy : ((b \to c) \to c) \to d} \qquad \dfrac{\dfrac{u : b \to c \quad z : b}{uz : c}}{\lambda u.uz : (b \to c) \to c}}{xy(\lambda u.uz) : d}}{\lambda z.xy(u.uz) : b \to d}}{\lambda yz.xy(\lambda u.uz) : a \to b \to d}}{\lambda xyz.xy(\lambda u.uz) : (a \to ((b \to c) \to c) \to d) \to a \to b \to d}$$

Then consider a substitution $\theta = [a := e, c := e]$ and the above TA-figure applied this substitution. Note that e is relevant to $\alpha\theta$ and that e occurs four times in $\alpha\theta$. Therefore $\alpha\theta$ is not relevantly balanced. Follow the TA-figure upwards. Then we see that M is the unique BCK-λ-term in $\beta\eta$- normal form having $\alpha\theta$ as its type-scheme.

Theorem 4 means the uniqueness of normal proof for relevantly balanced formulas. However the uniqueness among BCK-proofs does not always imply the uniqueness among LJ-proofs in general. The previous example $(e \to ((b \to e) \to e) \to d) \to e \to b \to d$ has the unique BCK-proof as we saw above. But $\lambda xyz.zy(\lambda u.y)$ is another non-BCK-λ-term which has the type as its type-scheme.

With respect to relevantly balanced formulas, we conjecture that the uniqueness holds not only among BCK-proofs but also among LJ-proofs. Moreover we conjecture that any β-normal form proof for a relevantly balanced formula is a BCK-proof. If this is the case, we can extend a theorem in [9] which states that normal form proofs for balanced formulas are BCK-proofs.

Acknowledgements

The author would like to thank to Y. Akama, R. Kashima, Y.Komori, H.Ono and M. Takahashi for their discussion on the subject. This work was partially supported by a Grant-in-Aid for Encouragement of Young Scientists No.02740115 of the Ministry of Education.

References

[1] Babaev, A.A., Solov'ev, S.V., A coherence theorem for canonical morphism in cartesian closed categories, *Zapiski nauchnykh Seminarov Lenigradskogo Otdeleniya matematichskogo Instituta im. V.A. Steklova An SSSR* **88** (1979) 3-29.

[2] Girard, J.-Y., Linear logic, *Theoret. Comput. Sci.* **50** (1987) 1-101.

[3] Hindley,J.R., Seldin, J.P., Introduction to Combinators and Lambda-Calculus (Cambridge University Press, London, 1986)

[4] Hindley, J.R., BCK-combinators and linear lambda-terms have types, *Theoret. Comput. Sci.* **64** (1989) 97-105.

[5] Hindley, J.R., BCK and BCI logics, condensed detachment and the 2-property, a summary, Report, University of Wolongon, Aug 1990.

[6] Hirokawa, S., Principal types of BCK-lambda terms, submitted.

[7] Hirokawa, S., Principal type assignment to lambda terms, submitted.

[8] Hirokawa, S., Converse principal-type-scheme theorem in lambda-calculus, Studia Logica (to appear).

[9] Hirokawa, S., Linear lambda-terms and coherence theorem, preprint.

[10] Howard, W.A., The formulae-as-types notion of construction, in: Hindley and Seldin Ed., *To H.B. Curry, Essays on Combinatory Logic, Lambda Calculus and Formalism* (Academic Press, 1980) 479-490.

[11] Jaskowski, S., Über Tautologien, in welchen keine Variable mehr Als zweimal vorkommt, *Zeitchrift für Math. Logic* 9 (1963) 219-228.

[12] Komori, Y., BCK algebras and lambda calculus, in: Proc. 10th Symp. on Semigroups, Sakado 1986, (Josai University, Sakado 1987) 5-11.

[13] Mints, G.E., A simple proof of the coherence theorem for cartesian closed categories, Manuscript 1982.

[14] Lambek,J., Scott,P.J., Introduction to higher order categorical logic, (Cambridge University Press, 1986)

[15] Ono,H., Komori,Y., Logics without the contraction rule, *J. Symbolic Logic* **50** (1985) 169-201.

[16] Tatsuta, M., Uniqueness of normal proofs in implicational logic, Manuscript, Sep 1988.

[17] Wronski, A., On Bunder and Meyer's theorem, Manuscript, Aug 1987.

Proof Nets and Coherence Theorems

Richard Blute
University of Pennsylvania
Department of Mathematics
Philadelphia, PA 19104-6395
email address:rblute@pennsas.upenn.edu

ABSTRACT

Proof nets, a natural deduction system for multiplicative linear logic (mLL), see [G], are used to obtain general coherence theorems for various closed categories. The main definition is the Autonomous Deductive System,(ADS), a general framework for defining theories of autonomous categories, such as compact closed and *-autonomous categories. These theories are viewed as theories over mLL. Necessary and sufficient conditions are then given for the theory to have the coherence property. The criterion is essentially whether each of the axioms of the theory is correct in mLL with the MIX rule added. The Girard notion of short trip is shown to be related to Kelly and Mac Lane's notion of incompatibility of graphs, see [KM].

INTRODUCTION

Logical methods have been used in the study of coherence questions since the work of Lambek, see [L1]. The method used is that the morphisms specified by the theory, such as the theory of symmetric monoidal closed (hereafter called autonomous) categories, are viewed as deductions in a deductive system. Techniques analogous to cut elimination can then be put to use. This was done by Kelly and Mac Lane, see [KM] and [M], to establish a partial coherence result for autonomous categories. They show that under certain conditions, and with some restrictions, all diagrams of allowable morphisms commute. Allowable morphisms are the morphisms specified by the theory. The restriction involves the notion of a graph. A graph is a pairing of the variables which satisfies a certain variance condition. It can be shown that every allowable morphism has a graph. The restriction is that only diagrams yielding morphisms of the same graph are considered. One of the results of this paper is that these graphs can be interpreted as axiom links from Girard's notion of proof net, see [G]. The basic information on graphs is contained in APPENDIX 1.

Previous work, such as [KM], [M], and [L1], has used a sequent calculus approach to the cut elimination result. Since, from a logical viewpoint, coherence conditions amount to an equivalence relation on proofs, it seems possible that natural deduction may be a better framework for proving coherence theorems. This is suggested by [G1], where Girard looks at the question, "What is a proof?". Natural deduction is seen as a more intrinsic structure, because sequent proofs which only differ by irrelevant choices in the order of rules used are identified in natural deduction. In multiplicative linear logic, the natural deduction system provided by Girard is especially well suited for the study of coherence theorems.

Girard's proof nets are a natural deduction system for mLL. One of the advantages of proof nets is that the cut elimination procedure is confluent and strongly normalizing. Thus there is a unique normal form for proofs, the cut free proof net. This normal form result will simplify coherence proofs.

We will actually need a generalized version of proof nets, due to Danos and Regnier, see [DR] and [D]. They are described in APPENDIX 2. We need these because they can be extended to the theory mLL+(MIX) which is a proper extension of mLL. For our purposes the mix rule can be written as follows:

$$\frac{\Gamma \vdash \qquad\qquad \vdash \Delta}{\Gamma \vdash \Delta}$$

The mix rule is discussed in [G1], [G] as well as [DR].

For our purposes, the mix rule is important for the following reason. The Danos and Regnier correctness criterion concerns the existence of a cycle in certain graphs. A proof structure is a graph as described in APPENDIX 2. Please note that this notion of graph is different from the Kelly-Mac Lane notion. To each proof structure is associated a family of subgraphs. The correctness criterion is then that each subgraph must be acyclic and connected. It is the acyclicity which is important. We will show that it is closely related to an idea in the original Kelly-Mac Lane paper, [KM]. One of their key definitions is that of incompatibility of allowable graphs. Two graphs are incompatible if, when trying to compose them, loops are formed. Kelly and Mac Lane show that for the theory of autonomous categories, composable morphisms always have compatible graphs.(We will show that this is not true for arbitrary theories.) Also it will be shown that there is a close analogy between the acyclicity of proof structures and the compatibility of graphs. However, the connectedness condition is stronger than we need. Thus we choose to work in mLL+(MIX) where the correctness criterion is that the subgraphs only be acyclic. This is the correct notion for our purposes.

One of the main goals of this work is to define a general framework for specifying the theories of various closed categories. That framework is the autonomous deductive system, defined as follows.

Autonomous Deductive Systems

Autonomous Deductive Systems, (ADS) are a general framework for defining the theories of various closed categories. They are an extension of Lambek's notion of deductive system, see [L1],[L2], to linear logic sequents. As in any deductive system, there are deductive rules and equality rules. We impose additional restrictions on these rules which make the system behave more like linear logic. There are two types of ADS, the intuitionistic and the classical. Their syntax is defined as follows.

Intuitionistic Case

Connectives are: $\otimes, \multimap$

Classical Case

Connectives are: $\otimes, \wp$

In the classical case, the operator $(-)^{\perp}$ is built into the syntax using Girard's notion of De Morgan's laws, see [G]. So then variables are of the form **A** or $\mathbf{A}^{\perp}$. Negation is then extended to nonatomic formulas using the De Morgan rules. However, in this paper, we will often write expressions such as $(\mathbf{A}^{\perp} \otimes \mathbf{B}^{\perp})^{\perp}$ with the obvious meaning. (In previous papers studying closed categories and coherence issues, the primitive connectives were always $\otimes, (-)^{\perp}$).

Propositions are built up from variables and the connectives. Sequents are of the form:

Intuitionistic- $\Gamma \vdash \mathbf{A}$

Classical- $\Gamma \vdash \Delta$

Γ, Δ are finite lists. **A** is a single proposition.

It is further stipulated that to each sequent is associated a Kelly-Mac Lane graph. Each rule of inference must come equipped with a rule for assigning a graph to the conclusion, given the graphs of the premisses. The inference rules are simply those of the appropriate fragment of linear logic. These rules may be found in the classical case in [G] and in the intuitionistic case in [GL]. It is easy to see how they operate on graphs. For example, the cut rule acts on graphs by graph composition, which is described in APPENDIX 1. To model the right tensor rule, take the disjoint union of the graphs of the premisses, as in the example:

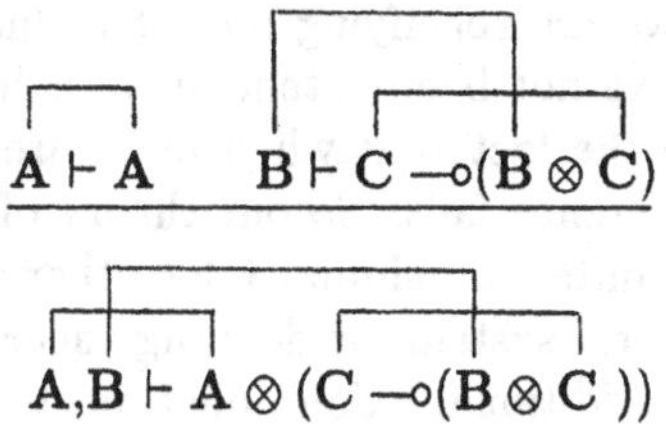

Equality Rules

We wish to allow that certain deductions can be equated, but with some restrictions. Two deductions are equatable if they are deductions of the same sequent and satisfy the following graph condition. First the two deductions assign the same graph to the sequent, and second, the two deductions have the same number of instances of incompatibility. This means that when trying to compose the underlying graphs, the same number of loops are formed. Remember that the composition of graphs occurs when applying the cut rule, see above. So, in particular, a deduction without an instance of incompatibility cannot be equated with a deduction with an instance of incompatibility. As an example of when two deductions cannot be equated, consider the ADS with the following two additional axioms(each with the obvious graph):

$$\mathbf{A} \vdash \mathbf{A} \otimes \mathbf{B} \otimes \mathbf{B}^{\perp}$$

$$\mathbf{A} \otimes \mathbf{B} \otimes \mathbf{B}^{\perp} \vdash \mathbf{A}$$

In this ADS, there is a deduction of the following form:

$$\frac{A \vdash A \otimes B \otimes B^{\perp} \qquad A \otimes B \otimes B^{\perp} \vdash A}{A \vdash A}$$

This deduction cannot be equated with the identity deduction of $A \vdash A$, because the above deduction has an instance of incompatibility in the underlying graphs. That the graphs of the two nonlogical axioms above are incompatible is shown by the following representation, imagining the formulas written vertically:

There are a large number of equality rules, of three basic types.

-Rules for making the connectives functorial

-Rules for making the ADS a multi\polycategory

-Rules specified by the theory of closed categories.

A complete list is contained in my thesis, see [T].

Note that the theories we are specifying do not include units. As pointed out in APPENDIX 2, proof nets have not been extended to include the multiplicative units. Intuitively, this is related to the fact that when units are added to the theory of autonomous categories, full coherence fails. So our theory of autonomous categories will be the usual theory without units, and similarly for other categories.

Since this is to be a general system for defining theories, one is allowed to add in additional axioms. But each additional axiom must meet the following condition.

Restriction on Additional Axioms

Any added nonlogical axiom must come equipped with a graph, and any axiom which has the same variable appearing more than twice must be part of an axiom scheme in which each variable appears exactly twice. For example, if one wishes to have an added axiom of the form:

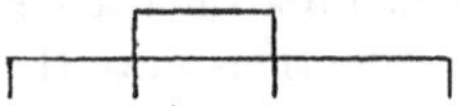

$$A \otimes (A^{\perp} \otimes A) \longrightarrow A$$

then it must appear as a substitution instance of the more general axiom:

$$B \otimes (A^{\perp} \otimes A) \longrightarrow B$$

This restriction is intended to match the intuition that if two variables are not linked by the axiom's graph, then it should be possible to substitute different terms for them. Perhaps it is best to think of the links themselves as the real variable, and the letters **A**, etc. as placeholders

One is also allowed to introduce new equalities between deductions. But again, this is only allowed when the two sequents satisfy the above criterion. What theories can be specified in this way?

EXAMPLES

1) In the intuitionistic case, with no additional axioms or equations, the result is the theory of autonomous categories.

2) In the classical case, with no additional axioms or equations, the result is the theory of *-autonomous categories, due to Barr, see [*].

3) In the classical case, add in two additional axioms, each with the evident graph:

$$(\mathbf{A} \otimes \mathbf{B})^{\perp} \vdash \mathbf{A}^{\perp} \otimes \mathbf{B}^{\perp}$$

$$\mathbf{A}^{\perp} \otimes \mathbf{B}^{\perp} \vdash (\mathbf{A} \otimes \mathbf{B})^{\perp}$$

Also, add the equations that make the two inverse to each other. Then, the result is the theory of compact closed categories, due to Kelly, see [K] and [KL].

4) Only adding the second of the two axioms in 3) gives a theory which, as far as I know, has not been studied previously. Girard's category of coherence spaces, see [G2], and [P&T], provides a model. One way to see this axiom is as a form of weakening. It is a consequence of the MIX rule mentioned in the introduction. So this is an example of a sequent which is correct in mLL+(MIX), but not in mLL.

We now recall what it means for an ADS to satisfy composability and compatibility. Compatibility of graphs is defined in APPENDIX 1, and in [KM].

Definition. *Two sequents,* $\Gamma \vdash \Delta$ *and* $\Gamma' \vdash \Delta'$ *are said to be* **composable** *if the following holds.*

If we are in an ADS without the mix rule and $\Gamma', \Delta \neq \emptyset$ (if either is empty, they are not composable), then the sequents are composable if $\bigotimes \Gamma' = \wp\Delta$. Of course, for these two formulas to be equal, either Γ or Δ must have only one formula.

If the ADS does have the mix rule, then the sequents are composable if either the above holds or both Γ', Δ are empty.

The intention of this definition is that when viewed categorically the morphisms interpreting the deduction should be composable. However the definition of model of an ADS turns out to be complicated by the fact that we do not have units. The way to model an ADS is with autonomous polycategories. We hope to explore this in a later paper.

Definition. *An ADS satisfies compatibility, or is* **compatible**, *if any two composable sequents have underlying graphs which are compatible.*

Note. *The following theorems are all for classical ADS's only. They can be modified to include the intuitionistic case. We will mention this modification later in the paper.*

Also note that the word **graph** is being used in two different ways; the Kelly-Mac Lane notion of graph, which corresponds to the axiom links, and the Danos-Regnier notion which corresponds to the entire proof structure. There should be no confusion, but we will refer to Kelly and Mac Lane's notion as simply "graphs", and the Danos-Regnier notion as "proof structures".

Theorem 1. *An ADS satisfies compatibility iff each axiom has a proof structure which is a generalized proof net. This means that it is a correct sequent in mLL+(MIX) and that its graph corresponds to the axiom links of a generalized proof net.*

The idea behind the proof is that any loop formed when attempting to compose the graphs will be a short trip, and conversely.

Proof of Theorem

$\Leftarrow$

Suppose we have two sequents, $\Gamma \vdash \Delta$ and $\Gamma' \vdash \Delta'$, which have the correct graphs and which are correct in mLL+(MIX). Suppose also that they are composable and incompatible. If they are incompatible, it follows that both Δ and Γ' are nonempty. So replace Δ with $\wp\Delta$ and Γ' with $\bigotimes(\Gamma')$. Since the two sequents are composable, these two formulas must be equal. Call this formula S. This direction of the theorem follows from the following lemma.

Lemma. *Given sequents of the form* $\Gamma \vdash \mathrm{S}$ *and* $\mathrm{S} \vdash \Delta'$, *then one of the following two deductions is correct iff the other is.*

$$\frac{\Gamma \vdash \mathrm{S} \qquad \mathrm{S} \vdash \Delta'}{\Gamma \vdash \Delta'}$$

$$\frac{\Gamma \vdash \mathrm{S} \qquad \mathrm{S} \vdash \Delta'}{\Gamma \vdash \mathrm{S} \otimes \mathrm{S}^{\perp}, \Delta'}$$

This lemma was proved for ordinary proof nets by Girard in [G3]. The situation here is analogous. The reason this lemma holds is that geometrically the cut rule behaves like a tensor link.

The proof of this direction now follows. If we have two sequents which are correct and incompatible, then it is easily seen that the proof structure of the following deduction would have a cycle, contradicting the lemma.

$$\frac{\Gamma \vdash \mathrm{S} \qquad \mathrm{S} \vdash \Delta'}{\Gamma \vdash \mathrm{S} \otimes \mathrm{S}^{\perp}, \Delta'}$$

$\Rightarrow$

We prove the other direction by contradiction. Suppose that $\Gamma \vdash \Delta$ has a proof structure which is not a generalized proof net. Then we must construct two composable sequents, $\Gamma'' \vdash \Delta''$ and $\Gamma' \vdash \Delta'$, with incompatible graphs. We proceed as follows.

Since $\Gamma \vdash \Delta$ fails to be a generalized proof net, it must have at least one cycle in the proof structure. List all such cycles, then choose the one which has the least number of formulas appearing in it. List all the formulas in the order they appear in the cycle, $\mathbf{T}_1, \mathbf{T}_2, \ldots, \mathbf{T}_n$. Start with the formula which appears farthest to the left.

There may be some $\mathbf{T}_i$'s which appear as subformulas of formulas in Γ. Bring any such formulas to the other side of the sequent using linear negation. So $\Gamma \vdash \Delta$ has been replaced with $\Gamma'' \vdash \Delta''$. Clearly, Δ'' is not empty so it can be replaced by $\wp(\Delta'')$. Call this formula S. This will be the first of the two sequents. We now construct a second sequent $\mathbf{S} \vdash \Delta'$ which is incompatible with it.

Take the original sequence of formulas $\mathbf{T}_1 \ldots \mathbf{T}_n$. Then take the subsequence of just the propositional atoms.

$$\Sigma = \mathbf{A}_1, \mathbf{A}_1^{\perp}, \ldots, \mathbf{A}_m, \mathbf{A}_m^{\perp}$$

It is easily verified that Σ must be of this form. We now make the following substitution in S.

$$\textit{Let } \mathbf{A} = \mathbf{A}_1 = \mathbf{A}_2 = \ldots = \mathbf{A}_m.$$

Note that A must be a fresh variable. Also, this forces the following equalities.

$$\mathbf{A}^{\perp} = \mathbf{A}_1^{\perp} \ldots = \mathbf{A}_m^{\perp}$$

We now construct the sequent $\mathbf{S} \vdash \Delta'$. Take the formula S (after the above substitution) and consider the tree determined by its subformulas, ordered by containment.

EXAMPLE

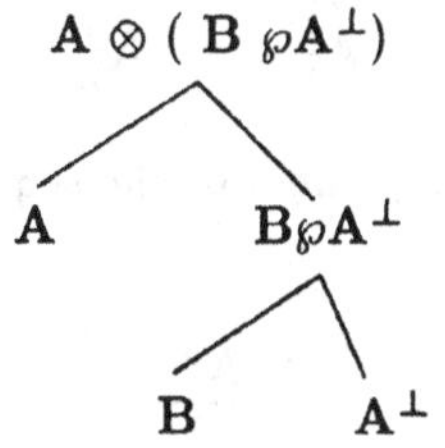

We induct on the height of the tree. The proof will use the following two constructions. If the top node is a $\otimes$, then the construction is as follows.

$$\dfrac{\dfrac{\mathbf{T}_1 \vdash \mathbf{T}_2 \qquad \mathbf{T}_3 \vdash \mathbf{T}_4}{\mathbf{T}_1, \mathbf{T}_3 \vdash \mathbf{T}_2, \mathbf{T}_4}}{\mathbf{T}_1 \otimes \mathbf{T}_3 \vdash \mathbf{T}_2, \mathbf{T}_4}$$

This deduction uses the mix rule followed by the left $\otimes$ rule.

If the top link is a par, then the construction is as follows.

$$\dfrac{\mathbf{T}_1 \vdash \mathbf{T}_2 \qquad \mathbf{T}_3 \vdash \mathbf{T}_4}{\mathbf{T}_1 \wp \mathbf{T}_3 \vdash \mathbf{T}_2, \mathbf{T}_4}$$

To get the induction started, we need the following lemma.

Lemma. *Given* $\mathbf{A}_i$ *in* Σ, *there is an* $\mathbf{A}_j^\perp$ *in* Σ *and a path in the tree from* $\mathbf{A}_i$ *to* $\mathbf{A}_j^\perp$ *such that the top node is a tensor link.*

Proof. *It is easy to see that* $\mathbf{A}_i$ *would not be part of the cycle, if this were not the case.*

We begin the construction at the variable $\mathbf{A}_1$. Choose the first $\mathbf{A}_j^\perp$ in Σ satisfying the property of the above lemma. In the sequent, $\mathbf{S} \vdash \Delta'$, these two variables will be connected via an axiom link. This can be done since all the $\mathbf{A}_i$'s have been equated. These two variables will be introduced into the sequent by the axiom:

$$\mathbf{A},\mathbf{A}^\perp \vdash$$

Repeat this process for the variable $\mathbf{A}_2$, choosing the first of the remaining $\mathbf{A}_j^\perp$'s as its pair, and so on until all of the variables in Σ have been paired off. One can check that this pairing off process always works. We thus have a list of m instances of axiom links.

We introduce them into the sequent $\mathbf{S} \vdash \Delta'$ by the following inductive procedure.

If $\mathbf{A}_i$ and $\mathbf{A}_j^\perp$ are connected in $\mathbf{S}$ this way:

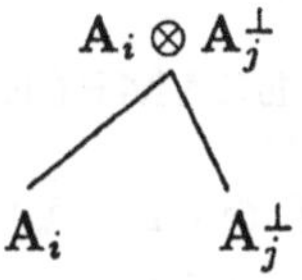

then we proceed by introducing them to the sequent using this deduction.

$$\frac{\mathbf{A}, \mathbf{A}^\perp \vdash}{\mathbf{A} \otimes \mathbf{A}^\perp \vdash}$$

Remember that all of the $\mathbf{A}_i$'s have been equated.

But, if there is an intermediate formula, as in $\mathbf{A} \otimes (\mathbf{B} \wp \mathbf{A}^\perp)$, then use the following procedure.

$$\frac{\dfrac{\mathbf{A}, \mathbf{A}^\perp \vdash \qquad \mathbf{B} \vdash \mathbf{B}}{\mathbf{A}, \mathbf{A}^\perp \wp \mathbf{B} \vdash \mathbf{B}}}{\mathbf{A} \otimes (\mathbf{A}^\perp \wp \mathbf{B}) \vdash \mathbf{B}}$$

So, such formulas are introduced into the deduction via the axiom $\mathbf{B} \vdash \mathbf{B}$.

The construction of the sequent $\mathbf{S} \vdash \Delta'$ now proceeds inductively, moving up the subformula tree. If the next connective is a tensor, then use the first of the two constructions mentioned above. If the connective is a par, then use the second.

So the sequent $S \vdash \Delta'$ can be built up inductively in this way. We now claim that $\Gamma'' \vdash S$ and $S \vdash \Delta'$ are incompatible. This means that we must give a sequence of atoms which fail the Kelly-Mac Lane definition of compatibility. The sequence will be Σ. The pairings for $\Gamma \vdash S$ are as follows. A_1 is paired to $A_1^\perp$, and so on. For $S \vdash \Delta'$, A_1 is paired to the $A_j^\perp$ specified by the above lemma, and so on. That a loop is formed follows from the fact that the original proof structure had a cycle.

End of Proof

Coherence for an ADS

Definition. *An ADS satisfies* **coherence** *if any two deductions of* $\Gamma \vdash \Delta$ *with the same graph are equal.*

Theorem 2. *An ADS satisfies coherence iff each axiom corresponds to a generalized proof net.*

In other words, coherence=compatibility.

This theorem is a consequence of the Church-Rosser property of proof nets. A cut-free proof net can be thought of as the normal form of a proof. Thus the coherence theorem says that every proof has a unique normal form.

Proof of Theorem

$\Leftarrow$

Suppose that we have an ADS in which each axiom corresponds to a generalized proof net.

Given a set of deductions, each of the same sequent and with the same graph, we define a reduction system on the deductions which is confluent and strongly normalizing.

The reduction system is simply cut elimination for Danos-Regnier proof nets, as described in APPENDIX 2. Several things have to be checked, the most important being that the reduction system preserves the deduction. In other words, after the application of a reduction procedure one still has the same deduction, up to equivalence. But this is built into the equality rules. From the point of view of allowable morphisms, this follows from the fact that any morphism is the composition of a sequence of basic morphisms. And composition is modelled in the deductive system as the cut rule. This should be compared to Kelly and Mac Lane's cut elimination theorem, which they call the "allowable=constructible lemma" in [KM].

It is shown in several places that this system is confluent and strongly normalizing, such as in the Danos-Regnier paper and Danos's Thesis, see [DR] and [D].

It is straightforward to check that, given a correct sequent in mLL or mLL+(MIX) together with its axiom links, then the cut free proof net is uniquely determined. This, combined with the confluence and strong normalization of the above reduction system, implies that every deduction of this sequent converges to the unique cut free proof net. This direction of the theorem now follows. Since there is a unique normal form which every deduction of the given sequent converges to, they must all be equal.

$\Rightarrow$

We prove this direction by contradiction. By Theorem 1, it is sufficient to show that if an ADS has an instance of incompatibility, then it is not coherent. We first need the following lemma.

Lemma. *Suppose we have two sequents, $\Gamma \vdash \Delta$ and $\Gamma' \vdash \Delta'$, which are composable and incompatible, and let* **A** *be a variable which appears in one of the loops. Then, the above two sequents can be replaced by two new sequents which are still composable and incompatible and in which the variable* **A** *only appears in the instances which are part of the loop.*

This is proved by induction and relies on the restriction on new axioms in the definition of ADS.

The proof of the proposition now proceeds as follows. Suppose that we have two sequents which are composable and incompatible. Let **A** be a variable appearing in the loop. If **A** has any occurrences in either sequent other than those appearing in the loop, use the above lemma to replace the two original sequents by two new sequents for which this is not the case. Now, choose two new variables, not appearing in either sequent, say **C** and **D**. Replace **A** by the term $\mathbf{C} \otimes \mathbf{D}$. Then it is easily verified that this new pair of sequents has at least one more instance of incompatibility than the previous pair of deductions. In other words, when attempting to compose the underlying graphs, in the second pair of sequents, one more loop is formed than in the first pair. The result is two different deductions of the same sequent and with the same graph. This last claim holds because when composing two graphs, you ignore any loops which may be formed, see APPENDIX 1.

However, the two deductions cannot be set equal because of the convention defined in the equality rules. Thus there are two unequal sequents of the same graph. So the ADS is not coherent.

End of Proof

Intuitionistic Case

For the intuitionistic case, the theorems need to be modified slightly. Since sequents are of the form $\Gamma \vdash \mathbf{A}$, where **A** is a single formula, the MIX rule cannot even be stated in Intuitionistic mLL, (ImLL). Since the construction in the proof of the compatibility theorem explicitly uses the MIX rule, it is incorrect for ImLL.

Definition. *An ImLL sequent is* **strongly incompatible** *if the following holds. When the sequent is viewed as a classical sequent, (so change $\mathbf{A} \multimap \mathbf{B}$ to $\mathbf{A}^{\perp} \wp \mathbf{B}$), the construction in the proof of the compatibility theorem does not require the MIX rule.*

In my thesis, [T], there is an intrinsic characterization of strongly incompatible sequents. Theorems 1 and 2 above become four theorems for intuitionistic ADS, as follows.

Theorem 3. *For an intuitionistic ADS, if all of the axioms, when viewed as classical sequents, are correct in mLL+(MIX), then the ADS satisfies compatibility.*

Theorem 4. *If the ADS contains a strongly incompatible sequent, then the ADS is incompatible.*

Theorem 5. *If all of the axioms of the ADS are correct in ImLL, then the ADS is coherent.*

Theorem 6. *If the ADS is coherent, it is compatible.*

Proofs. *Same as in the classical case.*

Corollary. *The following chart is a corollary of the above theorems.*

THEORY	*COMPATIBILITY*	*COHERENCE*
autonomous	*yes*	*yes*
***-autonomous**	*yes*	*yes*
compact closed	*no*	*no*
example 4	*yes*	*yes*

The proof is by examination. In the compact closed case, the first of the two axioms fails in mLL+(MIX). This is easiest to see using proof nets. For example 4, it can be verified that this axiom is a consequence of the mix rule. So the mix rule implies that $\otimes$ implies $\wp$, which can be thought of as a form of weakening. The deduction is as follows:

$$\cfrac{\cfrac{\mathbf{A} \vdash \mathbf{A} \qquad \mathbf{B} \vdash \mathbf{B}}{\mathbf{A}, \mathbf{B} \vdash \mathbf{A} \wp \mathbf{B}}}{\mathbf{A} \otimes \mathbf{B} \vdash \mathbf{A} \wp \mathbf{B}}$$

The first of the above four cases is analogous to the Kelly-Mac Lane result, but not quite as strong. What the Kelly-Mac Lane result says is that coherence holds in the theory of autonomous categories for all allowable morphisms which do not contain subshapes of the form $\mathbf{R} \multimap \mathbf{I}$, see [KM] and APPENDIX 1. This corollary says coherence holds in the theory of autonomous categories without units. However, this approach has the advantage that it allows one to study coherence for autonomous categories with additional structure, i.e. additional morphisms or equations.

CONCLUSION

As mentioned before, we hope to study the model theory of an ADS in a later paper, using autonomous polycategories. It is then possible to define a structure-preserving, faithful functor to ordinary categories such that the result will be the ordinary autonomous categories, *-autonomous categories, etc.

There are many possible extensions to this work. First it is possible to use the previous results to explore the issue of dinaturality. In particular, similar conditions to the above can be given which ensure that the dinatural transformations which arise from an ADS compose. Thus it is possible to model polymorphism, in a similar manner to the work in [FP] and [GSS].

Also, it is possible to define the notion of autonomous fibration, which is a fibered category for which each fiber is a model of an ADS. Also each morphism in the base category has both adjoints. Then using Girard's paper [G4], in which he extended proof nets to quantifiers, it should be possible to get a coherence result for these. Also, Chirimar has recently given a version of Danos-Regnier type proof nets for mLL with quantifiers, see [Ch].

Finally, it has been shown in several places that petri nets can be viewed as theories in mLL, see [MM] and [GG]. It would be interesting to study what types of petri nets can be specified as an ADS, and what the coherence property translates to as a property of petri nets.

APPENDIX 1
Overview of Kelly-Mac Lane

In studying autonomous categories, Kelly and Mac Lane begin by defining the allowable morphisms. These are defined inductively by beginning with the basic morphisms specified by the theory and closing under the operations of $\otimes, \Rightarrow$, and composition.

The next step is to assign to each allowable morphism a graph, as follows. First, define **shapes** to be the formal objects of the theory, built up from variables, $I, \otimes$, and $\Rightarrow$. The allowable morphisms are then seen as morphisms between shapes.

Assign to each shape a variable set, as follows. (A variable set is a sequence of +'s and –'s.) The variable set is denoted v(T).

$v(I)=\emptyset$

$v(X)=\{+\}$

$v(T\otimes S)=v(T)\coprod v(S)$

$v(T\Rightarrow S)=v(T)^{op}\coprod v(S)$

In the above, $v(T)^{op}$ is v(T) with +'s changed to –'s and vise versa. $\coprod$ stands for concatenation of lists. A graph is then defined to be a fixed-point free involution on $v(T)\coprod v(S)$ such that paired elements have opposite variance in $v(T)^{op}\coprod v(S)$.

EXAMPLES

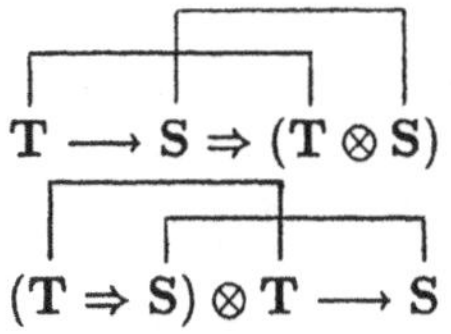

It can then be shown that every allowable morphism has a graph.

Two graphs $f{:}T\to S$ and $g{:}S\to R$ are **compatible** if there does not exist a sequence $t_1, t_2, \ldots, t_{2r}$ such that t_{2i-1} and t_{2i} are paired under f and t_{2i}, t_{2i+1} are paired by g, and t_{2r}, t_1 are also paired by g. The following are examples of incompatible graphs. Imagine that the domain and codomain are written vertically, this makes the loops easier to see.

EXAMPLES

Intuitively, the definition of compatibility says that no loops are formed when you try to compose the graphs.

Note, the graphs can still be composed by ignoring any loops that do arise. In this way, we get the category of Kelly-Mac Lane graphs, denoted by **G**.

The two main theorems of the Kelly-Mac Lane paper are as follows.

Theorem. *If* $f : \mathbf{T} \to \mathbf{S}$ *and* $g : \mathbf{S} \to \mathbf{R}$ *are allowable, then they are compatible.*

Theorem. *If* $f, g : \mathbf{T} \to \mathbf{S}$ *are allowable and have the same graph, and the shapes* $\mathbf{T}, \mathbf{S}$ *do not contain subshapes of the form* $\mathbf{R} \multimap \mathbf{I}$, *then* $f = g$.

From the viewpoint of Linear Logic, shapes are propositions in the intuitionistic fragment of Multiplicative Linear Logic, (mLL). Allowable morphisms are correct deductions, coherence conditions form an equivalence relation on deductions of the same graph. The linkings of Kelly-Mac Lane graphs correspond to the axiom links of a proof structure. Incompatibility of graphs corresponds to a failure of the long trip condition of Girard's proof nets. In terms of Danos-Regnier proof nets the analogy is to a cycle.

APPENDIX 2
DANOS-REGNIER PROOF NETS

Girard, in [G], introduced proof nets. They are the appropriate version of natural deduction for mLL. See also [P&T]. Danos and Regnier, in [DR], give an alternate characterization of proof-nets which is the version we need here. The reason we need this notion of proof net is that the correctness criterion can be altered to handle the theory mLL+(MIX).(It also has the advantage that a polynomial time correctness criterion can be given for mLL. This criterion and related issues are in Danos's thesis [D].)

The construction proceeds in the following way. A **proof structure** is built inductively from 3 types of links, tensor links, par links, and axiom links:

AXIOM LINKS

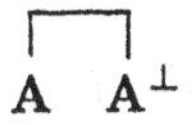

TENSOR LINKS

PAR LINKS

So a proof structure is a graph.(Not as in the Kelly-Mac Lane sense, but the ordinary sense.) The correctness criteria for mLL and mLL+(MIX) are given by associating to a proof structure a family of subgraphs. This family is obtained by removing one of the two edges from each par link. Thus there are 2^n subgraphs where n is the number of par links. The criterion for mLL is then that, each subgraph must be acyclic and connected. For mLL+(MIX), the criterion is that each subgraph must be acyclic.

Proof structures which are correct in mLL+(MIX) will be called generalized proof nets.

Note that the one portion of the multiplicative fragment which is not incorporated in proof nets is the units.

Cut elimination can be described in a way analogous to Girard's method in [G], or in [P&T]. It can also be shown that cut elimination is confluent and strongly normalizing for both systems.

REFERENCES

[FP] E. Bainbridge, P. Freyd, A. Scedrov, P. Scott, Functorial Polymorphism, TCS (1990) 70:35-64

[*] M. Barr, *-Autonomous Categories, Springer Lecture Notes in Mathematics Vol 752, (1980)

[T] R. Blute, Linear Logic, Coherence and Dinaturality, Thesis,1991

[D] V. Danos, La Logique Lineaire Appliquee a L'etude de Divers Processus de Normalisation et Pricipalement du λ-calcul, thesis, 1990

[DR] V. Danos, L. Regnier, The Structure of Multiplicatives, Arch. Math. Logic (1989) 28:181-203

[GG] V. Gehlot, C. Gunter, Normal Process Representatives, Proceedings of Logic in Computer Science, IEEE, (1990)

[G] J. Y. Girard, Linear Logic, TCS(1987) 50:1-102

[G1] J. Y. Girard, Towards a Geometry of Interaction, Proceedings AMS Conference on Categories in Computer Science and Logic (1990)

[G2] J. Y. Girard, The System F of Variable Types, 15 Years Later (1986) 45:159-192

[G3] J. Y. Girard, Multiplicatives, Rendiconti Semin. Univ. Polit. Torino (1988)

[G4] J. Y. Girard, Quantifiers in Linear Logic, preprint, 1989

[GL] J. Y. Girard, Y. Lafont, Linear Logic and Lazy Computation, SLNCS 250

[P&T] J. Y. Girard, Y. Lafont, P. Taylor, Proofs and Types, Cambridge University Press

[GSS] J. Y. Girard, A. Scedrov, P. Scott, Normal Forms and Cut Free Proofs as Natural Transformations, preprint (1990)

[KM] G. M. Kelly, S. Mac Lane, Coherence in Closed Categories, JPAA (1971) 1:97-140

[KL] G. M. Kelly, M. La Plaza, Coherence for Compact Closed Categories, JPAA (1980) 19:193-213

[K] G. M. Kelly, An Abstract Approach to Coherence, Springer Lecture Notes in Mathematics 281

[L1] J. Lambek, Deductive Systems and Categories II, Springer Lecture Notes in Mathematics 87, (1969)

[L2] J. Lambek, Multicategories Revisited, Proceedings AMS Conference on Categories in Computer Science and Logic, (1990)

[MM] N. Marti-Oliet, J. Meseguer, From Petri Nets to Linear Logic, Springer Lecture Notes in Computer Science 389, (1989)

[M] G. Mints, Closed Categories and the Theory of Proofs, Journal of Soviet Math. (1981) 15:45-62

A Modular Approach to Denotational Semantics

Eugenio Moggi
Dipartimento di Matematica
Universitá di Genova
via L.B. Alberti 4
16132 Genova, ITALY
moggi@igecuniv.bitnet

We propose an incremental approach to the denotational semantics of complex programming languages based on the idea of *monad transformer*.

The traditional way of giving denotational semantics to a programming language is to translate it into a *metalanguage* ML with a fixed intended interpretation in the category *Cpo* of cpos (or some variant of it). We depart from this approach by translating a programming language PL in a metalanguage $ML(\Sigma)$, where some constants do not have a fixed intended interpretation in *Cpo*. These constants, specified in the signature Σ, include a unary type constructor T and a collection of (polymorphic) operations for constructing terms of type TA. A key property of the translation is that programs of type A are translated into terms of type $T(A^{\circ})$, where A° is the translation of A.

This approach does not yield any new semantics for PL, since eventually one has to interpret the constants in Σ, e.g. by translating them into ML. However, it is an integral part of the incremental approach to be able to change the interpretation of constants in Σ (without invalidating *adequacy* of the denotational semantics w.r.t. some given operational semantics), when extending the programming language. Suppose that PL is obtained from PL_0 by a sequence of simple extensions $PL_i \subset PL_{i+1}$ and that we have a semantics for PL_0, how can we built a semantics for PL? In terms of signatures for metalanguages the problem can be rephrased as follows:

> given an increasing sequence of signatures $\Sigma_i \subset \Sigma_{i+1}$ and an interpretation for Σ_0 (the initial signature), we want to construct an interpretation for Σ (the final signature).

This can be achieved *incrementally*, provided at each step $\Sigma_i \subset \Sigma_{i+1}$ we can construct an interpretation for Σ_{i+1} from an interpretation for Σ_i. At first, one is tempted to extend the interpretation for Σ_i to the whole of Σ_{i+1}, but in general this is not possible. Instead, one has first to redefining the interpretation of T (and therefore of all operations in Σ_i), and then extend it to Σ_{i+1}.

We will show that for a wide range of signature extensions $\Sigma_i \subset \Sigma_{i+1}$, there is a *uniform way* of redefining an interpretation for Σ_i and to extend it to Σ_{i+1}. We will illustrate the incremental approach with few examples, such as:

- a parallel language with side-effects, where

$$TA = \mu X.\mathcal{P}((A + X) \times S)^S$$

- a sequential language with side-effects and exceptions, where

$$TA = ((A + E) \times S)_\perp^S$$

The two interpretations of T, given above, can be constructed starting from some very simple monads by applying few basic monad transformers, i.e. endofunctors on the category of monads (over *Cpo*). For instance, T for the parallel language with side-effects is given by $Res(Sef(\mathcal{P}))$, where

- $\mathcal{P}$ is some powerdomain monad
- Sef is the monad transformer for adding side-effects in S

$$Sef(T)A = T(A \times S)^S$$

- *Res* is the monad transformer for adding resumptions

$$Res(T)A = \mu X.T(A + X)$$

For each monad transformer, we will explain what kind of operation can be redefined (and how), and which new operations can be added. For instance, in the case of Sef we can redefine any polymorphic operation $op_X: A \times (TX)^B \to TX$ and add operations *lookup*: $L \to TU$ and *update*: $L \times U \to T1$ (provided $S = U^L$). We will also mention problems and limitations of such an incremental approach.

PROGRAMS IN PARTIAL ALGEBRAS - A CATEGORICAL APPROACH

Grzegorz Jarzembski

Institute of Mathematics, N.Copernicus University,
Toruń 87-100, ul.Chopina 12, Poland

The theory of partial algebras seems to be a natural framework for a study of algebraic properties of programs. However, this theory is not the most explored part of universal algebra and it may happen that it does not always offer tools supporting fruitfully these investigations.

In our categorical researches we have developed special methods and techniques introduced mainly for investigation of "categories of mixed structures (i.e. neither purely algebraic, nor purely topological). Some classes of partial algebras, called weak varieties, are within the scope of the categorical theory developed. In this note we are going to present applications of these categorical methods and techniques in studying algebraic properties of programs.

An intuitive meaning of a program in a class of algebraic systems (or partial algebras) is clear; we start with fundamental operations and term operations and then we construct programs using "branches", "loops", "restrictions" and composition of programs.

In this note, however, we do not follow this scheme of construction of programs. Instead, we proceed as follows: first we define the notion of an implicit operation in a weak variety which is, from semantical point of view, a natural generalization of the concept of a term operation. We give a structural description of implicit operations referring to the categorical concept of a free spectrum of a weak variety. Next, using this structural description, we define the class of "programs" as a subclass of implicit operations.

The defined class is greater then the class of programs in the intuitive sense. However, we prefer this approach because the defined class has a nice structure which makes possible a precise analysis of relationship between programs. Namely, we show that programs with a fixed arity form a sheaf, uniquely determined by the class of partial algebras considered.

I. BASIC CONCEPTS

Let Ω denote an arbitrary but fixed finitary type. By PalgΩ we denote the category of partial Ω-algebras and their homomorphisms.

Recall that a homomorphism of partial Ω-algebras $h:(A,(q^A))\longrightarrow(B,(q^B))$ is *strong* provided for every operation symbol q and $a\in A^n$, whenever $q^B(ha)$ is defined then $q^A(a)$ is defined (and then $hq^A(a) = q^B(ha)$) ([3]).

Definition 1.([6]) A class V of partial Ω-algebras (as well as the full subcategory of PalgΩ it determines) is called a weak variety provided it satisfies the following:

i.) whenever $e:A\longrightarrow B$ is surjective and strong and $A\in V$, then $B\in V$,

ii.) if $(m^i:A\longrightarrow A^i:\ i\in I)$ is a jointly monomorphic family containing at least one strong homomorphism and every $A^i\in V$, then $A\in V$.

We call a weak variety *elementary* iff it is closed under formation of ultraproducts. For a first order description of elementary weak varieties we refer the reader to [6].

If a weak variety V consists of total Ω-algebras only, then V is a variety (i.e. an equationally definable class). If V is closed under products, then V is an *ece-variety* in the sense of P.Burmeister ([1]).

Every relation may be considered as a partial projection: for a given relation $r\subseteq A^n$ we define $p_r:A^n\longrightarrow A$, where $\mathrm{dom}(p_r) = r$ and p_r is a restriction of the first-component projection. Thus relational systems as well as algebraic systems may be regarded as partial algebras. In particular, every universal class of algebraic systems may be regarded as a weak variety of partial algebras of a suitably defined type ([6]).

Also many sorted (partial) algebras may be regarded as partial algebras.

EXAMPLE. Let $\hat{\Omega}$ be a two sorted signature with one binary operation symbol q of the scheme (0,1;1)([4]). Then the category of all total $\hat{\Omega}$-algebras is isomorphic to the weak variety V of partial Ω-algebras where Ω consists of a binary symbol q and two unary symbols s,r, while V consists of partial Ω-algebras satisfying the following (universal) formulas:

$$\exists r(x) \vee \exists s(x)\ , \qquad \neg(\exists r(x)\wedge \exists s(x)),$$
$$\exists r(x)\wedge \exists s(y) \Leftrightarrow \exists q(x,y), \qquad \exists q(x,y) \Rightarrow \exists s(q(x,y)).$$

(Here and in what follows, for a given term t, the symbol $\exists t(x_1,..,x_n)$ is an abbreviation of a first order formula such that for any partial algebra A and a valuation h of variables $x_1,..,x_n$ in A, A satisfies $\exists t(x_1,..,x_n)$ at h ($A \models \exists t(x_1,..,x_n)[h]$ in symbols) iff the term $t(hx_1,..,hx_n)$ is defined in A ([6]). These formulas are called *existential atomic formulas)*.

We hope these remarks show that the concept of a weak variety covers a wide class of categories of algebraic structures investigated in the algebraic theory of programs.

Definition 2. By an n-ary implicit operation in a weak variety $V\subseteq$ PalgΩ we mean a family ϕ of partial functions,

$\phi = \{\phi_A : A^n \longrightarrow A:\ A \in V\}$

such that for every homomorphism $h: A \longrightarrow B$ in V, $h \cdot \phi_A \le \phi_B \cdot h^n$ (here and in what follows "$\le$" denotes the usual partial order on sets of parallel partial functions), and moreover, $h \cdot \phi_A = \phi_B \cdot h^n$, provided h is strong.

Roughly speaking, an implicit operation is a family of partial functions indexed by V-algebras and "compatible" with homomorphisms.

Clearly, the family of term operations determined by a fixed Ω-term t is an implicit operation in any weak variety $V \subseteq \mathrm{Palg}\Omega$. We shall show later that the only implicit operations in PalgΩ are "restrictions of term operations". For weak varieties the situation is much more interesting.

EXAMPLE. Let $\Omega = \Omega_1 = \{p,q,r,s\}$ and $V = \mathrm{Mod}(\exists s(x) \wedge \exists r(x) \Rightarrow (p(x)=q(x))$.
The family of unary partial functions $\phi = (\phi_A : A \in V)$ given by:
for every $A \in V$ and $a \in A$, $\phi_A(a)$ is defined iff $s^A(a)$ or $r^A(a)$ is defined and then

$$\phi_A(a) = \begin{cases} p^A(a) & \text{if } r^A(a) \text{ is defined,} \\ \\ q^A(a) & \text{if } s^A(a) \text{ is defined} \end{cases}$$

is an implicit operation which cannot be described by a single term.

"Programs" in a weak variety form a subclass of implicit operations:

TERM OPERATIONS $\subset$ "PROGRAMS" $\subset$ IMPLICIT OPERATIONS

II. FREE SPECTRA OF A WEAK VARIETY

Usually by a "syntax of a program" we mean its logical description. But it is not the most convenient tool if one wants to investigate the structure of the class of programs. In order to get a more appropriate tool, we shall make use of a categorical properties of weak varieties.

Every weak variety considered as a concrete category over sets is a *partially monadic category* ([5]). We shall not need here the detailed definition of a partially monadic category - it may be found in [5]. We focus our attention on some characteristic features of partially monadic categories which are useful for our purposes.

For a given weak variety V, let $U: V \longrightarrow \mathrm{Set}$ denote the obvious underlying functor into sets.

Definition 3. By a free spectrum of $U:V \longrightarrow Set$ over a set X we mean an ordered family of "arrows" $(\eta_i^X : X \longrightarrow J_X(i) = UJ_X(i);\ i \in SX)$ with the following property:
for every $A \in V$ and a function $h:X \longrightarrow UA$ there exists a unique $i \in SX$ and a unique strong homomorphism $\tilde{h}:J_X(i) \longrightarrow A$ such that $U\tilde{h}\cdot\eta_i^X = h$ and, moreover, for every homomorphism $g:J_X(j) \longrightarrow A$, whenever $Ug\cdot\eta_j^X = h$, then $j \le i$ in SX and $g = \tilde{h}\cdot J_X(j \le i)$.

More formally: the spectrum is a triple consisting of a poset SX, a functor $J_X:SX \longrightarrow V$ and a natural transformation $\eta:X \longrightarrow UJ_X = J_X$ satisfying the condition above.

The existence and uniqueness of free spectra in weak varieties has been proved in [6]. This is a one of the characteristic feature of partially monadic categories.

CONSTRUCTION OF SPECTRA

1. Spectra of $U^\Omega:Palg\Omega \longrightarrow Set$. By an *initial Ω-segment* over a given set X we mean every subset X_i of the set ΩX of all Ω-terms such that $X \subseteq X_i$ and together with any term t, X_i contains every subterm of t. Every initial segment X_i carries a unique structure of a partial Ω-algebra X_i being an initial subalgebra of the (total) algebra of Ω-terms ΩX.
The spectrum of U^Ω over X (with respect to strong homomorphisms) consists of all algebras of the form X_i, where X_i is an initial segment. The arrows $\eta_i^X:X \longrightarrow X_i = U^\Omega X_i$ are identity embeddings (compare [9]).

2. Let $V \subseteq Palg\Omega$ be a weak variety. We construct the required spectrum $(S^V X, J_X^V, \eta^X)$ as follows: $S^V X$ is a subset of the set $S^\Omega X$ of all initial segments over X such that for every $X_i \in S^\Omega X$:

$X_i \in S^V X$ iff there is a strong homomorphism $h:X_i \longrightarrow A$ with $A \in V$.

For every $X_i \in S^V X$, $J_X^V(X_i)$ is a strong epimorphic image of the algebra X_i obtained as a central object of the (strong epi, monosource) factorization of the source

$V(X_i) = (h_i:X_i \longrightarrow A^h:\ A^h \in V)$

For the details of this construction we refer the reader to [6].

Notational convention. For a natural number n by **n** we denote the set {0,1,..,n-1}. For notational simplicity, we denote elements of $S^V\mathbf{n}$ by small letters i,j,d...etc. and we will write $J_n^V(i)$ instead of $UJ_n^V(X_i)$.

The following theorem explains the role of spectra in the investigation of implicit operations.

Theorem 4. Let ϕ be an n-ary implicit operation in a weak variety V. Then there exists a uniquely determined increasing subset $D \subseteq S^V\mathbf{n}$ and an element $r = (r_d:d \in D) \in \lim(J_n^V(d):d \in D)$ such that the following holds true:

for every A in **V** and $h{:}n \longrightarrow UA = A$ (i.e. $h \in A^n$) with the strong extension $\tilde{h}{:}J_n^V(i) \longrightarrow A$ of (h,A),

$\phi_A(h)$ is defined iff $i \in D$ and then $\phi_A(h) = \tilde{h}(r_i)$.

Proof. Put

$D = \{i \in S^V n{:}\ \phi$ is defined in $J_n^V(i)$ on the n-tuple $\eta_i^n{:}n \longrightarrow J_n^V(i)\}$.

If $i \in D$ and $i \le j$, then there is a homomorphism $J_n^V(i{\le}j)$ such that $J_n^V(i{\le}j)(\eta_i^n) = \eta_j^n$. From this it follows that $j \in D$, too, i.e. D is an increasing set. Next, let $r = (r_d{:}\ d \in D)$ be a sequence such that every r_d is the value of the operation ϕ in the algebra $J_n^V(d)$ on the n-tuple η_d^n.

It is easily checked that $r \in \lim(J_n^V(d){:}d \in D)$.

We call the pair (D_ϕ,r_ϕ) a *structural description* of the operation ϕ the more that the converse of the theorem above holds true:

Proposition 5. Let D be an increasing subset of $S^V n$ and let $r = (r_d) \in \lim(J_n^V(d){:}\ d \in D)$. Then there exists a unique n-ary implicit operation ϕ such that $(D,r) = (D_\phi,r_\phi)$.

Proof. Define ϕ as follows: for A in V and $h{:}n \longrightarrow A$, with the strong extension $\tilde{h}{:}J_n^V(i) \longrightarrow A$, $\phi_A(h)$ is defined iff $i \in D$ and then $\phi_A(h) = \tilde{h}(r_d)$.

An implicit n-ary operation ϕ in V is said to be *irreducible* iff for any family $\{\gamma^i{:}i \in I\}$ of n-ary implicit operations, whenever $\phi \le \bigcup\{\gamma^i{:}i \in I\}$, then $\phi \le \gamma^j$ for some $j \in I$. ($\phi \le \gamma$ means that $\phi_A \le \gamma_A$ for every A in V.)

Corollary 6.

i.) Every n-ary irreducible operation is uniquely determined (in the sense of Theorem 4) by a pair $(i \in Sn, r \in J_n^V(i))$.

ii.) every n-ary implicit operation is a join of irreducible operations.

A straightforward consequence of the given structural description of implicit operation is the following

III. LOGICAL DESCRIPTION OF IMPLICIT OPERATIONS

By Theorem 4 and the description of free spectra in PalgΩ, implicit operations in PalgΩ are simply "restrictions of term operations", i.e. every n-ary operation ϕ is uniquely determined by a pair $(t,(p^i{:}i\in I))$, where t and every p^i are Ω-terms with variables $x_1,..,x_n$ in the sense that for every A in **V** and a valuation $h{:}\{x_1,..x_n\} \longrightarrow A$,

$\phi_A(h)$ is defined iff $A \models \Lambda(p^i{:}i\in I)[h]$ and then

$\phi_A(h)$ is the value of t at the valuation h.

Thus the pair $(t,(p^i{:}i\in I))$ may be treated as a *logical description* of the operation ϕ.

Assume now that $\mathbf{V} \subseteq \mathrm{Palg}\Omega$ is a weak variety. We need more detailed analysis of free spectra in **V**.

Let $n_i \in S^V n \subseteq S^\Omega n$ and let $\beta^n_i{:}n_i \longrightarrow J^V_n(i)$ be the strong surjective epimorphism being the first component of the (strong epi, monosource) factorization of the source $V(n_i)$ (section 1). Thus every element of $J^V_n(i)$ is an equivalence class of some Ω-term with n variables.

Notice that for any $t,r\in n_i$, $\beta^n_i(t) = \beta^n_i(r)$ provided the formula

$\Lambda(\exists p{:}p\in n_i) \Rightarrow (t = r)$

is valid in **V** ([6]).

Lemma 7. Let ϕ be an irreducible n-ary implicit operation in **V**. Then:

i. there exists a pair $(t,(p^i{:}i\in I))$ where t and every p^i are terms with variables $x_1,..,x_n$ which describe the operation ϕ in the sense above,

ii. two such pairs $(t,(p^i{:}i\in I))$, $(r,(s^k{:}k\in K))$ describe the same implicit operation provided every A in **V** satisfies the formulas:

$\Lambda(\exists p^i{:}i\in I) \Leftrightarrow \Lambda(\exists s^k{:}k\in K)$,

$\Lambda\exists p^i{:}i\in I) \Rightarrow (t = r)$.

Proof. This is a straightforward consequence of Corollary 6.i.and the description of the sets $J^V_n(i)$ given above.

As a consequence of Lemma 7 and Corollary 6.ii. we obtain

Corollary 8. Every n-ary implicit operation ϕ in **V** has a logical description, i.e. there exists a pair of sequences $((t^i{:}i\in I),(\Phi^i{:}i\in I))$ such that:

every t^i is an Ω-term with variables in $n = \{0,1,..,n-1\}$,

every Φ^i is a set of Ω-terms with variables in n such that the formula

$\Lambda\Phi^i \wedge \Lambda\Phi^j \Rightarrow (t^i = t^j)$

is valid in **V**, and

for every A in **V** and a valuation of variables $h{:}n \longrightarrow A$,

$\phi_A(h)$ is defined iff $A \models \bigvee(\bigwedge\Phi^i : i\in I)[h]$

and then

$\phi_A(h) = t^i(hx_1,..,hx_n)$ (the value of the term t^i at the valuation h)

provided $A \models \bigwedge\Phi^i$ [h].

Notice the following:

in contrary to the structural description given in Theorem 4, a logical description of an implicit operation need not be unique,

in order to give a logical description of an implicit operation we need not only first order formulas but also infinite conjunctions and infinite disjunctions.

It is natural to distinguish the following subclass of implicit operations:

Definition 9. We call an implicit operation ϕ *finitely definable* provided it has a first order description, i.e., there exists a pair of finite sequences $((t_1,...,t_m),(\Phi^1,..,\Phi^m))$ such that every Φ^i is a finite set and this pair describes ϕ in the sense of Corollary 8.

One may ask the following question:

how we can describe the structure of finitely definable operations using the concept of the free spectrum?

We are going to answer this question for elementary weak varieties.

IV. SHEAVES OF PROGRAMS

We show that for elementary weak varieties free spectra over finite sets are endowed not only with an ordered structure but also with a more subtle topological structure. This topological structure is then used in order to distinguish structural descriptions of finitely definable operations.

Let us start with the category PalgΩ of all partial Ω-algebras. By Corollary 6, finitely definable operations in PalgΩ may be described as follows:

- every restriction of a term operation, where the restriction is described by a finite conjunction of existential atomic formulas, is an irreducible finitely definable operation,

- every finitely definable operation in PalgΩ is a finite join of such restrictions.

Thus the structural description (D_ϕ, r_ϕ) of a finitely definable operation ϕ in PalgΩ is such that:

$D_\phi = \uparrow n_1 \cup ... \cup \uparrow n_k$, where every n_i is a finite initial segment over **n** while

$\uparrow n_i = \{ n_j \in S^\Omega n : n_i \leq n_j \}$,

and $\quad r_\phi \in \lim(n_k : n_k \in D_\phi) = \bigcap(n_j:\ j = 1,2,..k)$

Since $S^\Omega n$ is an algebraic lattice, it may be endowed with the Lawson topology $\lambda(S^\Omega n)$ with an open subbase consisting of sets $\uparrow n_k$ and $S^\Omega n \setminus \uparrow n_k$ where n_k is finite (i.e. a compact element of $S^\Omega n$)([2]).

Notice that clopen increasing sets in $S^\Omega n$ are finite joins of sets of the form $\uparrow n_k$ for a finite initial segment n_k.

Let $\lambda(S^\Omega n)^+$ denote the subtopology of $\lambda(S^\Omega n)$ generated by these sets. Define a sheaf ([8]) $\bar{J}^\Omega_n : (\lambda(S^\Omega n)^+)^{op} \longrightarrow \mathbf{Set}$ by:

$\bar{J}^\Omega_n(\uparrow n_1 \cup ... \cup \uparrow n_k) = \bigcap(n_j:\ j = 1,2,..k)$

for every clopen increasing set $D = \uparrow n_1 \cup ... \cup \uparrow n_k$. Then we extend $\bar{J}^\Omega_n$ to all $\lambda(S^\Omega n)^+$-open sets in the usual way.

Notice that every initial segment is a stalk of the sheaf considered.

From above it follows that the structural description of finitely definable operations in PalgΩ may be given in the following form:

Corollary 10. For every finitely definable n-ary operation ϕ there exists a clopen increasing set D and $r \in \bar{J}^\Omega_n(D)$ such that for every $A \in$ PalgΩ and $h: n \longrightarrow A$ with the strong extension $\tilde{h}: n_i \longrightarrow A$,

$\phi_A(h)$ is defined iff $i \in D$ and then $\phi_A(h) = \tilde{h} \cdot \psi_{Di}(r)$,

(where $\psi_{Di} : \bar{J}^\Omega_n(D) \longrightarrow n_i$ denotes the stalk-embedding).

We are going to show that finitely definable operations in any elementary weak variety V can be described in the same way, for some suitably defined topology on its free spectra and suitably defined sheaves.

Let $V \subseteq$ PalgΩ be an elementary weak variety. Since V is closed under formation of ultraproducts, the spectrum $S^V n$ is a closed subset of $S^\Omega n$ with respect to the topology $\lambda(S^\Omega n)$ ([7]). Let $\mathcal{T}_n$ denote the induced topology on $S^V n$, while $\mathcal{T}^+_n$ the subtopology of $\mathcal{T}_n$ generated by $\mathcal{T}$-clopen increasing sets.

It has been proved in [7] that the sets of the form $V_k = \uparrow n_k \cap S^V n$, where n_k is compact (i.e. finite) in $S^\Omega n$, form a base of $\mathcal{T}^+_n$. Let $[k] = \inf V_k$ in $S^\Omega n$.

Consider the (epi,monosource)-factorization of the source

$$(n_{[k]} \longrightarrow n_j \xrightarrow{\beta^n_j} J^V_n(j) \ :\ n_j \in V_k\ ).$$

Let $\hat{J}^V_n(V_k)$ be the central object of this factorization.

It is easy to observe that $\hat{J}^V_n$ may be extended to a presheaf

$\hat{J}^V_n: (\mathcal{T}^+)^{op} \longrightarrow Set.$

Finally, let $\bar{J}^V_n$ be the sheaf associated to the presheaf $\hat{J}^V_n$ ([8]). Observe that for every clopen increasing set D, $\bar{J}^V_n(D)$ is a subset of $\lim(J^V_n(i): i\in D)$.

Theorem 11. Let ϕ be an n-ary implicit operation in an elementary weak variety V with the structural description (D,r). ϕ is finitely definable iff D is a clopen increasing set and $r\in \bar{J}^V_n(D)$ and then for every A in V and $h: n \longrightarrow A$ with the strong extension $\tilde{h}: J^V_n(i) \longrightarrow A$,

$\phi_A(h)$ is defined iff $i\in D$ and then $\phi_A(h) = \tilde{h}\cdot\psi_{Di}(r)$,

(where $\psi^V_{Di}: \bar{J}^V_n(D) \longrightarrow J^V_n(i)$ denotes the stalk-embedding).

Proof. Observe that the set $\hat{J}^V_n(V_k)$ constructed above is a quotient of $n_{[k]}$. Two terms $t,p\in n_{[k]}$ have the same image in $\hat{J}^V_n(V_k)$ provided the first order formula:

$\bigwedge(\exists r: r\in n_k\ ) \Rightarrow (t = p)$

is valid in V. Hence elements of $\hat{J}^V_n(V_k)$ represents finitely definable operations having the logical description of the form

$(t\in n_{[k]}, \Phi = \{\exists r: r\in n_k\})$.

It is clear that finitely definable operations are precisely finite joins of such operations.

Now take $r\in \bar{J}^V_n(D)$, where D is a clopen increasing set. This means that there exists a covering of D, $D = \bigcup(V_{k(i)}: i\in I)$ and a "compatible" family $(r^i: r^i\in \hat{J}^V_n(V_{k(i)})$ which "represents" r. Since the Lawson space $S^\Omega n$ is compact, $S^V n$ and D are closed subsets (hence compact), every $\mathcal{T}^+_n$-open covering of D may be reduced to a finite covering of the form

$D = V_{k(1)} \cup \ldots \cup V_{k(m)}$, where $V_{k(i)} = \uparrow n_i \cap S^V n$ and n_i is finite.

Hence every element $r\in \bar{J}^V_n(D)$ is representable by a compatible finite sequence $(r_i\in \hat{J}^V_n(V_{k(i)})$, $i = 1,..,m)$ for some finite covering of D of the form described above. This means that the implicit operation corresponding to the pair $(r\in \bar{J}^V_n(D), D)$, where D is clopen and increasing, is a finite join of "finite restrictions of term operations" i.e. the pair $(r\in \bar{J}^V_n(D), D)$ is the structural description of a finitely definable operation.

The opposite implication is obvious.

Up to now, discussing finitely definable operations, we considered only the restrictions of the associated sheaves $\bar{J}^V_n$ to clopen increasing sets in $S^V n$. The global analysis of sheaves $\bar{J}^V_n$ leads to the following definition.

Definition 12. An n-ary implicit operation ϕ in an elementary weak variety $V \subseteq Palg\Omega$ is called a program iff there exist a $D \in \mathcal{T}_n^+$ and an $r \in \bar{J}_n^V(D)$ such that ϕ is described by the pair (D,r) in the sense of Theorem 11.

Thus, taking into account the construction of the sheaf $\bar{J}_n^V$ we may say that programs are *locally finitely definable implicit operations* i.e. every such operation is a direct sum of finitely definable operations. In other words: for any program ϕ , an algebra $A \in V$ and $h: n \longrightarrow A$, whenever $\phi_A(h)$ is defined then there exists a finitely definable operation ψ such that $\psi \leq \phi$ and $\psi_A(h)$ is defined (and then ,of course, $\psi_A(h) = \phi_A(h)$).

Notice that every $J_n^V(i)$ is the stalk of the sheaf $\bar{J}_n^V$ at $i \in S^V n$. This means that for every irreducible operation ϕ, every A in V and h: $\longrightarrow UA$, there exists a program ψ such that $\psi_A(h)$ is defined and $\psi_A(h) = \phi_A(h)$, i.e. programs form a "dense" subset of the set of implicit operations.

V. OPERATIONS ON PROGRAMS

In this last section we describe briefly operations which may be considered on the class of programs defined in Definition 12.

Throughout $V \subseteq Palg\Omega$ denotes an arbitrary but fixed elementary weak variety.

1. ***RESTRICTION.*** Given n-ary programs ϕ and ψ by a *restriction* of ϕ determined by ψ we mean a program $\phi_{|\psi}$ such that for every A in V, $dom(\phi_{|\psi})_A = dom\phi_A \cap dom\psi_A$ and for every $h \in dom(\phi_{|\psi})_A$, $(\phi_{|\psi})_A(h) = \phi_A(h)$.

In other words: if (D_ϕ, r_ϕ) and (D_ψ, r_ψ) are structural descriptions of ϕ and ψ, resp. in the sense of Theorem 11, then the structural description of $\phi_{|\psi}$ has the form $(D_\phi \cap D_\psi, (r_\phi)_{|D_\psi})$, where $(r_\phi)_{|D_\psi} = \bar{J}_n^V(D_\phi \cap D_\psi \subseteq D_\phi)(r_\phi)$ (the restriction of r to $D_\phi \cap D_\psi$).

2. ***COMPOSITION.*** Given a sequence of m-ary programs $\phi_1, .., \phi_n$ and an n-ary program ψ we may define a composition $\psi(\phi_1, .. \phi_n)$ in the obvious way. It is easily checked that a composition of programs is a program.

The structural description of composition may be derived from structural descriptions of its components too. However it needs more detailed analysis of connections between spectra over different finite sets which was omitted here. We refer on this subject to [7].

3. ***JOIN.*** Given n-ary programs ϕ and ψ such that for every A in V, and $h \in A^n$, whenever both $\phi_A(h)$ and $\psi_A(h)$ are defined then they are equal,we may define a join $\phi \vee \psi$ as follows: for every $A \in V$, $dom(\phi \vee \psi)_A = dom\phi_A \cup dom\psi_A$ and whenever $(\phi \vee \psi)_A(h)$ is defined, then

$$(\phi\vee\psi)_A(h) = \begin{cases} \phi_A(h) & \text{if } \phi_A(H) \text{ is defined,} \\ \psi_A(h) & \text{if } \psi_A(h) \text{ is defined.} \end{cases}$$

Given structural descriptions (D_ϕ,r_ϕ) and (D_ψ,r_ψ) of ϕ and ψ, resp. (in the sense of Theorem 11), the structural description of $\phi\vee\psi$ is the pair $(D_\phi\cup D_\psi,r)$, where $r\in \bar{J}^V_n(D_\phi\cup D_\psi)$ is a unique element such that $r_{|D_\phi} = r_\phi$ and $r_{|D_\psi} = r_\psi$.

In a similar way one can define an infinite join of programs.

4. *LOOP* Every program derived from other programs using "loops" may be semantically represented as a directed join of programs. Thus the defined class is also closed under formation of "loops".

Note however, that we do not claim that every directed join of programs may be derived from finitely definable programs as a result of a finite sequence of operations decribed above. Thus in fact the considered class as greater then the class of "programs" in the intuitive sense. But, as we have mentioned in the introduction, the sheaf of programs defined here seems to be a proper framework for investigation of algebraic properties of "real programs".

REFERENCES

[1] P.Burmeister, *A Model Theoretic Oriented Approach to Partial Algebras (Introduction to Theory and Applications of Partial Algebras - Part I)*, Math.Research, vol.32, Akademie-Verlag, Berlin, 1986.

[2] G.Gierz, K.H.Hoffmann, K.Keimel, J.D.Lawson, M.Mislove, D.S.Scott, *A Compendium of Continuous lattices*, Springer-Verlag, 1980.

[3] G.Gratzer, *Universal Algebra, 2nd ed.* Springer-Verlag, 1979.

[4] P.J.Higgins, *Algebras with a scheme of operators*, Math.Nach.27, 115-132, 1963.

[5] G.Jarzembski, *Finitary spectral algebraic theories*, J.Pure and Appl.Algebra 52, 31-50, 1988.

[6] G.Jarzembski, *Weak varieties of partial algebras*, Alg.Univ.25, 247-262, 1988.

[7] G.Jarzembski, *Sheaves of finitely definable operations in weak varieties*, to appear in Alg.Univ.

[8] P.Johnstone, *Topos Theory*, Academic Press, 1977.

[9] J.Schmidt, *A homomorphism theorem for partial algebras*, Coll. Math.21, 5-21, 1970.

Tail Recursion from Universal Invariants

C. Barry Jay *
University of Edinburgh
University of Ottawa

June 28, 1991

Abstract

The categorical account of lists is usually given in terms of initial algebras, i.e. head recursion. But it is also possible to define them by interpreting tail recursion by means of the colimit of a loop diagram, i.e. its universal invariant. Parametrised initial algebras always have universal invariants, while the converse holds in the presence of equalisers.

Consequences include categorical descriptions of vectors and matrices, which allow definitions of inner products, transposes and matrix multiplication.

*Research supported by The Royal Society of Edinburgh/BP, and NSERC operating grant OGPIN 016.

1 Introduction

When the object of lists in a category is defined as an initial algebra (a parametrised (initial) stack), then head recursion, represented by the operation of foldright [?], becomes the focus of interest. The resulting mathematics is familiar and useful, but suffers from the defect that much post-hoc manipulation is required to obtain efficient descriptions of many operations of interest. An alternative is now available, namely convergent stacks, defined in terms of tail recursion, with basic operation foldleft.

Tail recursion (and while-loops etc.) can be expressed in terms of an extremely elementary, but neglected class of categorical diagrams which have exactly one object and one morphism, namely the loops.

The limit of a loop $f : D \rightarrow D$ is its fixpoints $\mathtt{fix}(f) : \mathtt{Fix}(f) \rightarrow D$ while the colimit is its *universal invariant* denoted $\mathtt{inv}(f) : D \rightarrow Inv(f)$. A loop *converges* if we can choose $\mathtt{Fix}(f) = \mathtt{Inv}(f)$ with $\mathtt{inv}(f) \circ \mathtt{fix}(f) = \mathtt{id}$. A special case of convergence (in the presence of a natural numbers object) is *termination* where every sequence obtained by iterating the loop reaches a fixpoint after a specified number of steps. A *terminating stack* (*pre-recursive* in [?]) is a convergent stack for which `tail` terminates with fixpoints given by `nil`.

An appropriate setting for the study of stacks (solutions to the domain equation for lists) is a *polynomial category* i.e. a category which has all finite products and sums, with the products distributing over the sums. In such a category the various concepts of list above are closely related. Parametrised stacks are convergent stacks; the converse holds in the presence of equalisers. Here is given a slightly weaker result, namely the existence of a parametrised natural numbers object is assumed to show that parametrised stacks are terminating stacks (and hence convergent). The more general result can be found in [?]. At this stage it appears that foldright is good for mathematical constructions, foldleft is good for computation (and is suggestive for other datatypes, e.g. trees) while contractions are best for proofs of list properties.

A *distributive category* [?] is a finitely complete polynomial category where the sums are disjoint and universal. Then having all parametrised stacks is equivalent to having all convergent stacks, which are unambiguously called list objects. Such a distributive category is called a *locos.*

Examples of list properties are that the list functor preserves connected limits [?]. This yields the *zip* operation which maps pairs of lists of equal length (a pullback) to lists of pairs. Another new result is that many of the natural transformations associated with lists (e.g. flatten) have their naturality square being a pullback.

More importantly, from lists can be constructed vectors, whose operations must be parametric in their length. From the vectors can be extracted the list objects in slices of the category, which yields a new proof that a slice of a locos is a locos.

Vectors of vectors are matrices. Transposition of matrices is defined using

of matrices can be defined, and is the composition of an internal category. The power of the approach can be illustrated by the construction of the Discrete and Fast Fourier Transforms, together with a proof of their equivalence [?].

2 Preliminaries

A polynomial category is one which has all finite products and sums, with the products distributing over the sums. Examples include **Sets**, and **Sp** the category of topological spaces and continuous functions. The structure is inherited by many sub-categories of **Sp** such as that of bottomless c.p.o.'s, or **Met** the category of metric spaces (with possibly infinite distances) and distance-decreasing maps. In the presence of finite sums, cartesian closure suffices since the functor $A\times-$ is then a left adjoint, and so preserves all colimits. It is not essential, however, as demonstrated by **Sp**. Let us establish some notation.

The pairing of $f : C\to A$ and $g : C\to B$ is denoted $\langle f,g\rangle : C\to A\times B$ with projections $\pi_{A,B} : A\times B\to A$ and $\pi'_{A,B} : A\times B\to B$ and symmetry $c_{A,B} : A\times B\to B\times A$. Unless they aid comprehension the subscripts on natural transformations will be suppressed.

The cases morphism for $f : A\to C$ and $g : B\to C$ is $[f,g] : A+B\to C$ with inclusions $\iota_{A,B} : A\to A+B$ and $\iota'_{A,B} : B\to A+B$ and symmetry $\mathtt{switch}_{A,B} = [\iota',\iota] : A+B\to B+A$.

Distributivity of products over binary sums asserts that

$$(A\times B)+(A\times C)\xrightarrow{[1\times\iota,1\times\iota']}A\times(B+C) \tag{1}$$

is an isomorphism (with inverse denoted $d_{A,B,C}$). Similar calculations for the empty sum shows that

$$A\times 0\cong 0 \tag{2}$$

from which it follows that 0 is strict, i.e. any morphism $A\to 0$ is an isomorphism.

Fix a polynomial category C for the rest of this abstract.

3 Loops

Given a category $\mathcal{A}$ define $\mathcal{A}^c$ to be the category of loop diagrams in $\mathcal{A}$. Its objects are the endomorphisms $f : D\to D$ of $\mathcal{A}$ thought of as diagrams with *one* object and one arrow. The *loop morphisms* from f to $g : E\to E$ are the

usual diagram morphisms,

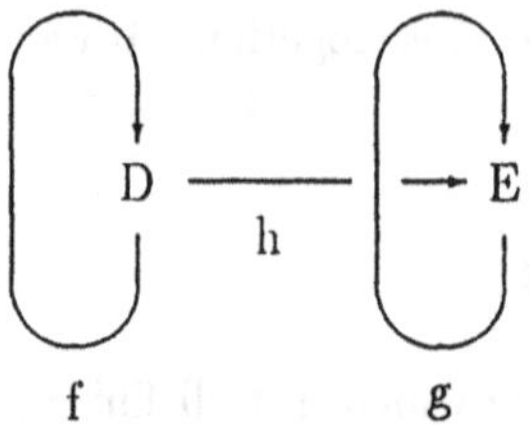

i.e. $h : D \rightarrow E$ such that $h \circ f = g \circ h : D \rightarrow E$. Note that $\mathcal{A}^c$ is *not* a full subcategory of $\mathcal{A}^{\rightarrow}$ the arrow category of $\mathcal{A}$.

Such loops may be used to represent the conditional of a while-loop or the unfolding of a fixpoint combinator [?]. The (categorical) limit of such a loop f on D is its *fix object* denoted $\mathtt{Fix}(f)$ with inclusion $\mathtt{fix}(f) : \mathtt{Fix}(f) \rightarrow D$. Thus any morphism $x : X \rightarrow D$ which is fixed by f (that is $f \circ x = x$) factors through $\mathtt{fix}(f)$.

An *invariant* of the loop f above is a morphism $g : D \rightarrow X$ such that $g \circ f = g$. Alternatively, g is a loop morphism from f to $\mathtt{id}_Q$ (a *cocone* for the loop diagram [?]). The colimit of the loop, if it exists, is its ***universal invariant***, denoted $\mathtt{inv}(f) : D \rightarrow \mathtt{Inv}(f)$. Then the invariant g factors through $\mathtt{inv}(f)$ in a unique way. Of course the universal invariant is only determined up to isomorphism.

Definition 3.1 *The loop f on D in $\mathcal{A}$* converges *if it has a universal invariant* $\mathtt{inv}(f) : D \rightarrow \mathtt{Inv}(f)$ *whose object also serves as the fixobject of f via some* $\mathtt{fix}(f) : \mathtt{Inv}(f) \rightarrow D$ *such that* $\mathtt{inv}(f) \circ \mathtt{fix}(f) = 1 : \mathtt{Inv}(f) \rightarrow \mathtt{Inv}(f)$.

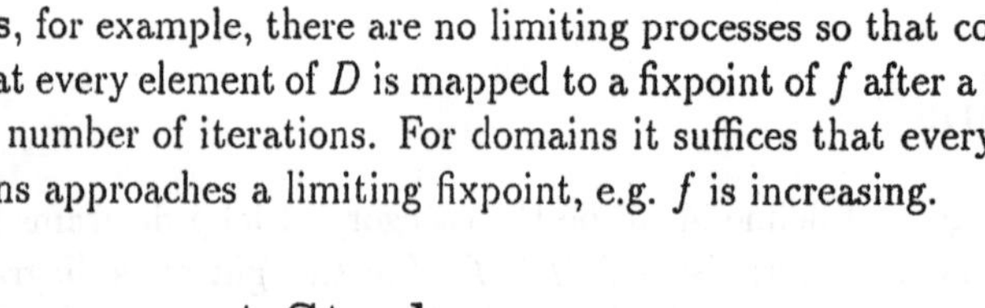

In **Sets**, for example, there are no limiting processes so that convergence implies that every element of D is mapped to a fixpoint of f after a finite (but unknown) number of iterations. For domains it suffices that every sequence of iterations approaches a limiting fixpoint, e.g. f is increasing.

4 Convergent Stacks

Let A be an object in $\mathcal{C}$. Consider a solution $(L, \mathtt{empty}, \mathtt{push})$ for the domain equation for stacks on A [?]

$[\mathtt{empty}, \mathtt{push}]$

The inverse is called pop : $L \to 1 + (A \times L)$. When the stack is thought of as a list then `empty` and `push` may be denoted `nil` and `cons` respectively.

Given a right A-action $\alpha : C \times A \to C$ on an object C we may then define the loop $\mathtt{shunt}(\alpha) : C \times L \to C \times L$ by

$$C \times L \xrightarrow{d \circ (1 \times \mathtt{pop})} (C \times 1) + (C \times A \times L) \xrightarrow{[1 \times \mathtt{nil}, \alpha \times 1]} C \times L$$

When $C = 1$ with its unique A-action then **shunt** corresponds to $\mathtt{tail} = [\mathtt{nil}, \pi'] \circ \mathtt{pop} : L \to L$.

Definition 4.1 *$(L, \mathtt{nil}, \mathtt{cons})$ is a* convergent stack *for A if $\mathtt{shunt}(\alpha)$ converges with fixpoints given by $\langle 1, \mathtt{nil}\rangle : C \to C \times L$ for all right A-actions α. The corresponding universal invariant is $\mathtt{foldl}(\alpha) : C \times L \to C$ called* foldleft *of α.*

In **Sets** if $\alpha(x, a)$ is denoted by $x \oplus a$ then we have

$$\mathtt{foldl}(\alpha)(x, [a_i]) = (\ldots((x \oplus a_0) \oplus a_1)\ldots a_{n-1})$$

which is the usual operation *foldleft* on lists.

Define

$$\begin{aligned} \mathtt{rev} &= \mathtt{foldl}(\mathtt{cons} \circ \mathtt{c}) \circ \langle \mathtt{nil}, 1\rangle : L \to L^2 \to L \\ \mathtt{snoc} &= \mathtt{rev} \circ \mathtt{cons} \circ (1 \times \mathtt{rev}) \circ \mathtt{c} : L \times A \to L \\ @ &= \mathtt{foldl}(\mathtt{snoc}) : L^2 \to L \end{aligned}$$

rev reverses a stack, **snoc** is like **cons** except that it attaches the new entry to the tail of the list, rather than the head, and @ is the append function.

Lemma 4.2 *The following diagram commutes (for $\alpha = \mathtt{cons} \circ \mathtt{c} : L \times A \to L$).*

$$\begin{array}{ccc} L^2 & \xrightarrow{\mathtt{foldl}(\alpha)} & L \\ {\scriptstyle \mathtt{c}}\big\downarrow & & \big\downarrow{\scriptstyle \mathtt{rev}} \\ L^2 & \xrightarrow[\mathtt{foldl}(\alpha)]{} & L \end{array} \qquad (4)$$

Thus $\mathtt{rev} \circ \mathtt{rev} = \mathtt{id}$.

Proof It suffices to show that $\mathtt{foldl}(\alpha) \circ \mathtt{c}$ is an invariant for the loop $\mathtt{shunt}(\alpha)$ on L^2 since then it equals $h \circ \mathtt{foldl}(\alpha)$ for some $h : L \to L$. Now composing with $\langle 1, \mathtt{nil}\rangle$ shows that $h = \mathtt{rev}$. Stacking two of these squares together yields the second result. □

Another example of the proof techniques available is given by

Lemma 4.3 *Append is associative, i.e.* $@ \circ (@\times 1) = @ \circ (1\times @) : L^3 \to L$.
Proof Both sides are the colimit of the diagram with one object L^3 and two loops given by $\mathtt{shunt(snoc)}\times 1$ and $1\times\mathtt{shunt(snoc)}$. This shows that @ is associative up to isomorphism, but this isomorphism must be the identity since the two colimits have a common splitting, namely $1\times\mathtt{nil}\times\mathtt{nil}$. □

5 Parametrised Stacks

Given an object A in C define the functor $F : C \to C$ by $F(-) = 1 + (A\times -)$. An *initial stack* for A is then an initial algebra for F. More usefully, consider the functor $G : C^2 \to C$ defined by $G(-, B) = B + (A\times -)$. A *parametrised initial stack* for A (or *parametrised stack* for short) is an F-algebra $(L, \mathtt{nil}, \mathtt{cons})$ such that, for each B $(L\times B, \langle\mathtt{nil}, 1\rangle, \mathtt{cons}\times 1)$ is an initial algebra for $G(-, B)$. That is, given objects B, C and morphisms $x : B \to C$ and $\alpha : A\times C \to C$ there is a unique morphism $\mathtt{foldr}(x, \alpha) : L\times B \to C$ called *foldright* of x and α, making the following diagram commute

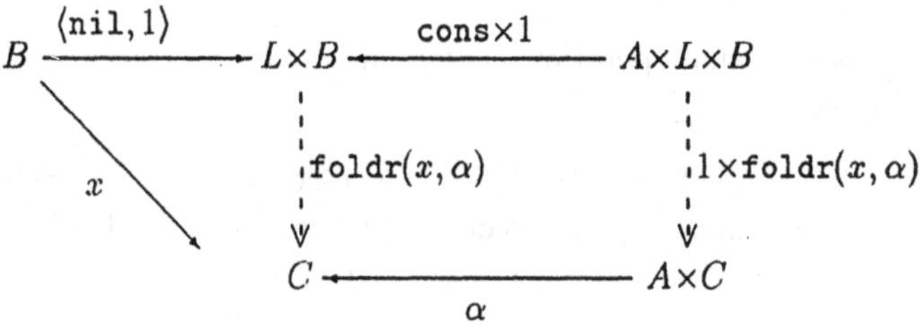

If $x = 1_C$ then it is denoted by $\mathtt{foldr}(\alpha)$. In **Sets** if $x = 1$ and $\alpha(a, y)$ is denoted by $a \oplus y$ then we have

$$\mathtt{foldr}(\alpha)([a_i], y) = a_0 \oplus (a_1 \oplus (\ldots (a_{n-1} \oplus y) \ldots))$$

which is the usual operation *foldright* on lists.

The parameter B is necessary since C need not be cartesian closed. It is also more natural. For example, append can be defined here as $@ = \mathtt{foldr(cons)} : L^2 \to L$ whereas the usual definition of lists as an initial algebra for F forces us to first define the transpose $L \to (L \to L)$ of append, which is rather messy, and then transpose back.

A parametrised stack N for the terminal object is exactly a parametrised natural numbers object [?, ?, ?, ?] with zero $0 = \mathtt{nil} : 1 \to N$ and successor $S = \mathtt{cons} \circ l^{-1} : N \to 1\times N \to N$. The (parametric) universal property requires that given $g : B \to C$ and $h : C \to C$ there is a unique map $\mathtt{It}(g, h) : N\times B \to C$

making the following diagram commute

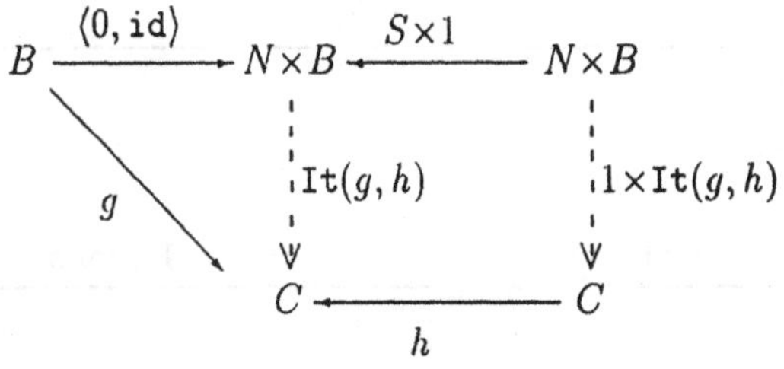

Of course, $\mathtt{It}(g, h) = \mathtt{foldr}(g, h \circ l)$ which may be abbreviated to $\mathtt{It}(h)$ if $g = \mathtt{id}$.

If there is a natural numbers object N then a parametrised stack L for A has an associated length morphism

$$\# = \mathtt{foldr}(0, S \circ \pi') \circ r^{-1} : L \to L \times 1 \to N$$

Given N then a loop $f : D \to D$ is a *bound* if there a *bounding map* $\beta : D \to N$ such that $\mathtt{foldr}(f)^{\beta} = \mathtt{foldr}(f) \circ \langle \beta, \mathtt{id} \rangle$ is fixed by f. If it actually factors through a morphism $m : F \to C$ fixed by f then f *terminates.* Informally, this says that β-many applications of f yields a fixpoint. It follows [?] that terminating loops are all convergent.

A stack whose `tail` terminates with fixpoints given by `nil` is called a *terminating stack.*

Theorem 5.1 *Let C be a polynomial category with a parametrised natural numbers object and let* $\mathbf{L} = (L, \mathtt{nil}, \mathtt{cons})$ *be a stack on A. The following are equivalent:*

(i) **L** *is a convergent stack.*

(ii) **L** *is a parametrised stack.*

(iii) **L** *is a terminating stack.*

Proof (i)⇒(ii): Given $\alpha : A \times C \to C$ define

$$\mathtt{foldr}(\alpha) = \mathtt{foldl}(\alpha \circ c) \circ (1 \times \mathtt{rev}) \circ c : L \times C \to C$$

Then with $x : B \to C$ define

$$\mathtt{foldr}(x, \alpha) = \mathtt{foldr}(\alpha) \circ (1 \times x) : L \times B \to C$$

The compatability of $\mathtt{foldr}(x, \alpha)$ with `nil` is straightforward. For `cons`

consider the following diagram

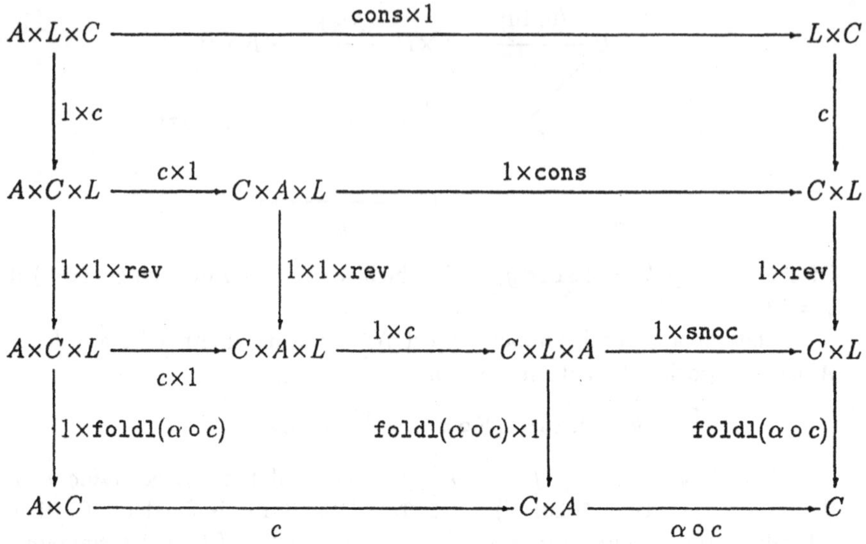

The bottom right-hand square commutes because $\texttt{foldl}(\alpha \circ c) \circ (1 \times \texttt{snoc})$ is an invariant for $\texttt{shunt}(\alpha \circ c) \times 1$ (use the associativity of @) and so factors through $\texttt{foldl}(\alpha \circ c)$ by some morphism, which is revealed to be $\alpha \circ c$ on composing with $1 \times \texttt{nil} \times 1$.

It remains to show that $\texttt{foldr}(x, \alpha)$ is the unique map making the desired diagram commute. If there are two such maps then their equaliser is a subobject $m : R \to B \times L$ compatible with the parametrised action. This induces a left inverse to m which forces it to be an isomorphism. Hence the two maps are equal.

(ii)⇒(iii) The bounding map for `tail` is the length $\#$ of the list. Now show that $\texttt{tail}^{\#} = \texttt{nil}$ (see [?]).

(iii)⇒ (ii) If `tail` terminates with bounding map β then so does $\texttt{shunt}(\alpha)$: $C \times L \to C \times L$ with bounding map given by $\# \circ \pi'$. The fixpoints can be found by observing that

$$\pi' \circ \texttt{shunt}(\alpha)^{\beta \circ \pi'} = \texttt{tail}^{\beta} \circ \pi'$$

□

6 Locoses

A polynomial category is *distributive* if it is finitely complete and its finite sums are disjoint and universal. Recall that disjointness asserts that inclusions to sums are monomorphisms whose intersection is the initial object. Universality asserts that sum diagrams are preserved by pullbacks. A *locos* is a distributive category in which every object has a parametric (equivalently, convergent) stack LA, now unambiguously called its *list object*. The action

form natural transformations, and will be subscripted where appropriate. Assume that $\mathcal{C}$ is a locos for the rest of the abstract.

It is easy to prove that (L, η, μ, τ) is a strong monad on $\mathcal{C}$ where $\eta =$ **cons** $\circ \langle 1, \texttt{nil}\rangle : A \rightarrow LA$ is listification, $\mu = \texttt{foldr(nil,@)} : L^2A \rightarrow LA$ is flattening, and $\tau : LA \times B \rightarrow L(A \times B)$ is the operation 'pairwith' which pairs up each entry in the list of A's with the same element of B. Thus it is an example of a computational monad [?].

The functor L doesn't preserve products, since, for example, $L1 = N$ describes the possible shapes that a list may have, i.e. its length, which is a natural number. However, we do have the following

Theorem 6.1 (Cockett) *L preserves any connected limits that exist in $\mathcal{C}$, e.g. pullbacks and equalisers.*

Proof The proof is given for pullbacks. The general result follows similarly, but will not be required below. Let

$$\begin{array}{ccc} Q & \longrightarrow & LA \\ \big\downarrow & \lrcorner & \big\downarrow Lf \\ LB & \xrightarrow[Lg]{} & LC \end{array}$$

be a pullback. It suffices to show that Q is a contractible list object for the pullback P of f along g. Stability of sums under pullbacks shows that it is a solution of the domain equation for list on P. Fixpoints of its `tail` are given by the pullback of the fixpoints of the other tails (i.e. 1) and the length morphism on LA (or LB) provides a bounding map. □

For example, every product $A \times B$ is a pullback of $!_A$ along $!_B$. Let $LA \times_{\#} LB$ be the pullback of $\#_A$ along $\#_B$. Then the canonical morphism

$$L(A \times B) \xrightarrow{\langle L\pi, L\pi' \rangle} LA \times_{\#} LB$$

is an isomorphism, whose inverse, denoted $zip : LA \times_{\#} LB \rightarrow L(A \times B)$ maps a pair of lists of equal length to a list of pairs. If $\oplus : A \times B \rightarrow C$ is a morphism of $\mathcal{C}$ then the usual primitive operation $zip(\oplus)$ can be interpreted by $L \oplus \circ zip : LA \times_{\#} LB \rightarrow LC$.

Consequently, if $(A, +, 0, *, 1)$ is a ring object in $\mathcal{C}$ then the *inner product* $p : LA \times_{\#} LA \rightarrow A$ is given by

$$LA \times_{\#} LA \xrightarrow{zip} L(A \times A) \xrightarrow{L*} LA \xrightarrow{\Sigma} A$$

where $\Sigma = \texttt{foldr(+)} : LA \rightarrow A$ is the monad algebra corresponding to $+$.

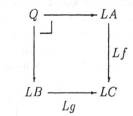

It is also worth pointing out here that many of the natural transformations associated to L have the property that the squares which assert their naturality are all pullbacks, as is the case with flattening:

$$\begin{array}{ccc} L^2A & \xrightarrow{\mu} & LA \\ {\scriptstyle L^2 f}\downarrow & \lrcorner & \downarrow{\scriptstyle Lf} \\ L^2B & \xrightarrow[\mu]{} & LB \end{array}$$

These results arise because the shape of a list (its length) is independent of the type of its entries. (Similar results hold for trees, etc.)

7 Vectors

An object $\alpha : A \to I$ of the arrow category $\mathcal{C}^{\to}$ can be thought of as a datatype A (e.g. of lists) with α computing some static data or compile-time information of type I (e.g. the length of the list). Morphisms can then interpret programs between the datatypes for which the static data of the output is determined by that of the input. Define $V\alpha$ by the following pullback

$$\begin{array}{ccc} VA & \xrightarrow{v_A} & LA \\ {\scriptstyle V\alpha}\downarrow & \lrcorner & \downarrow{\scriptstyle L\,\alpha} \\ N \times I & \xrightarrow[\copyright]{} & LI \end{array}$$

where $\copyright = Ll \circ \tau : L1 \times I \to L(1 \times I) \to LI$.

VA should be thought of as the object of those lists of A's, whose entries all take the same value under α. Then $V\alpha$ is the length of the list paired with this common value in I. When $I = 1$ then $VA \to N$ is the usual object of vectors on A equipped with their lengths, and $V^2A \to N^2$ represents vectors of vectors that all have the same length, i.e. the *matrices* of A.

These intuitions are reinforced by the following

Theorem 7.1 *Given $\alpha : A \to I$ then $\pi' \circ V\alpha : VA \to I$ is a list object for α in the slice category $\mathcal{C}/I$.*

Proof Show that VA satisfies the domain equation $VA \cong I + (A \times_I VA)$ for lists in $\mathcal{C}/I$ by pulling back v_A along `nil` and `cons`. That the fixpoints for `tail` are given by `nil` follows upon showing that they must all have length 0. The bounding map for `tail` is $\# \circ v$. □

We will name the list operations on VA as usual, even when in $\mathcal{C}^{\to}$.

Corollarv 7.2 (Cockett) *Slices of a locos are locoses*

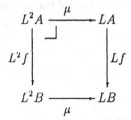

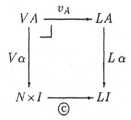

Theorem 7.3 *V extends to a strong monad on $C^{\rightarrow}$.*

Proof This can be proved elementarily if somewhat tediously. For an elegant 2-categorical proof see [?]. □

8 Matrices

For simplicity, let the object of $C^{\rightarrow}$ of interest be $A \rightarrow 1$. The explicit use of lengths means that matrices of dimensions $0 \times n$ and $m \times 0$ are well-defined, and all distinct. Thus $\texttt{nil} : N \rightarrow V^2A$ produces matrices of the first kind with 0 rows while $V\texttt{nil} : N \rightarrow V^2A$ produces a matrix with 0 columns. Now transposition of matrices is defined using the list structure in C/N by

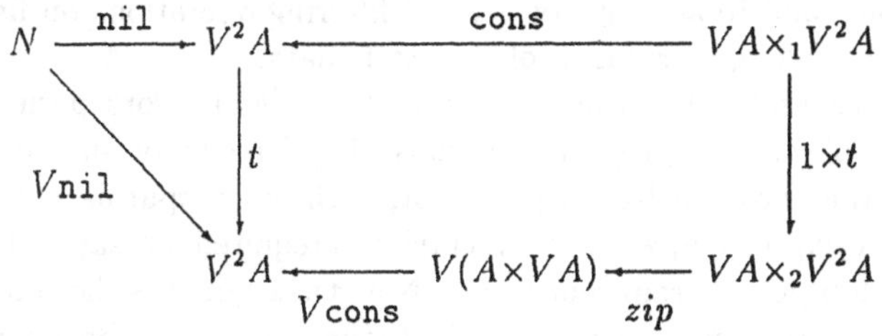

where $\times_i$ indicates that the pullback is along the lengths $\#_i$ given by ith application of V (if there is a crhoice). Alternatively, it is given by $t = \texttt{foldl}(V\texttt{cons} \circ zip) \circ \langle \texttt{nil}, \texttt{rev} \rangle$.

Now matrix multiplication can be defined as follows. Composable pairs of matrices are given by the pullback

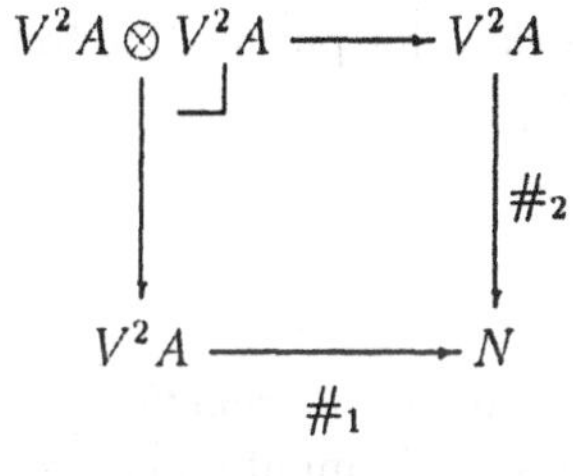

whose composition is given by

$$V^2A \otimes V^2A \xrightarrow{1 \times t} V^2A \times_2 V^2A \xrightarrow{V(\tau \circ c) \circ \tau} V^2(VA \times_{\#} VA) \xrightarrow{V^2 p} V^2A$$

where p is the inner product defined above. The ring axioms imply that the composition is associative and unitary as usual.

Theorem 8.1 *If A is a ring object in a locos C then C has an internal category whose objects are the natural numbers and whose morphisms are the matrices. Domain and codomain are given by $\#_2, \#_1 : V^2A \rightarrow N$ respectively.*

When C = **Sets** this is the usual category of matrices.

9 Further and Related Work

These results arose in an attempt to understand the Squiggol language as used by Jones [?] to prove the correctness of the Fast Fourier Transform in a language which expresses the computational steps required for its implementation. Matrices are there coded as lists of lists which must satisfy a side condition which is here described by a pullback. The correctness can also be proved categorically, but full generality requires some further results about vector morphisms which take the length as a parameter e.g. pointwise operators [?].

Other possible applications of this work are to a general static analysis of programs, where static properties of the input are used to determine those of the output, and to account for general filtering operations on lists in terms of pullbacks along morphisms of the static data.

More generally still, this type of analysis should work perfectly well for trees and other datatype constructions. Head recursion now works for operations that proceed from tip to root, such as computing the value of an arithmetic expression, while tail recursion is required for, say, pattern matching, which proceeds from root to tip. Now `tail` removes the root of the tree to yield a forest, and so is properly thought of as a loop on forests. Finally search algorithms can proceed in either direction, to yield depth-first, or breadth-first search.

10 Acknowledgements

My thanks go to Robin Cockett for pushing me to think about these things, and for locoses.

References

[1] R. Bird and P. Wadler, *Introduction to Functional Programming* International Series in Computer Science, ed: C.A.R. Hoare (Prentice Hall, 1988).

[2] J.R.B. Cockett, List-arithmetic distributive categories: locoi, J. Pure and Appl. Alg. 66 (1990) 1–29.

[3] J.R.B. Cockett, Distributive Theories, in: G. Birtwistle (ed), IV Higher Order Workshop, Banff, 1990, (Springer, 1991).

[4] A note on natural numbers objects in monoidal categories, Studia Logica 48(3) (1989) 389–393.

[5] C.B. Jay, Fixpoint and loop constructions as colimits, preprint.

[7] C.B. Jay, Matrices, monads and the Fast Fourier Transform, in preparation.

[8] G. Jones, Calculating the Fast Fourier Transform, in: G. Birtwistle (ed), IV Higher Order Workshop, Banff, 1990, (Springer, 1991).

[9] J. Lambek and P. Scott, Introduction to higher order categorical logic, Cambridge studies in advanced mathematics 7, Cam. Univ. Press (1986).

[10] S. Mac Lane, *Categories for the Working Mathematician* (Springer- Verlag, 1971).

[11] E. Moggi, Computational lambda-calculus and monads, Proceedings Fourth Annual Symposium on Logic in Computer Science (1989) 14–23.

[12] R. Paré and L. Roman, Monoidal categories with natural numbers object, Studia Logica 48(3) (1989).

[13] L. Roman, Cartesian categories with natural numbers object, J. of Pure and Appl. Alg. 58 (1989) 267–278.

[14] R.F.C. Walters, Datatypes in distributive categories, Bull. Australian Math. Soc. 40 (1989) 79–82.

A DIRECT PROOF OF THE INTUITIONISTIC RAMSEY THEOREM

THIERRY COQUAND

Introduction.

In Veldman 1991, Wim Veldman presents an intuitionistic version of Ramsey Theorem, and a proof of it, based on Brouwer's thesis 1927. In his argument, he uses also as a lemma the finite version of Ramsey theorem. We give here a direct proof of Veldman's intuitionistic Ramsey Theorem, using only Brouwer's thesis 1927, and not the finite version of Ramsey Theorem. As in Veldman 1991, our argument is both valid intuitionistically and classically. It can be read as a new proof of even the finite version of Ramsey Theorem.

We are using mainly this theorem as a concrete example that illustrates a possible formalisation of the logic of non-deterministic program as the logic of sheaves over a locale.

1. Almost Full Relations.

We let Ω be the set of all strictly increasing sequence α over N. Let a n-ary relation R over N be **almost full** iff for all $\alpha \in \Omega$, there exist $i_1 < \cdots < i_n$ such that $R(\alpha(i_1), \ldots, \alpha(i_n))$.

The goal of this paper is to give a proof of the following theorem, that can be seen as an intuitionistic version of Ramsey Theorem.

Theorem. *(Veldman) If R and S are two almost full n-ary relation, then so is $R \cap S$.*

We want to emphasize the fact that it is essential for the applications of this theorem (and for the proof we are giving) that the relations are not supposed decidable (however, they will all be "open" or "obervable" relations, in the terminology of Abramsky 1991).

2. A Reformulation of Brouwer's Thesis.

In this section, we reformulate first in topological terms the problem we want to solve. We give then an inductive description of a system of basic open sets of the topological set we consider. We look at it as a **transition system**, i.e. a set with a binary relation $\rightarrow$, where $U \rightarrow V$ implies that U contains V. Since transition systems are used to describe processes, this will give a more "dynamic" interpretation of the notion of almost full relations. Our reformulation of Brouwer's Thesis is a way to relate a statement about points to a statement about this transition system of open sets.

Typeset by $\mathcal{A}_{\mathcal{M}}\mathcal{S}$-TEX

The set Ω of all strictly increasing sequences will be considered as a topological space for the product topology, N being considered with the discrete topology. Any finite subset $p = \{i_1, \ldots, i_k\}$, with $i_1 < \cdots < i_k$, defines a basic open (and closed) subset U_p which is the set of all sequence α such that $\alpha(1) = i_1, \ldots, \alpha(k) = i_k$.

We shall view the set $\mathcal{F}$ of all finite subsets of N as a transition system, letting $p \rightarrow q$ mean that q has exactly one element more than p, which is bigger than all elements of p.

Each subset B of $\mathcal{F}$ defines an open subset U_B of Ω, which is the union of all U_p, $p \in B$.

Before stating our reformulation of Brouwer's thesis, we recall first some definitions of Coquand 1991, which provides a convenient terminology (used in the logic of processes) in which to state this reformulation.

Let P be a transition system, i.e. a set with a binary relation $\rightarrow$. For $x \in P$, we let $S(x)$ be $\{y \in P \mid x \rightarrow y\}$. If A is a subset of P, we let $\Box A$ be the set $\{p \in P \mid S(p) \subseteq A\}$. Evidently, the $\Box$ operator preserves inclusions and commutes with arbitrary intersections. A subset A of P is **hereditary** iff $\Box A \subseteq A$. It is **monotone** iff $A \subseteq \Box A$.

In order to simplify some definitions and discussions, we will consider only transition systems P such that, for all $p \in P$, the set $S(p)$ is inhabited.

If A is a subset of P, we let E_A be the least subset X of P such that $A \cup \Box X = X$. Thus, E_A is the intersection of all hereditary subsets of X that contain A.

Let B be a monotone subset of P. Let us say B is **ample** in P iff $E_B = P$, and that B is **unavoidable** iff, for any sequence $p_n \in P$ such that $p_n \rightarrow p_{n+1}$ for all n, there is a k such that $p_k \in B$. Both notions can be used a priori for formalising the intuitive idea that "B holds eventually." Notice that the notion of ampleness makes only references to observables, in opposition to the notion of unavoidable, which refers to the non observable notion of complete sequence of experiments.

We can now state Brouwer's thesis 1927 as follows.

Thesis. *(Brouwer) Let B be a monotone subset of $\mathcal{F}$. Then B is ample iff it is unavoidable.*

It is true directly that if B is ample then it is unavoidable. The converse holds classically for a non necessarilly monotone B modulo the axiom of dependent choices. An example of Kleene 1967, corollary 9.9, page 113, shows that the converse does not hold in general. Thus, if we work constructively (that is, following the terminology of Bell 1988, we want a result that holds in any local set theory), the choice of the formalisation of the notion of eventuality does matter.

We will see later that allowing ourselves to work in any local set theory, and to be able to change the "local set theory reference" seems essential in the formalisation of the idea of "arbitrary sequence", but in a first step, we will assume Brouwer's thesis, and prove Veldman's intuitionistic Ramsey's Theorem from it.

A first remark is that if we assume Brouwer's thesis, then the identification of ample and unavoidable holds for other transition systems than $\mathcal{F}$. For instance, given $p \in \mathcal{F}$, let $\mathcal{F}_p$ be the transition system of subsets q of which p is an initial segment, and where $q \rightarrow r$ holds iff it holds in $\mathcal{F}$. $L(\mathcal{F}_p)$ is then a locale corresponding to the open subset U_p of Ω. Brouwer's thesis implies that $B \subseteq \mathcal{F}_p$ is ample iff it is unavoidable. Indeed, it is directly shown that, if A is the set of $q \in \mathcal{F}$ incompatible with p (that is, such that p and q have no common extension), B is ample (resp. unavoidable) in $\mathcal{F}_p$ iff $B \cup A$ is ample (resp. unavoidable) in $\mathcal{F}$. Hence the result.

Any n-ary relation R over N defines a monotone subset $B(R)$ of P which is the set of all p such that there exist $i_1 < \cdots < i_n$ in p such that $R(i_1, \ldots, i_n)$, and an open subset $U(R) = U_{B(R)}$ which is the set of all sequences $\alpha \in \Omega$ such that there exist $i_1 < \cdots < i_n$ satisfying $R(\alpha(i_1), \ldots, \alpha(i_n))$. The relation R is almost full iff $B(R)$ is unavoidable.

3. Some remarks about transition systems.

A transition system will be thought of as the representation of possible experiments on a "process", or non determinitic program, by an observer. This process is intuitively like a "black box" for this observer. Each element $p \in P$ represents a "stage of knowledge" (an "observable") that has the observer about the process. At each stage p, the observer can choose to do an experiment, knowing a priori that once the experiment will be finished, he will be in one stage of knowledge $q \in S(p)$. We require furthermore that the process must then give the required information, which results in the evolution from the knowledge $p \in P$ to a knowledge $q \in S(p)$ that the observer got about the process by doing the experiment.

This interpretation of transition system is reminiscent of the informal explanation of Beth models by Kripke 1965. The author does not know how "standard" this interpretation of transition system is however.

As an example, the transition system $\mathcal{F}$ represents a non-deterministic program that produces strictly increasing integers, each time it is required to do so. The "process" that the transition system $\mathcal{F}$ represents is an infinite subset of N, and each partial knowledge $K \in \mathcal{F}$ about this subset is one of its finite initial segment. The observer, in a given stage of knowledge, when doing an experiment, gets to know the next element of this subset.

Let P be a transition system. We let $L(P)$ be the following locale (or point-free space, see Vickers 1989): the elements of $L(P)$ are the monotone subset of P, and we say that a family B_i of elements of $L(P)$ covers $B \in L(P)$ iff B is a subset of $E_{\bigcup B_i}$. The locale $L(P)$ is to be thought of as the space of non deterministic program defined by the transition system P. This definition is a formal summary of our interpretation of transition systems. In the case of the transition system $\mathcal{F}$, the locale $L(\mathcal{F})$ corresponds to the topological space Ω.

We recall also one simple lemma of Coquand 1991. Given two transition systems P_1, P_2, we let $P_1 \times P_2$ be the following transition system: the elements are pairs (p_1, p_2) with $p_1 \in P_1$ and $p_2 \in P_2$, and $(p_1, p_2) \rightarrow (q_1, q_2)$ means that $p_1 = q_1$ and $p_2 \rightarrow q_2$, or that $p_1 \rightarrow q_1$ and $p_2 = q_2$, or that both $p_1 \rightarrow q_1$ and $p_2 \rightarrow q_2$.

Lemma. *If $B_i \subseteq P_i$ is ample in P_i, then so is $B_1 \times P_2 \cup P_1 \times B_2$ in $P_1 \times P_2$.*

Proof. Let $X \subseteq P_1 \times P_2$ be such that $B_1 \times P_2 \cup P_1 \times B_2 \cup \Box X \subseteq X$. We let X_1 be the set of $p_1 \in P_1$ such that $(p_1, p_2) \in X$ for all $p_2 \in E_{B_2}$. We have $B_1 \subseteq X_1$.

Let us show that X_1 is hereditary. For $p_1 \in \Box X_1$, let X_2 be the set of $p_2 \in P_2$ such that $(p_1, p_2) \in X$. Then, $B_2 \subseteq X_2$ and $\Box X_2 \subseteq X_2$. Hence, $X_2 = P_2$ and $p_1 \in X_1$. □

We don't know if the lemma holds with "unavoidable" instead of "ample", if we don't assume the identification of ample and unavoidable.

We will use this lemma as follows. If $B_i \subseteq P_i$ is such that $E_{B_i} = P_i$, then for showing that a property $\phi(p_1, p_2)$ holds for all $p_1 \in P_1$, $p_2 \in P_2$, it is enough to show that $\phi(p_1, p_2)$ holds if $p_1 \in B_1$ or $p_2 \in B_2$ and that $\phi(p_1, p_2)$ holds whenever $\phi(q_1, p_2)$ holds for all q_1 such that $p_1 \rightarrow q_1$ and $\phi(p_1, q_2)$ holds for all q_2 such that $p_2 \rightarrow q_2$.

4. A direct Proof of the Intuitionistic Ramsey Theorem.

We show the intuitionistic Ramsey Theorem by induction on n. The case $n = 0$ is clear, because in this case, an almost full n-ary relation is a true proposition. We suppose then that $n > 0$, that the theorem holds for $n-1$, and we let R and S be two n-ary almost full relations.

Let p and q be two disjoint elements of $\mathcal{F}$. For any $\beta \in U_{p \cup q}$, we let $\psi(p, q, \beta)$ be the assertion that there exists $\{i_1, \ldots, i_n\}$ in the range of β meeting p but not q such that $i_1 < \cdots < i_n$ and $R(i_1, \ldots, i_n)$ or there exists $\{i_1, \ldots, i_n\}$ in the range of β meeting q but not p such that $i_1 < \cdots < i_n$ and $S(i_1, \ldots, i_n)$.

Let p and q be two disjoint elements of $\mathcal{F}$. We let $\phi(p, q)$ be the assertion that for all $\beta \in U_{p \cup q}$, $\psi(p, q, \beta)$ or $\beta \in U_{R \cap S}$. Notice that, if both p and q are empty, then $\phi(p, q)$ is the assertion we want to prove, namely that $R \cap S$ is almost full, because in this case, $\psi(p, q, \beta)$ never holds.

We show that $\phi(p, q)$ holds for all disjoints p, q in $\mathcal{F}$, using the lemma of the previous section and Brouwer's thesis. It is clear that $\phi(p, q)$ holds if p is in $B(R)$ or if q is in $B(S)$. Next, we suppose that p and q are disjoint elements of $\mathcal{F}$, that $\phi(p_1, q)$ holds whenever $p \rightarrow p_1$, p_1 and q disjoint, and that $\phi(p, q_1)$ holds whenever $q \rightarrow q_1$, q_1 and p disjoint, and we show that $\phi(p, q)$ holds. It follows then from the lemma that the assertion "if p and q are disjoints, then $\phi(p, q)$" holds for all p and q in $\mathcal{F}$.

Let x be a natural number bigger than all elements in p and q. We show that $\psi(p, q, \beta)$ or $\beta \in U_{R \cap S}$ for all $\beta \in U_{p \cup q \cup \{x\}}$. For this, we fix $\beta \in U_{p \cup q \cup \{x\}}$ and let k be such that

$\beta(k) = x$. We define then two $(n-1)$-ary relations on N. We let $R'(i_1,\ldots,i_{n-1})$ mean that $\psi(p,q,\beta)$ or $\beta \in U_{R\cap S}$ or $R(x,\beta(k+i_1),\ldots,\beta(k+i_{n-1}))$. Symmetrically, we let $S'(i_1,\ldots,i_{n-1})$ mean that $\psi(p,q,\beta)$ or $\beta \in U_{R\cap S}$ or $S(x,\beta(k+i_1),\ldots,\beta(k+i_{n-1}))$.

Notice that R' and S' may be not decidable, even if R and S are decidable.

We claim that both R' and S' are almost full. Let us show that R' is almost full, the proof for S' being similar. Let γ be an element of Ω. We show that γ is in $U_{R'}$.

We define α by $\alpha(i) = \beta(i)$ for $i \leq k$ and $\alpha(i) = \beta(k+\gamma(i-k))$ for $k < i$. We have $\alpha \in U_{p_1 \cup q}$, where p_1 is $p \cup \{x\}$. Since $\phi(p_1,q)$ holds by induction hypothesis, we know $\alpha \in U_{R\cap S}$ or $\psi(p_1,q,\alpha)$. If $\alpha \in U_{R\cap S}$, then $\beta \in U_{R\cap S}$ because α is a subsequence of β, and hence $U_{R'} = \Omega$. If $\psi(p_1,q,\alpha)$, then there are two cases:

1. There exist $\{i_1,\ldots,i_n\}$ in the range of α meeting p_1 andnot q such that $i_1 < \cdots < i_n$ and $R(i_1,\ldots,i_n)$. If $i_1 < x$, this implies $\psi(p,q,\beta)$, and thus $U_{R'} = \Omega$. If $i_1 = x$, this implies $\gamma \in U_{R'}$.

2. There exist $\{i_1,\ldots,i_n\}$ in the range of α meeting q and not p_1 such that $i_1 < \cdots < i_n$ and $S(i_1,\ldots,i_n)$. Then $\{i_1,\ldots,i_n\}$ is in the range of β, meets q and not $p \subset p_1$ and is such that $i_1 < \cdots < i_n$ and $S(i_1,\ldots,i_n)$. Hence $\psi(p,q,\beta)$, and thus $U_{R'} = \Omega$.

In all cases, we have $\gamma \in U_{R'}$, and this proves that R' is almost full.

By induction hypothesis, $R' \cap S'$ is almost full. But $(R' \cap S')(i_1,\ldots,i_{n-1})$ mean that $\psi(p,q,\beta)$ or $\beta \in U_{R\cap S}$ or $(R' \cap S')(x,\beta(k+i_1),\ldots,\beta(k+i_{n-1}))$, which is equivalent to $\psi(p,q,\beta)$ or $\beta \in U_{R\cap S}$. We deduce that $\psi(p,q,\beta)$ or $\beta \in U_{R\cap S}$ holds, hence the result. □

5. Ample versus unavoidable.

The previous argument is valid classically and also intuitionistically, but only modulo Brouwer's thesis. We would like now to have a proof that can be said "constructive", that is valid in every local set theory (Following Bell 1988 terminology). We shall give a statement that uses only the notion of inductive definitions corresponding to the intuitionistic Ramsey Theorem.

First, we reformulate the statement itself of Veldman's intuitionistic Ramsey Theorem, eliminating any references to Brouwer's thesis 1926. The idea behind this reformulation is the notion of point-free topology (see for instance, Vickers 1989; for other examples of this point-free technique can be found in Coquand 1991 and de Bruijn, Van Der Maiden 1968). We want to make references only to directly observations that we can make on elements of the space Ω, that is we want to describe this space in term of the transition system $\mathcal{F}$ uniquely, and not in term of its points.

We thus redefine the notion of "almost full" as follows. We say that a n-ary relation R is almost full iff the subset $B(R)$ of $\mathcal{F}$ is such that $E_{B(R)} = \mathcal{F}$, that is $B(R)$ is ample in the transition system $\mathcal{F}$. This definition mentions only the elements of the transition system $\mathcal{F}$ (the "observable" elements). It is equivalent to our previous definition (which can be formulated as "$B(R)$ is unavoidable in $\mathcal{F}$") only if we admit Brouwer's thesis 1927.

We are now going to prove, using only the notion of inductive definitions, that the intersection of two n-ary almost full relation is almost full, for this new notion of almost full.

6. A sheaf-theoretic reformulation.

We want to follow as much as possible the previous proof. It is easy to reformulate the first part of the proofs, avoiding references to Brouwer's thesis 1927. The only difference is that we have used systematically "ample" instead of unavoidable. At one point, we have introduced an arbitrary, but fixed $\beta \in U_{p \cup q \cup \{r\}}$ and defined some relations in term of it. This was crucial for the proof. The main problem is then to represent in constructive terms this introduction of an arbitrary $\beta \in U_{p \cup q \cup \{r\}}$.

The difficulty of this problem is emphasized by the example of Kleene 1967, corollary 9.9, page 113, that shows that it will be wrong in general to represent such a β by an infinite sequence (if we don't assume Brouwer's thesis).

The importance of this question is clear when we generalise it: for an arbitrary transition system P, what is a possible formalisation of the logic of an arbitrary non deterministic program represented by P?

Our proposal for such a representation is the following: the introduction of an arbitrary point β of $L(P)$ corresponds to moving from the current local set theory to the local set theory of **sheaves** over the locale $L(P)$ (see Bell 1989, for instance for an explanation of the notion of sheaves).

The intuition behind this proposal is the following. To work in the local set theory of sheaves over the locale $L(P)$ corresponds to adding a new "generic" point β to the locale $L(P)$. This generic point β has, by construction, the property that it belongs to an open U, defined in the local set theory we start with, iff $U = L(P)$.

For instance, in the topos of sheaves over the locale $L(\mathcal{F})$, we have a generic infinite subset of N, whose partial description at stage $K \in \mathcal{F}$ is the finite subset K itself. It is possible to prove that internally, this defines an infinite subset of N. By changing the base topos, it is possible to add such a generic subset of N, which otherwise could only be done in general by using dependent choices and classical logic.

Since adding a generic point corresponds to moving in the topos of sheaves, we arrive at the following reformulation of the intuitionistic version of Ramsey's theorem, which is this time valid constructively.

Theorem. *Let L be an arbitrary locale. If R and S are two almost full n-ary relation in the topos of sheaves over L, then so is $R \cap S$.*

Intuitively, the introduction of the locale L corresponds to the fact that the relation R and S may depend to a non-deterministic process (itself may be composed of several subprocesses). This is what happens when we use the induction hypothesis.

With this strenghtening of the theorem, we can follow the previous argument almost verbatim. Each call to the induction hypothesis corresponds to a change of the locale L. The introduction of the arbitrary, but fixed, $\beta \in L(\mathcal{F}_r)$, with $r = p \cup q \cup \{x\}$ corresponds to working over the locale $L \times L(\mathcal{F}_r)$. When we introduce the element γ, we are then in the topos of sheaves over the locale $L \times L(\mathcal{F}_r) \times L(\mathcal{F})$.

Remark that, even in the case $n = 1$ and with a trivial locale L, we need in the proof to consider the locale $L(\mathcal{F}_r) \times L(\mathcal{F})$. In general, the locale $L(P_1) \times L(P_2)$ corresponds to the observation of two independent processes described respectively by P_1 and P_2. This locale differs from $L(P_1 \times P_2)$. Even for this proof, we need to consider locales more general than the ones given by a transition system. However, we can limit ourselves to locales that are given explicitely by a site (see Vickers 1988, page 47).

For the convenience of the reader, let us explicit what it means for a subset B of a transition system P to be ample over the topos of sheaves over a locale L. First, we have to explicit what is a monotone subset of P ("ordinary" set) in the topos of sheaves over L. This can be seen as a subset B of $L \times L(P)$. To say that this subset is ample is then to say that this subset covers the product space $L \times L(P)$.

If L is given as a site, i.e. a inf semilattice with a covering relation, we can explicit this a little more. This means that, if $R \subseteq L \times P$ contains B and is such that $(x,p) \in R$ whenever $(x,q) \in R$ for all $q \in S(p)$, or $(y,p) \in R$ for all $y \in A$, where A covers x, then R is equal to $L \times P$.

Using this remark, it can be seen that for the proof of the sheaf-theoretic version of the intuitionistic Ramsey theorem, we do not need all the strength of topos theory, but only the notion of inductive definitions.

7. Application to Ramsey Theorem.

In this section, we present an application of our sheaf-theoretic version of the intuitionistic Ramsey Theorem. The main idea behind this application is due to Wim Veldman. Our contribution here is to show that we do not need Brouwer's thesis (or the fan theorem) for this application if we use the sheaf-theoretic version.

If $A \subseteq \mathbb{N}$ we let $\mathcal{F}_n(A)$ be the set of finite subsets of A of cardinal n. We introduce the space $\Omega(n)$ of n-set colouring over $\mathbb{N}$. An element of $\Omega(n)$ is an application $\mathcal{F}_n(\mathbb{N}) \to \{0,1\}$. We let $\mathcal{F}(n)$ be the following transition system. Its elements are maps $\sigma : \mathcal{F}_n([1,k]) \to \{0,1\}$, i.e. finite n-set colourings of an initial segment $[1,k]$ of $\mathbb{N}$. We write then $k = h(\sigma)$. We let $\sigma_1 \to \sigma_2$ mean that σ_2 extends σ_1, and $h(\sigma_2) = h(\sigma_1) + 1$.

Notice that $L(\mathcal{F}(n))$ is a localic description of $\Omega(n)$, and that $\mathcal{F}(n)$ is finitely branching, so that the locale $L(\mathcal{F}(n))$ is compact. This implies directly that if $A \subseteq \mathcal{F}(n)$ is ample, then there exists k such that $\sigma \in A$ whenever $k \leq h(\sigma)$.

If B is a subset of $\mathcal{F}$, we let $B^{(n)}$ be the set of elements σ of $\mathcal{F}(n)$ such that $[1, h(\sigma)] \in B$. Our goal here is to show that the following theorem is a consequence of the intuitionistic Ramsey theorem.

Theorem. *If $B \subseteq \mathcal{F}$ is ample, then $B^{(n)} \subseteq \mathcal{F}(n)$ is ample.*

We instance the sheaf-theoretic version intuitionistic Ramsey Theorem on the locale $L(\mathcal{F}(n))$. Let α be the generic n-set colouring of N in the topos of sheaves over $L(\mathcal{F}(n))$. We define two n-ary relations R_0 and R_1 over N, depending of α. We let $R_i(l_1, \dots, l_n)$ mean that, if $l_1 < \dots < l_n$, then $\alpha(\{l_1, \dots, l_n\}) = i$ or there exists $K \in B$ such that α is monochromatic for K. Then R_0 and R_1 are almost full.

Indeed, if β is the generic element of $L(\mathcal{F})$ (moving now over the topos of sheaves over the product $L(\mathcal{F}(n)) \times L(\mathcal{F})$), we have $K \in B$ included in the range of β, because B is ample by hypothesis. Then, since K is finite, and we can suppose it is of cardinal bigger than n, either K is monochromatic for α, or there exists $k_1 < \dots < k_n$ such that $\alpha(\{\beta(k_1), \dots, \beta(k_n)\}) = i$. This implies that there exists $k_1 < \dots < k_n$ such that $R_i(\beta(k_1), \dots, \beta(k_n))$, hence that R_i is almost full.

It follows from the intuitionistic Ramsey Theorem that $R_0 \cap R_1$ is almost full, and that means that there exists k such that the restriction of α to $[1, k]$ is in $B^{(n)}$, hence the result. □

Notice that it is essential for this proof; to work over the topos of sheaves over $L(\mathcal{F}(n))$.

Let now k be a fixed integer. If we take in particular for B the set of finite subsets of N of cardinal greater than k, then we get the finite Ramsey Theorem. If we take for B the set of finite subsets of N that contains $n_1, \dots, n_p$ such that $k < n_1 < \dots < n_p$ and $n_1 < p$, we get the theorem that Paris and Harrington 1977 have shown to be not provable in Peano Arithmetic.

This provides thus a mathematical example of the use of generalised inductive definitions in intuitionistic mathematics.

Conclusion.

We have given here an alternative proof of the intuitionistic Ramsey theorem, using first Brouwer's thesis 1927. We have explained then how to formulate this proof in a framework of inductive definitions without assuming Brouwer's thesis 1927. The statement we prove then is the relativised version of the statement over an arbitrary locale.

This example, which can be seen as a concrete illustration of some ideas in Fourman 1984, suggests a possible way of presenting the logic of a process described by a transition system, which consists in working in the topos of sheaves over a locale canonically associated to this transition system. From the logical point of view, this has the advantage of working with points (and actually, generic points), without having to introduce any further axioms.

Acknwledgements.

The idea of this proof comes from an analysis of the proof in Richman and Stolzenberg 1990 that the products of two wqos is a wqo. Email exchanges with Wim Veldman and Gabriel Stolzenberg have been essential for this work.

REFERENCES

1. Brouwer, L. E. J., *Uber definitionsbereiche von Funktionen*, Math. Ann. 96 (1927), 60 - 75.
2. Martin-Löf, P., *Notes on Constructive Mathematics*, Almqvist & Wiksell, 1970.
3. Richman F. and Stolzenberg G., *Well Quasi-Ordered Sets*, To appear in Advances in Mathematics (1991).
4. Paris J. and Harrington L., *A Mathematical Incompleteness in Peano Arithmetic*, in the Hanbook of Mathematical Logic (1977), J. Barwise, editor, North-Holland.
5. Veldman W., *Ramsey's Theorem and the Pigeonhole principle in Intuitionistic Mathematics*, Report 9017, Department of Mathematics, catholic University, Toernooiveld, 6525 ED Nijmegen,.
6. Ramsey, F.P., *On a Problem of Formal Logic*, Proc. London Math. Soc. 48 (1928), 122 - 160.
7. Coquand, Th., *An analysis of Ramsey's Theorem*, Submitted to Information and Computation (1991).
8. Kleene, S.C. and Vesley, R.E., *The Foundations of Intuitionistic Mathematics*, North-Holland Publishing Company, 1965.
9. Abramsky, S., *Domain Theory in Logical Form*, Annals of Pure and Applied Logic 51 (1991), 1 - 77.
10. Fourman, M. P., *Continuous Truth I*, in Logic Colloquium'82 (1984), North-Holland.
11. Bell, J.L., *Toposes and Local Set Theories, An Introduction*, Oxford Science Publications, 1988.
12. Coquand, Th., *Constructive Topology and Combinatorics*, to appear in the proceeding of Constructivity in Computer Science, Trinity University, San Antonio, Texas.
13. Vickers, S., *Topology via Logic*, Cambridge Tracts in Theoretical Computer Science 5, 1989.
14. De Bruijn, N. G. and Van Der Meiden, W., *Notes on Gelfand's theory*, Indagationes 31 (1968), 467–474.
15. Kripke, S., *Semantics analysis of intuitionistic logic I, in: Formal Systems and Recursive Functions*, North-Holland, 1965, pp. 92 - 130.

Constructions and Predicates

Duško Pavlović

Zevenwouden 223, Utrecht, The Netherlands

Abstract

In this paper, the *theory of constructions* is reinterpreted as a type theory of "sets" and "predicates". Following some set-theoretical intuitions, it is modified at two points: (1) a simple new operation is added – to represent a constructive version of the *comprehension principle*; (2) a restriction on contexts is imposed – "sets" must not depend on "proofs" of "predicates". The resulting theory is called *theory of predicates*. Sufficiently constructive arguments from naive set theory can be directly written down in it. On the other hand, modification (2) is relevant from a computational point of view, since it corresponds to a necessary condition of the modular approach to programming.

Our main result tells that, despite (2), the theory of predicates is as powerful as the theory of constructions: the constructions obstructed by (2) can be recovered in another form using (1). In fact, the theory of constructions is equivalent with a special case of the theory of predicates.

1. Introduction

The foundational role of type theory in computer science is comparable with the foundational role of set theory in mathematics. But the "set-theoretical" type theory of Russell and Church seems to have been less influential than the "logical" conception of *formulæ-as-types*, due to Curry and Howard (and traceable back to the Brouwer-Heyting-Kolmogorov interpretation of *proofs-as-constructions*). On the other hand, the experience of topos theory shows that the crucial set-theoretical notions can be given an elegant type-theoretical presentation (cf. Lambek-Scott 1986). So it seems worth-while to better explore the conceptual area in the intersection of type theory and set theory.

This paper reports on an effort to understand the *theory of constructions* (Coquand-Huet 1986, 1988, Hyland-Pitts 1989, Coquand 1990) as a strongly constructive theory of sets and propositions. With a similar idea, Ehrhard (1989) has argued that the categorical counterpart of the theory of constructions generalizes the notion of topos. Rather than semantically, we shall here approach the theory of constructions from another type theory, the *theory of predicates*.

Both these theories recognize two sorts of types, which can be understood as sets and propositions. So there are two *universes*. The universe of propositions is a type in the universe of sets; propositions appear as terms of this type. Terms in the universe of sets represent *elements*; terms in the universe of propositions are *proofs*. Viewed in this way, a family of propositions $\alpha(X)$ indexed by the elements of a set K is of course a *predicate* on K.

The theory of predicates starts from the idea that every predicate $\alpha(X)$ should be *comprehended* in the universe of sets by something like $\{X \in K / \alpha(X)\}$. An element of $\{X \in K / \alpha(X)\}$ would be a pair $\langle k, a \rangle$, where a is a proof of $\alpha(k)$. There may be many different constructive proofs of $\alpha(k)$ (i.e. many terms of this type) and the set $\{X \in K / \alpha(X)\}$, viewed constructively, may not be a subset of K.

Furthermore, indexing of a family of sets by proofs of a proposition will be forbidden in the theory of predicates. Philosophical justifications for this restriction (in the style: "all the elements must be created before proofs of propositions about them") become superfluous in the light of the main result of this paper, which tells that it really makes no difference – provided that predicates are comprehended among sets. We shall prove that the theory of predicates has slightly greater expressive power than the theory of constructions (although the latter theory imposes no special restrictions on indexing). In fact, the theory of constructions is equivalent (modulo a translation) to the *strict* theory of predicates, the one which satisfies a version of the ω-rule, well known from the untyped λ-calculus. Another characteristic of the strict theory of predicates is that every predicate $\alpha(X)$ in it can be recovered from (or even identified with) the set $\{X \in K / \alpha(X)\}$. In my thesis (1990) it was described how the theory of predicates corresponds to some small categories with small sums and products, while the theory of constructions and the strict theory of predicates correspond to those such categories which are (fully) generated by the terminal object.

And while conceptually nothing is lost by dumping the sets which depend on proofs, it seems that a lot can be gained. Some gains are technical: the imposed restriction reduces contexts to two layers (first sets, and then propositions), and many constructions – e.g. term models – become essentially simpler. But recent papers by Moggi (1990) and by Harper-Mitchell-Moggi

(1990) display this restriction as a *sine qua non* of the modular programming. Roughly speaking, Moggi understands as *programs* what I here call propositions, and my sets are for him *data types*. Clearly, a modular approach to programming can be effective only if the type-checking can be performed at compile time, before running actual programs. In other words, no type must depend on output of programs. This is called *phase distinction* between the compile-time and the run-time. Or between sets and propositions. It is amusing to think that this analogy of computational and foundational concepts is not accidental.[1]

2. Type theories

Keywords. We shall consider three kinds of *expressions*:

- *terms*, here denoted by metavariables p, q, r, s, t,
- *types*, denoted by P, Q, R, S , and
- *universes*, for which we use the letter $\mathcal{U}$.

The common name for terms and types is *constructions*; while *range* denotes a type or a universe. And now these expressions form two kinds of *judgements* (or *statements*):

- *equations*, or *conversion* judgements $T=T'$ between constructions T,T', and
- *formation* judgements $T:U$, meaning "the construction T has the range U".

The metavariable J will denote a judgement. The range of a construction can sometimes be indicated by a superscript: T^U.

The *variables* are special atomic terms. We use the letters X, Y, Z for them. If the terms are understood as programs, the variables are the input operations. Each term is represented by an expression $p(X_0,...X_n)$, in which the variables indicate the input gates. To supply input means to *substitute* a term $q(Y_0,...Y_m)$ for a variable X_i.

$$p(X_0,...,X_i,...,X_n)[q/X_i] := p(X_0,...,X_{i-1},q(Y_0,...Y_m),...,X_n).$$

Of course, q must have the same type as X_i. The type of a term-as-program is the type of its output data. Note that a data type may also vary, i.e. it may need some input before it is evaluated. A term must vary with its type. The universes will always remain constant – no variables can occur in them.

The actual objects of study in type theory are *sequents*

$$X_0:P_0,...,X_n:P_n \Rightarrow J \qquad (n\in\omega).$$

[1]Added in proof: Other such analogies can be found in Meseguer 1989.

An array $X_0{:}P_0,\ldots,X_n{:}P_n$ is called *context* and abbreviated by letters Γ or Δ. It can be understood as the list of declarations of the data used for constructions in J. All the variables occurring in these constructions must, of course, be declared. But a variable occurring only in the context of a construction, and not in the expression which actually names this construction, can not always be safely omitted. Intuitively, a program with some superfluous data among the declarations may change when this data is removed: if the superfluous data does not exist – if its type is empty –, a program containing it may never become executable.[2]

Sequents are derived using some *rules*, generally in the form

$$\frac{\Gamma, \Delta_0 \Rightarrow J_0 \qquad (\ldots) \qquad \Gamma, \Delta_n \Rightarrow J_n}{\Gamma \Rightarrow J}$$

The sequents above the line are *premises*, the one below is the *conclusion*. The variables from $\Delta_0,\ldots\Delta_n$ are said to be *bound* in the conclusion. Conventionally, we often omit the context Γ common to all the sequents in a rule. The rules by which the theories studied here are built up will be listed in Appendix I.

Derivations are trees built iteratively using the conclusions of some (instances of) rules as premises for other rules. This proces starts from *axioms*, which can be regarded as rules with empty set of premises.

A construction or a context is said to be *well-formed* (or *valid*, or *legal*) if it occurs in a derivable sequent. The name of a universe and the empty context are assumed to be well-formed.

A construction is *closed* when its context is not bigger than the context of its range. Thus, a closed type must have empty context (since the range of a type is a universe). A type is *inhabited* when it possesses a closed term.

If we allow not only the empty context, but also the "empty judgement", and assume the empty sequent $\Rightarrow$ (empty on both sides!) as an axiom, then we can show that a context Γ is well-formed iff the sequent $\Gamma\Rightarrow$ is derivable. (We can extend the notion of axiom to the rules with at most one premis, checking whether a context and a range are well-formed.)

In fact, the empty context and the empty judgement are a type – just as zero is a number. (This is essential for some proofs below.) In every universe $\mathcal{U}$ we shall assume a *unit* type $1{:}\,\mathcal{U}$,

[2]When is this the case and when not is a rather subtle matter: its categorical formulation leads into theory of *descent*. A forthcoming paper will explore this connection.

inhabited by a unique term $\phi{:}1$. The empty context and the empty judgement can now be written $\phi{:}1$ or $X{:}1$, which boils down to the same thing, since $X^1=\phi$.

Common to all type theories are also the *structural* rules, which govern manipulation with variables. The rules *Replacement* and *Typing* tell that equal constructions can replace each other: all the operations must preserve the convertibility relation (=). The rule *Assumption* tells that there is always a fresh variable of each well-formed type. Let me stress that this does *not* imply that each type must be inhabited (i.e. that data of each type must exist)!

To get an algebra from an algebraic theory, one can add some generators and equations (to the constants and equations included in the theory), and derive the well-formed expressions, which are then partitioned in the equivalence classes induced by equations. A type theory can similarly be *extended* by generators and additional equations. Generators must be given with well-formed contexts and ranges; equations may be imposed only on constructions with the same range and context. We call *system* the class of derivable formation sequents of an extended type theory; letters $\mathcal{M}$, $\mathcal{N}$ denote systems. (For convenience, we shall assume that a system also includes the names of universes.) Building a system is a dynamical process, since an atomic construction – a generator – can have a complex context and range, and can be thrown in only when they have been derived.

The convertibility relation (=) is extended from constructions to sequents in an obvous way – component-wise – modulo a renaming of variables (*α-rule*). Let us spell this out. By definition,

$$(X_0{:}P_0,\dots,X_m{:}P_m \Rightarrow T{:}U) = (X'_0{:}P'_0,\dots,X'_n{:}P'_n \Rightarrow T'{:}U')$$

<u>means</u> that

- $m=n$, and
- the following sequents are derivable

$$Y_0{:}P''_0,\dots,Y_j{:}P''_j \Rightarrow P_{j+1}[\vec{Y}/\vec{X}]=P'_{j+1}[\vec{Y}/\vec{X}'], \text{ for all } j<n;$$

$$Y_0{:}P''_0,\dots,Y_n{:}P''_n \Rightarrow U[\vec{Y}/\vec{X}]=U'[\vec{Y}/\vec{X}'];$$

$$Y_0{:}P''_0,\dots,Y_n{:}P''_n \Rightarrow T[\vec{Y}/\vec{X}]=T'[\vec{Y}/\vec{X}'];$$

where $P''_j{:=}P_j[\vec{Y}/\vec{X}]$, while $\vec{Y}:=(Y_0,\dots,Y_n)$ are fresh variables.

Partitioning a system of a type theory by the convertibility relation gives a *term model* for this theory. In the usual abuse of language, we often write T for whole sequent $\Gamma \Rightarrow T{:}U$, and even for its equivalence class; the context and range are meant to be kept implicite, and can be recovered by $CX(T)=\Gamma$ and $RG(T)=U$.[3]

The algebraic aspect of type theory is the study of the convertibility $(=) \subseteq \mathcal{M}\times\mathcal{M}$. Its proof-theoretical aspect concerns the relation of derivability $(\vdash) \subseteq \mathcal{M}^{*}\times\mathcal{M}$, transitive closure of all the instances of the given formation rules (together with the axioms and generators taken as rules), where $\mathcal{M}^{*} := \bigcup_{i\in\omega} \mathcal{M}^{i}$.

Theories of constructions and of predicates. The theory of constructions is a (Martin-Löf-style) type theory of sums and products – in two universes:

$\mathcal{S}$ – its types are called *sets*, its terms *elements* (or *functions*);

$\mathcal{P}$ – its types are *propositions*, terms are *proofs*.

Each of these universes is closed under all sums and products. Clearly, there are four possible kinds of indexing: $\mathcal{S}\Rightarrow\mathcal{S}$, $\mathcal{P}\Rightarrow\mathcal{P}$, $\mathcal{S}\Rightarrow\mathcal{P}$, $\mathcal{P}\Rightarrow\mathcal{S}$ – and four kinds of sums and products, two for each universe. The sums and products of propositions indexed over sets ($\mathcal{S}\Rightarrow\mathcal{P}$) are *quantifiers*. They will be written $\exists$ and $\forall$.

The axiom $\mathcal{P}:\mathcal{S}$ is assumed: "The universe of propositions is a set". It follows that every proposition is at the same time a type in $\mathcal{P}$ and a term in $\mathcal{S}$. So there are three levels of constructions:

$$\text{proofs } \begin{matrix} a,b,c \\ x,y,z \end{matrix} : \text{ propositions } \begin{matrix} \alpha,\beta,\gamma \\ \xi,\eta,\zeta \end{matrix} : \text{ sets } \mathcal{P}K := K \to \mathcal{P}.$$

Of course, sets which are not in the form $\mathcal{P}K$ may also be introduced. We denote by A,B,K sets in general, and their elements by f,g,k; the general element-variables remain X,Y,Z. We shall reserve $\phi:1$ for the *singlèton*, unit of $\mathcal{S}$; the *truth*, unit of $\mathcal{P}$, will be denoted by $*:\top$.

The intended meaning of the operation of *extent* ι is to assign to each proposition the set of its proofs. A constructive version of the *comprehension principle* should be captured in this way. The *selection operator* ι, which Alonzo Church introduced in his *simple theory of types* (1940), is the classical ancestor of our ι – though based on a quite different idea. On the other hand, one version of the calculus of constructions (Coquand 1990) contained an operation T, which was meant to replace a proposition by the set of its proofs. But a proposition in the calculus (or theory) of constructions is, in a sense, nothing *but* the set of its proofs. Conceptually, the operation T does not do much; it is actually a syntactical device, introduced

[3]This is a notational convention. In general, a construction need not determine a unique context and range.

to secure the uniqueness of derivations. If all the ι-rules (T had only the introduction rule) would be added in the theory of constructions, the extent operation would just switch a proposition from universe to universe.

This operation is more interesting when combined with the *phase distinction*, the requirement that sets and elements never depend on proofs. (I.e., the indexing $\mathcal{P}\Rightarrow\mathcal{S}$ is forbidden.) The (implicite) context in all the extent rules – listed in Appendix I – must now consist of sets only: otherwise, a proposition contained in the context of a proposition α would be passed in the context of the set $\iota\alpha$. Therefore, only a *predicate* – a proposition indexed only by sets – can have an extent. The elements of the extent $\iota\alpha$ now correspond to the *logically closed proofs* of α, i.e. to those proofs which do not depend on other proofs (and have only some element-variables in their contexts). – This combination of the extent operation and the phase distinction characterizes the *theory of predicates*.

The fragments obtained by removing the Σ-operations from type theories will be called *calculi* here. We shall abbreviate by COC the calculus of constructions, and by COP the calculus of predicates. TOC and TOP will be the theory of constructions and the theory of predicates.

3. What can be expressed by predicates?

Now we shall list some facts which might offer an impression of the power of predicates, and of questions arising from them. The proofs are omitted; they are beyond the scope and the intention of this section. (Some of them can be found in my thesis.)

The notations are explained in Appendix I (or in section 2). "$\models\alpha$" means that "α is inhabited".

31. For every pair of functions $f,g:A\to B$, and elements $h,h':A$, all in the same context, the following statements are true:

$\models \forall X{:}A.fX\equiv gX$ <u>iff</u> $f=g$;

$\models \exists Z{:}\{X{:}A/\ fX\equiv gX\}.h\equiv\pi_0 Z$ <u>iff</u> $fh=gh$

$\models \forall XX'{:}A.fX\equiv fX'\to X\equiv X'$ <u>iff</u> $fh=fh'$ <u>implies</u> $h=h'$

$\models \forall Y{:}B\exists X{:}A.gX\equiv Y$ <u>iff</u> g is a *quotient function*, i.e. for every $k:A\to K$, such that $\models \forall XX'{:}A.gX\equiv gX'\to kX\equiv kX'$ there is unique $\tilde{k}:B\to K$ such that $k=\tilde{k}\circ g$.

32. Writing $\wedge$ in place of $\times$, define

$\exists!X{:}K.\gamma(X) := \exists X{:}K.\gamma(X)\wedge\forall XY{:}K.(\gamma(X)\wedge\gamma(Y))\to X\equiv Y.$

Now consider the principle of *function coprehension:*

$\models \forall X{:}A\exists!Y{:}B.\alpha(X,Y)$ iff $\models \alpha(X,Y) \leftrightarrow fX \equiv Y$ for some f.

In other words, the functions may be identified with the total and single-valued relations, as in set theory. The if-direction of the function comprehension is true in TOP: the graph $fX \equiv Y$ of a function f is provably total and single-valued. The then-direction, however, requires an operation ιX which would *extract singletons*, in the sense that

whenever $\models \exists!X{:}K.\gamma(X)$, then there is $\iota X.\gamma(X) : K$ with $\models \gamma(\iota X.\gamma(X))$.

In Church's simple theory of types (1940), the operation ιX was derivable using the selector ι. (The logical systems of Frege, of Russell-Whitehead, of Hilbert-Bernays also contained operations like ιX.) Constructively, however, the function comprehension is independent from the set comprehension. It is not derivable in the theory of predicates[4], but it can be neatly introduced. For instance – by a slight intervention on the phase distinction:

Predicate γ can occur in the context of a set only if $\models \gamma(X)\wedge\gamma(X') \rightarrow X \equiv X'$

Given $P=S=K$, $Q=\gamma$ and closed proofs $b{:}\exists X{:}K.\gamma$ and $c{:}\gamma(X)\wedge\gamma(X')\rightarrow X\equiv X$, the term

$\iota X.\gamma(X) := \pi_0 b$

can now be formed by ΣE and proved to be independent of b and c. (We assume that the condition ($S{\leq}Q$) is omitted from ΣE in TOP. To introduce ιX in TOC, it is sufficient to strengthen ΣE by extending this condition to ($S{\leq}Q$ or $\models Q(X)\wedge Q(X')\rightarrow X\equiv X'$).)

33. Define

$\upsilon_A \;:=\; \lambda X^A Y^{\mathcal{P}A}.YX : A \rightarrow \mathcal{P}\mathcal{P}A$, and

$\mathcal{P}f \;:=\; \lambda Y^{\mathcal{P}B} X^A.\, Y(fX) : \mathcal{P}B \rightarrow \mathcal{P}A$, for an arbitrary function $f{:}A\rightarrow B$.

In ordinary set theory, for every set A there is a bijection

$A \simeq \{X \in \mathcal{P}\mathcal{P}A /\ \upsilon_{\mathcal{P}\mathcal{P}A} X \equiv \mathcal{P}(\mathcal{P}\upsilon_A)X\}$.

In TOP, we have a term u from left to right and – if the function comprehension is supported – a term n from right to left. They satisfy $n \circ u = id_A$, but not $u \circ n = id_{\{...\}}$. An intuitive explanation can be that the set on the right side contains not just the *principal filters* on $\mathcal{P}A$, but also the proofs that they are principal filters, and there can be many of those for each of them.

Similar phenomena are met in encoding other set-theoretical constructions in TOP. E.g., the *disjoint union* can be defined by:

---

[4]To see this, consider a Heyting algebra H as a model for the theory of predicates. The sets are interpreted as the members of H. For $a,b \in H$, the relation $a \leq b$ represents a function from a to b. The type $\mathcal{P}$ of propositions will be the unit 1 of H. (In terms of my thesis, we are looking at the category of predicates $id{:}H \rightarrow H$.) – The function comprehension fails in this model.

$$A_0{+}A_1 := \{X{:}\mathcal{P}(\mathcal{P}A_0{\times}\mathcal{P}A_1) /\ \upsilon_{\mathcal{P}(\mathcal{P}A_0\times\mathcal{P}A_1)}X \equiv \mathcal{P}(\mathcal{P}\upsilon_{A_0}\times\mathcal{P}\upsilon_{A_1})X\ \}$$

Of course, there are inclusions $\kappa_i{:}A_i{\to}A_0{+}A_1$ $(i{\in}2)$ and the operation $[\_,\_]$, which assigns to each pair of terms $f_i{:}A_i{\to}B$ $(i{\in}2)$ a term $[f_0,f_1]:A_0{+}A_1{\to}B$, such that $[f_0,f_1]\circ\kappa_i = f_i$. However, $[\kappa_0, \kappa_1] = id$ need not be true.

Yet another example: If, except the powersets, no other products of sets were given in our theory, we could define them using the extents of some equations, just as above, adapting the constructions from topos theory. However, the λ-abstraction obtained in this way would not satisfy the $\Pi\eta$-rule.[5]

Morale: The constructions with constructive extents are not extensional, because these extents are blown up by some constructive proofs.

4. Comparing theories: the conceptual part

What are we going to do? The starting point of our reduction of TOC is a simple observation, formulated in lemmas 21, Appendix II:

- the universe of propositions is embedded in the universe of sets by the operation $\_{\times}1{:}\mathcal{P}\to\mathcal{S}$ (and $\_{\times}T{:}\mathcal{S}\to\mathcal{P}$ is its reflection);
- this embedding preserves (up to isomorphism) all operations except the existential quantifier.

In particular, every sum or product over α is isomorphic with a sum resp. product over $\alpha{\times}1$. This means that the theory of constructions is sufficiently redundant that propositions occurring in contexts can be replaced by sets. If we restrict TOC by allowing only sets to occur in the contexts – call such a theory $TOC_{\mathcal{S}}$ – and translate TOC-constructions into $TOC_{\mathcal{S}}$-constructions:

$$\big(...x{:}\alpha...{\Rightarrow}T(x)\big) \quad\mapsto\quad \big(...X{:}\alpha{\times}1...{\Rightarrow}T(\pi_0X)\big)$$

– nothing will be lost, in the sense that an isomorphic copy of each TOC-type will still be generated in $TOC_{\mathcal{S}}$.

[5] The exponent $A{\to}B$ could be obtained as a subset of $\mathcal{P}(A{\times}\mathcal{P}B)$. The sum $A{+}B$ is a subset of $\mathcal{P}(\mathcal{P}A{\times}\mathcal{P}B)$. Note the resemblance with classical logic, where $(A{\to}B)\leftrightarrow\neg(A\wedge\neg B)$ and $(A\vee B)\leftrightarrow\neg(\neg A\wedge\neg B)$.

But now, TOC$_{\mathcal{S}}$ respects the phase distinction, and can be translated in TOP. So TOC can be translated in TOP. On the other hand, TOP can surely be translated in TOC, since the extent operation is definable there:

$$\iota\alpha \quad := \quad \alpha \times 1$$
$$\delta a \quad := \quad \langle a,\phi \rangle$$
$$\tau k \quad := \quad \pi_0 k.$$

By this translation, however, many types which were not isomorphic in TOP become isomorphic in TOC; the former theory has "more" types. (Out of seven isomorphisms "through the border of the universes", which can be extracted from lemma 212 for TOC, only two exist in TOP: those from lemmas 222 and 223.) To relate the theories precisely, we added in TOP the terms $x{:}\alpha \Rightarrow \delta^* x{:}\ \iota\alpha \times T$. They behave just like "$\langle \delta x, * \rangle$" would, if only δx could be formed. These terms force isomorphism of each predicate with (the reflection of) its extent (lemma 231). Consequently, the extent operation $\iota{:}\mathcal{P} \longrightarrow \mathcal{S}$ becomes an embedding, with the same preservation properties as $\_ \times 1{:}\mathcal{P} \longrightarrow \mathcal{S}$ in TOC (lemma 232).

In the *strict theory of predicates* (STOP) – the one with $\delta^* x$ – the sums and products over propositions can be reduced to the sums and products over sets, just like in TOC. So we have a subtheory STOP$_{\mathcal{S}} \subseteq$ STOP, just like TOC$_{\mathcal{S}} \subseteq$ TOC. Moreover, STOP$_{\mathcal{S}}$ and TOC$_{\mathcal{S}}$ are isomorphic. The conclusion that STOP and TOC are equivalent can now be made following the topological idea that

two spaces are homotopy equivalent iff they have isomorphic deformation retracts.

The next proposition shows the strict extents from another angle.

Proposition. **(In STOP.)** Let $T(x^\alpha)$ and $T'(x^\alpha)$ be arbitrary propositions, or proofs of the same proposition. The following rule is true:

ω   if $T(a)=T'(a)$ for all logically closed proofs $a{:}\alpha$,
then $T(x)=T'(x)$.

In the presence of δ^* and $\delta^*\eta$, the ω-rule implies $\delta^*\beta$.

Proof. If $T(a)=T'(a)$ for all logically closed $a{:}\alpha$, then it holds for $\tau X{:}\alpha$, i.e.

$$X{:}\iota\alpha \Rightarrow T\big(\nu(\langle X,*\rangle,(X,*).\tau X)\big) = T(\tau X) = T'(\tau X) = T'\big(\nu(\langle X,*\rangle,(X,*).\tau X)\big).$$

According to lemma 14 (still Appendix II!), this implies

$$z{:}\ \iota\alpha \times T \Rightarrow T\big(\nu(z,\ (X,*).\tau X)\big) = T'\big(\nu(z,\ (X,*).\tau X)\big).$$

Using $\delta^*\beta$, we get

$$x{:}\alpha \Rightarrow T(x) = T\big(\nu(\delta^* x,\ (X,*).\tau X)\big) = T'\big(\nu(\delta^* x,\ (X,*).\tau X)\big) = T'(x).$$

To derive $\delta^*\beta$ from ω, note that for

$$c(x) := \nu(\delta^* x, (X,*).b[\tau X/z]))$$

and for any logically closed $a{:}\alpha$ holds

$$c(a) = c(\tau(\delta a)) = \nu\big(\delta^* \tau(\delta a), (X,*).b[\tau X/z]\big) = \nu\big(\langle \delta a, * \rangle, (X,*).b[\tau X/z]\big) =$$

$$= b(\tau(\delta a)) = b(a).\bullet$$

Remarks. The last proposition is the type-theoretical version of the fact that the category $\mathcal{P}$ of propositions is generated by the terminal object in the models of TOC and STOP. This means that the operations

$$\beta(x) \quad \mapsto \quad \iota(\beta(\tau X))$$
$$b(x) \quad \mapsto \quad \delta(b(\tau X))$$

are injective. In fact, a TOP-system supports the strict extents iff the second operation induces a bijection between the sets of closed terms of type $\alpha \to \alpha'$ and of $\iota\alpha \to \iota\alpha'$. (This can be deduced from III.4.3 and IV.2.2 in Pavlović 1990).

The ω-rule owes its name to the fact that it is an infinitary rule (with infinitely many premises). In our setting, however, it can be equivalently expressed with just one premis:

$$\omega \qquad \frac{X{:}\iota\alpha \Rightarrow T(\tau X) = T'(\tau X)}{x{:}\alpha \Rightarrow T(x) = T'(x)}$$

5. Comparing theories: the technical part

Instanciation. Consider a construction $T(X)$ and terms p and q which can be substituted for X. If for every judgement $J_T(X)$, involving $T(X)$ and possibly some more occurrences of X,

$J_T(p)$ implies $J_T(q)$,

then we say that $T(q)$ is an *instance* of $T(p)$.

Usually, $T(p)$ is $T(X)$, and its instances are obtained by substitution. The example of the ω-rule shows, however, that this is not the only way to instanciate. (In the ω-rule, $T(q)$ is $T(x)$!) In the sequel, we shall actually use *instanciation* as the common name for the substitution and the ω-rule.

Equivalences. Let $\mathcal{M}$ and $\mathcal{N}$ be two systems. A *translation of systems* is a mapping $F{:}\mathcal{M} \to \mathcal{N}$ which preserves the derivability ($\vdash$) and the convertibility (=). Moreover, it should be *coherent*, in the sense that

$$\left.\begin{array}{lcl} F(\Gamma \Rightarrow T{:}U) & = & (\Gamma' \Rightarrow T'{:}U') \\ F(\Gamma \Rightarrow U{:}V) & = & (\Gamma'' \Rightarrow U''{:}V'') \end{array}\right\} \ \underline{\text{imply}} \ U' = U''.$$

Let M and N be two type theories. A *translation* $F{:}M \to N$ assigns to every M-system $\mathcal{M}$ an N-system $F\mathcal{M}$ and a translation of systems $F_{\mathcal{M}}{:}\mathcal{M} \to F\mathcal{M}$.

A subsystem $\mathcal{N} \subseteq \mathcal{M}$ is a *retract* of $\mathcal{M}$ if there is a translation $F{:}\mathcal{M} \to \mathcal{N}$, which restricts to the identity on $\mathcal{N}$; moreover, every type Q from $\mathcal{M}$ must be isomorphic with an instance of $F(Q)$. More precisely, there is a chain of instanciations Ξ, which brings $F(Q)$ in the context of Q, and

$$F(Q)[\Xi] \simeq Q.$$

A subtheory $N \subseteq M$ is a *retract* of M if there is a translation $F{:}M \to N$ such that every $F\mathcal{M}$ is a retract of $\mathcal{M}$ by $F_{\mathcal{M}}$.

Theories M and N are *equivalent* if there are translations $F{:}M \to N$ *and* $G{:}N \to M$, such that for every M-system $\mathcal{M}$ and N-system $\mathcal{N}$, $GF\mathcal{M}$ is a retract of $\mathcal{M}$ and $FG\mathcal{N}$ is a retract of $\mathcal{N}$.

Comments. Recall (from section 2) that a system is assumed to contain its universes, together with all "other" derivable formation sequents. The coherence requirement for translations applies therefore not only when U is a type, but also when it is a universe.

Usually, a subobject $\Gamma : \mathcal{N} \hookrightarrow \mathcal{M}$ is called retract of $\mathcal{M}$ when there is a map $F{:}\mathcal{M} \to \mathcal{N}$ such that $F \circ \Gamma = id_{\mathcal{N}}$. The above definition requires $\Gamma \circ F \simeq id_{\mathcal{M}}$ too. Because of this, $\mathcal{N}$ can be understood as a *deformation* retract of $\mathcal{M}$; and our notion of equivalence can be understood as the *homotopy* equivalence. Note that each deformation retract of a system is equivalent to that system.

The idea is that theories should be equivalent if they have the same class of models.[6] For instance, the theory of Boolean algebras is equivalent with that of Boolean rings. The theory of Boolean algebras with the signature $\langle \vee, \to, 0 \rangle$ is a retract of the one using $\langle \vee, \wedge, \to, \neg, 0, 1 \rangle$. The cut elimination is a retraction of a sequent calculus.

Eliminating redundancies from a theory is like removing synonyms from a natural language. It becomes harder to speak, but easier to understand – closer to semantics. E.g., the cut-elimination yields unnatural proofs, but offers a crucial insight into what is provable.

[6]The morphisms which they induce on this class can be different.

As far as type theory is concerned, we want to consider as synonymous exactly those isomorphic types that would be identified semantically. (A complete semantics for the theory of predicates has been given in Pavlović 1990.)

Theorem. The theory of constructions (TOC) and the strict theory of predicates (STOP) are equivalent.

Proof. As explained in section 4, we shall define the following translations

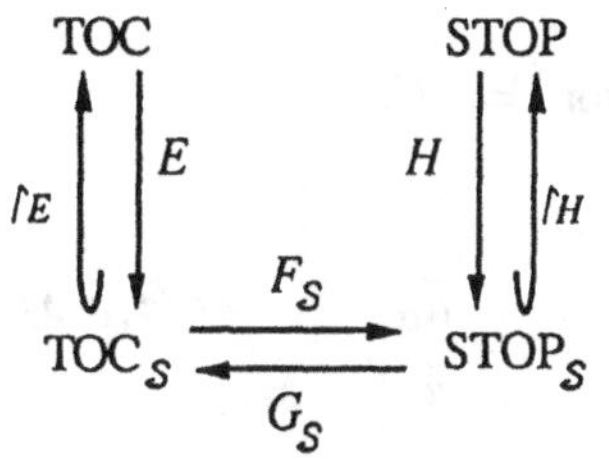

The subtheories which we consider are obtained from TOP resp. STOP by the restriction

TOC_S, $STOP_S$	*Only sets may occur in contexts.*

In TOC_S, however, a provision must be made for the operation $\_\times 1 : \mathcal{P} \to \mathcal{S}$

TOC_S	*The context of $1 : \mathcal{S}$ may contain propositions.*

Translation *E*. For an arbitrary TOC-system $\mathcal{M}$, we simultaneously define two translations, D and $E : \mathcal{M} \to \mathcal{M}$:

$$D(\ldots X{:}Q \ldots \Rightarrow T{:}U) \;:=\; (\ldots X{:}\lfloor Q \rfloor \ldots \Rightarrow \lfloor T \rfloor{:}\lfloor U \rfloor[d_Q X/X]),$$

$$E(\ldots X{:}Q \ldots \Rightarrow T{:}U) \;:=\; (\ldots X{:}\lfloor Q \rfloor \ldots \Rightarrow \lceil T \rceil{:}\lceil U \rceil[d_Q X/X])^{7},$$

where $\lceil\_\rceil$ and $\lfloor\_\rfloor$ translate expressions as follows. ¢ denotes an atom, and □ stands for Σ or Π.

$$
\begin{array}{lcl}
\lceil ¢ \rceil & := & ¢ \\
\lceil \square X{:}P.Q \rceil & := & \square X{:}\lfloor P \rfloor.\lceil Q \rceil \\
\lceil \lambda X.q \rceil & := & \lambda X.\lceil q \rceil \\
\lceil pq \rceil & := & \lceil p \rceil \lfloor q \rfloor \\
\lceil \langle p,q \rangle \rceil & := & \langle \lfloor p \rfloor, \lceil q \rceil \rangle \\
\lceil v(r,\ (X,Y).s) \rceil & := & v(\lceil r \rceil,\ (X,Y).\lceil s \rceil)
\end{array}
$$

---

[7]People who would prefer to change the name of a variable when translating it into a different type should assume a bookkeeping algorithm for variables here.

$$\lfloor \mathcal{P} \rfloor := \mathcal{S} \qquad \lfloor \mathcal{S} \rfloor := \mathcal{S}$$
$$\lfloor \alpha \rfloor := \lceil \alpha \rceil \times 1 \qquad \lfloor K \rfloor := \lceil K \rceil$$
$$\lfloor a \rfloor := \langle \lceil a \rceil, \emptyset \rangle \qquad \lfloor k \rfloor := \lceil k \rceil$$

Let us define the terms d_Q now. We want to substitute $d_Q X$ for $X{:}Q$ in order to replace $X{:}Q$ in a context by $X{:}D(Q)$. So we must have $d_Q{:}D(Q) \to Q[\Delta_Q]$, where Δ_Q is a sequence of substitutions of $d_P Y{:}D(P)$ for each $Y{:}P$ in the context of Q. In other words, Δ_Q brings Q in the context of $D(Q)$ and $E(Q)$.

Note that $E(¢) = ¢[\Delta_¢]$, for a generator $\Gamma \Rightarrow ¢{:}U$.

$d_Q : D(Q) \to Q[\Delta_Q]$

$$d_\alpha := e_\alpha \circ \pi_0 \qquad \tilde{d}_\alpha := \lambda x. \langle \tilde{e}_\alpha x, \emptyset \rangle$$
$$d_K := e_K \qquad \tilde{d}_K := \tilde{e}_K$$

$e_Q : E(Q) \to Q[\Delta_Q]$

$$e_¢ := id_{¢[\Delta_¢]} \qquad \tilde{e}_¢ := id_{¢[\Delta_¢]}$$
$$e_{\Box X:P.Q} := v_\Box \circ w \qquad \tilde{e}_{\Box X:P.Q} := \tilde{w} \circ \tilde{v}_\Box$$

$v_\Box : \big(\Box X{:}E(P).E(Q)\big) \to \big(\Box X{:}P.Q\big)$

$$v_\Pi := \lambda Z.\, e_Q \circ Z \circ \tilde{e}_P \qquad \tilde{v}_\Pi := \lambda Z.\, \tilde{e}_Q \circ Z \circ e_P$$
$$v_\Sigma := \nu(Z, (X,Y).\langle e_P X, e_Q Y \rangle) \qquad \tilde{v}_\Sigma := \nu(Z, (X,Y).\langle \tilde{e}_P X, \tilde{e}_Q Y \rangle)$$

$w : \big(\Box X{:}D(P).E(Q)\big) \to \big(\Box X{:}E(P).E(Q)\big)$ is the isomorphism from lemma 212; $\tilde{w}$ is its inverse.

This completes the definition of mappings D and E. Clearly, the substitution will be:

$$D(T[p/X]) := D(T)[D(p)/X]$$
$$E(T[p/X]) := E(T)[D(p)/X].$$

A straightforward inductive argument shows that E and D are translations. The image of E is a TOC$_\mathcal{S}$-subsystem of $\mathcal{M}$. Call this subsystem $E\mathcal{M}$. Since all d_Q are isomorphisms, there are substitutions Ξ_Q which bring $D(Q)$ and $E(Q)$ back in the context of Q. (Ξ_Q puts $\tilde{d}_P Y{:}P$ in place of $Y{:}D(P)$.) From the isomorphisms e_Q we get

$$e_Q[\Xi_Q] : E(Q)[\Xi_Q] \simeq Q$$

for every type Q from $\mathcal{M}$. Hence, $E\mathcal{M}$ is a retract of $\mathcal{M}$; TOC$_\mathcal{S}$ is a retract of TOC.

Translation *H*. The approach is completely the same: For an arbitrary STOP-system $\mathcal{N}$, we define two translations $I, H : \mathcal{N} \to \mathcal{N}$, using $\lceil \_ \rceil$ and $\lfloor \_ \rfloor$ just as above: write H in place of E, I in place of D, and i_Q in place of d_Q.

The definition of $\lceil\_\rceil = \lceil\_\rceil_H$ is the same as that of $\lceil\_\rceil_E$ above, plus:

$$\lceil \iota\alpha \rceil := \iota\lceil\alpha\rceil$$
$$\lceil \delta a \rceil := \delta\lceil a\rceil$$
$$\lceil \tau k \rceil := \tau\lceil k\rceil$$

$\lfloor\_\rfloor_H$ deviates from $\lfloor\_\rfloor_E$ a bit more:

$$\lfloor P \rfloor := S \qquad\qquad \lfloor S \rfloor := S$$
$$\lfloor \alpha \rfloor := \iota\lceil\alpha\rceil \qquad\qquad \lfloor K \rfloor := \lceil K\rceil$$
$$\lfloor a \rfloor := \delta\lceil a\rceil \qquad\qquad \lfloor k \rfloor := \lceil k\rceil$$

A real difference with respect to the situation in TOC is that there are no terms from propositions to sets in STOP – hence no isomorphisms between $I(\alpha)$ and α.

$i_Q : I(Q) \rightarrow Q[\Delta_Q]$

$$i_\alpha := h_\alpha \circ \tau$$
$$i_K := h_K$$

$h_Q : H(Q) \rightarrow Q[\Delta_Q]$

$$h_¢ := id_{¢[\Delta_¢]} \qquad\qquad \tilde{h}_¢ := id_{¢[\Delta_¢]}$$
$$h_{\iota\alpha} := \delta \circ h_\alpha \circ \tau \qquad\qquad \tilde{h}_{\iota\alpha} := \delta \circ \tilde{h}_\alpha \circ \tau$$
$$h_{\square X:P.Q} := v_\square \circ w \qquad\qquad \tilde{h}_{\square X:P.Q} := \tilde{w} \circ \tilde{v}_\square$$

$v_\square : (\square X{:}H(P).H(Q)) \rightarrow (\square X{:}P.Q)$ is defined exactly as in the E-part, but with h instead of e.

$w : (\square X{:}I(P).H(Q)) \rightarrow (\square X{:}H(P).H(Q))$ is the isomorphism from lemma 232.

By a substitution Δ_Q along the terms i_P (for P from the context of Q), each type Q is brought in the context of $H(Q)$. The question is now how to get $H(Q)$ back in the context of Q without any inverses of i_P?

Note that the variables $X{:}I(\alpha)$ occurring in the context of $H(Q)$ are substituted in $\lceil Q \rceil$ by $[i_\alpha X/x]$. But $i_\alpha X = h_\alpha(\tau X)$. We can now instanciate by the ω-rule, and replace τX by x. So we put in the context of $H(Q)$ the variable $x{:}H(\alpha)$ in place of $X{:}I(\alpha)$ $(=\iota H(\alpha))$; and now we substitute: $\lceil Q \rceil[h_\alpha x/x]$.

If this is done for all propositions α occurring in the context of $H(Q)$, a chain of instanciations Θ_Q is obtained, which brings the term $h_Q{:}H(Q) \rightarrow Q[\Delta_Q]$ in a context "parallel" with that of Q. The only difference between the two contexts is that instead of $Y{:}P \in CX(Q)$, the context of $h_Q[\Theta_Q]$ contains $Y{:}H(P)$.

The terms h_Q and $\bar{h}_Q$ remained, of course, inverse under the instanciation Θ_Q; hence $h_Q[\Theta_Q]:H(Q)[\Theta_Q] \simeq Q[\Delta_Q,\Theta_Q]$. To get these two terms back in the original context of Q, substitute now $\bar{h}_P[\Theta_Q]Y$ for each $Y:H(P)$ in their contexts. Denote this sequence of substitutions by Ξ_Q.

It is not hard to see that $Q[\Delta_Q,\Theta_Q,\Xi_Q]=Q$. Namely, Δ_Q substituted i_PY for $Y:P$; Θ_Q replaced i_PY with h_PY; Ξ_Q put $\bar{h}_PY$ in place of Y in h_PY; and $h_P(\bar{h}_PY) = Y$. Hence

$$h_Q[\Theta_Q,\Xi_Q] : H(Q)[\Theta_Q,\Xi_Q] \simeq Q$$

for every type Q from $\mathcal{N}$. $H\mathcal{N}$ is a retract of $\mathcal{N}$; STOP$_\mathcal{S}$ is a retract of STOP.

Translations *F* and *G*. The maps $F_\mathcal{S}$: TOC$_\mathcal{S}$ $\rightarrow$ STOP$_\mathcal{S}$ and $G_\mathcal{S}$: STOP$_\mathcal{S}$ $\rightarrow$ TOC$_\mathcal{S}$ are easy to guess. The latter rewrites all the expressions from a STOP$_\mathcal{S}$-system $\mathcal{N}_\mathcal{S}$, replacing only:

$$\iota a \mapsto a\times 1,$$
$$\delta a \mapsto \langle a,\phi\rangle,$$
$$\tau k \mapsto \pi_0 k;$$

the former goes the other way around. Note that the rules for ι and those for $\_\times 1$ are completely the same. So we have an isomorphism.

Given a TOC-system $\mathcal{M}$, define $F\mathcal{M}$ to be the smallest STOP-system containing the STOP$_\mathcal{S}$-system $F_\mathcal{S}\mathcal{M}_\mathcal{S}$. Given a STOP-system $\mathcal{N}$, let $G\mathcal{N}$ be the smallest TOC-system which contains $G_\mathcal{S}\mathcal{N}_\mathcal{S}$. Clearly, $GF\mathcal{M} \subseteq \mathcal{M}$ and $FG\mathcal{N} \subseteq \mathcal{N}$.

Further define for systems $\mathcal{M}$ and $\mathcal{N}$ the translations $F = F_\mathcal{M} : \mathcal{M} \rightarrow F\mathcal{M}$ and $G = G_\mathcal{N} : \mathcal{N} \rightarrow G\mathcal{N}$ as follows:

$$F := \bar{I}_H \circ F_\mathcal{S} \circ E \text{ and}$$
$$G := \bar{I}_E \circ G_\mathcal{S} \circ H.$$

Using $E \circ \bar{I}_E = id$, $H \circ \bar{I}_H = id$, $F_\mathcal{S} \circ G_\mathcal{S} = id$ and $G_\mathcal{S} \circ F_\mathcal{S} = id$, we get

$$G \circ F = \bar{I}_E \circ E \text{ and}$$
$$F \circ G = \bar{I}_H \circ H.$$

$F \circ G$ and $G \circ F$ are thus retractions, since E and H are.

Remark. The danger of working modulo isomorphisms is that whole groups (of automorphisms) can be swept away: reduced to an identity. This will not happen if *unique canonical* isomorphisms are used. The isomorphisms in the preceding theorem are clearly canonical, i.e. defined uniformly for all types. A curious reader will perhaps want to check that they are unique. (The assertions to be proved: For every canonical isomorphism $f_Q:E(Q) \rightarrow Q$, $D(f_Q)=id_{D(Q)}$ implies $f_Q=e_Q$; for every canonical $g_Q:H(Q) \rightarrow Q$, $I(g_Q)=id_{I(Q)}$ implies $g_Q=h_Q$.) –

For a full precision, the unicity requirement should be put in the definition of retracts. We refrained from this for the sake of simplicity.

6. How to compare calculi?

In the calculus of constructions, all operations can be reduced to those within the universe of sets: the exception from lemma 212 disappears. The restriction of the translation D on COC will therefore be a retraction. (Whole $D: \mathrm{TOC} \rightarrow \mathrm{TOC}_S$ is not a retraction because of the mentioned exception: $D(\exists X{:}K.\beta) \not\simeq \exists X{:}K.\beta$.) So we can translate expression-wise here: a D-image of a sequent is obtained by simply applying $\llcorner\lrcorner$ at each expression in it.

In the calculus of predicates, on the other hand, a new way of making extents strict must be invented, since the operation δ^* needs Σ. Two possibilities are suggested by proposition III.4.3 in my thesis. One is to force $\iota(\alpha \rightarrow \beta) \simeq \iota\alpha \rightarrow \iota\beta$ (by adding something like δ^*); otherwise force $\alpha \rightarrow \beta \simeq \forall X{:}\iota\alpha.\beta$. A proof of equivalence of the *strict calculus of predicates* – which contains these isomorphisms – and the calculus of constructions can be built along the same lines as the one presented above.

Appendix I

Rules

Structure (all type theories)

Assumption $\dfrac{\Gamma \Rightarrow P : \mathcal{U}}{\Gamma, X{:}P \Rightarrow X{:}P}$ $(X{:}P \notin \Gamma)$

Weakening $\dfrac{\Gamma, \Delta \Rightarrow J \qquad \Gamma \Rightarrow P : \mathcal{U}}{\Gamma, X{:}P, \Delta \Rightarrow J}$ $(X{:}P \notin \Gamma, \Delta)$

Substitution $\dfrac{\Gamma, X{:}P, \Delta \Rightarrow J \qquad \Gamma \Rightarrow p{:}P}{\Gamma, \Delta[p/X] \Rightarrow J[p/X]}$

Replacement $\dfrac{\Gamma, X{:}P, \Delta \Rightarrow T{:}U \qquad \Gamma \Rightarrow p{=}q}{\Gamma, \Delta[p/X] \Rightarrow T[p/X] = T[q/X]}$

Typing $\dfrac{\Gamma \Rightarrow p{:}P \qquad \Gamma \Rightarrow P{=}Q}{\Gamma \Rightarrow p{:}Q}$

Equality (all) $\dfrac{}{p=p}$ $\dfrac{p=q}{q=p}$ $\dfrac{p=q \quad q=r}{p=r}$

Unit (all) $\dfrac{}{1 : \mathcal{U}}$ $\dfrac{}{\phi : 1}$ $\dfrac{p : 1}{p = \phi}$

Universes (COC, COP, TOC, TOP, STOP)

$\dfrac{}{\mathcal{P} : \mathcal{S}}$

Products (COC, COP, TOC, TOP, STOP)

Π $$\frac{X{:}P \Rightarrow Q{:}\mathcal{U}}{\Pi X{:}P.Q{:}\mathcal{U}}$$

ΠI $$\frac{X{:}P \Rightarrow q{:}Q \qquad X{:}P \Rightarrow Q{:}\mathcal{U}}{\lambda X.q : \Pi X{:}P.Q}$$

ΠE $$\frac{r : \Pi X{:}P.Q \qquad p{:}P}{rp : Q[p/X]}$$

Πβ $$(\lambda X.\, q)p \;=\; q[p/X]$$

Πη $$\lambda X.\,(tX) \;=\; t \qquad (X \notin CX(t))$$

Sums (TOC, TOP, STOP)

Σ $$\frac{X{:}P \Rightarrow Q{:}\mathcal{U}}{\Sigma X{:}P.Q{:}\mathcal{U}}$$

ΣI $$\frac{p{:}P \qquad q{:}Q[p/X] \qquad X{:}P \Rightarrow Q{:}\mathcal{U}}{\langle p,q\rangle : \Sigma X{:}P.Q}$$

ΣE $$\frac{r{:}\Sigma X{:}P.Q \;\; X{:}P,Y{:}Q \Rightarrow s{:}S[\langle X,Y\rangle/Z] \;\; Z{:}\Sigma X{:}P.Q \Rightarrow S{:}\mathcal{U}}{\nu(r,(X,Y).s) : S[r/Z]} \; (S \leq Q)$$

Σβ $$\nu(\langle p,q\rangle, (X,Y).s) \;=\; s[p/X,\, q/Y]$$

Ση $$\nu(r, (X,Y).t[\langle X,Y\rangle/Z]) = t[r/Z] \qquad (X,Y \notin CX(t))$$

Comment. $S \leq Q$ <u>means</u> $RG(S){:}RG(Q)$ <u>or</u> $RG(S)=RG(Q)$. In other words, ΣE must not be applied when S is a set and Q a proposition. Due to the next rule, this cannot happen in (S)TOP at all; so that the condition can be omitted there.

Phase distinction (COP, TOP, STOP)

The context of a set or an element must contain no propositions.

Extent (COP, TOP, STOP)

ι $\dfrac{\alpha : \mathcal{P}}{\iota\alpha : \mathcal{S}}$

ιI $\dfrac{a : \alpha}{\delta a : \iota\alpha}$ ιE $\dfrac{k : \iota\alpha}{\tau k : \alpha}$

ιβ $\tau(\delta a) = a$ ιη $\delta(\tau k) = k$

ιT $\iota T = 1$

Strict extent (STOP)

δ* $\dfrac{a : \alpha}{\delta^* a : \iota\alpha \times T}$

δ*β $\nu(\delta^* a, (X,*).b[\tau X/z]) = b[a/z]$

δ*η $\delta^*(\tau k) = \langle k, * \rangle.$

Comment. Because of the phase distinction, the (implicite) context in all the extent rules may contain only sets; $\iota\alpha$ and δa can be formed only in such a context.

Notations

$P \to Q := \Pi X{:}P.Q \quad (X \notin CX(Q))$

$id_P := \lambda X^P.\, X^P$

$\pi_0 := \lambda Z.\nu(Z,(X,Y).X)$

$\mathcal{P}K := K \to \mathcal{P}$

$X \underset{K}{\equiv} Y := \forall \xi{:}\mathcal{P}K.\xi X \leftrightarrow \xi Y$

$P \times Q := \Sigma X{:}P.Q \quad (X \notin CX(Q))$

$p \circ q := \lambda X.p(qX)$

$\pi_1 := \lambda Z.\nu(Z,(X,Y).Y)$

$\{X{:}K /\ \alpha(X)\} := \Sigma X{:}K.\iota\alpha(X)$

Appendix II

Lemmas

1. About Σ. Due to the restriction on ΣE (in TOC, or to the phase distinction in TOP) the projection $\pi_0 : \exists X{:}P.Q \to P$ cannot be formed when P is a set and Q a proposition. The other three combinations of P and Q (set-set, proposition-set, proposition-proposition) allow both projections. In these situations, ΣE can be replaced by the projection rules, as in Hyland-Pitts 1989. (The equivalence of the two presentations follows from 11-13.)

11. $\pi_i \langle X_0, X_1\rangle = X_i,\ i \in 2$

12. $\langle \pi_0 Z, \pi_1 Z\rangle = Z$

13. $s(\pi_0 Z, \pi_1 Z) = \nu(Z, (X_0,X_1).s)$

14. If $s(\langle X,Y\rangle) = t(\langle X,Y\rangle)$ then $s=t$.

15. In the case when P is a set and Q a proposition, the rule ΣE can be modified (following the idea of $\exists$-elimination) by removing $Z{:}\Sigma X{:}P.Q$ from the context of S . In the theories considered here, the full ΣE-rule is still derivable from this modified instance. (Cf. Pavlović 1990, I.1.52.)

2. Isomorphisms are of course terms $\Gamma \Rightarrow p{:}P' \to P$ and $\Gamma \Rightarrow p'{:}P \to P'$, such that $p \circ p' = id$ and $p' \circ p = id$. We write $p{:}P' \simeq P$ to denote that p is an isomorphism, and $P' \simeq P$ to say that an isomorphism exists.

21. In TOC.

211. $\alpha \times 1 \simeq \alpha$

212. The statement:

$$\text{if } p{:}P' \simeq P \text{ and } Q(X^P) \simeq Q'(X^P) \text{ then } \square X{:}P.\,Q(X) \simeq \square X'{:}P'.\,Q'[pX'/X]$$

holds for all types P, P', Q, Q' and for $\square \in \{\Sigma, \Pi\}$, with one exception:

$$A \simeq \alpha \text{ does not imply } \Sigma X{:}K.A \simeq \exists X{:}K.\alpha.$$

22. In TOP.

221. $\iota(\iota\alpha \times T) \simeq \iota\alpha$

222. $\Sigma X{:}\iota\alpha.\iota(\beta(\tau X)) \simeq \iota(\Sigma x{:}\alpha.\beta)$

223. $\Pi X{:}A.\iota\beta \simeq \iota(\forall X{:}A.\beta)$

23. In STOP.

231. $\iota\alpha \times T \simeq \alpha$

232. For $\square \in \{\Sigma, \Pi\}$ holds:

$$\iota(\square x{:}\alpha.\beta) \simeq \iota(\square X{:}\iota\alpha.\beta) \simeq \square X{:}\iota\alpha.\iota\beta$$

$$\square x{:}\alpha.\beta \simeq \square X{:}\iota\alpha.\beta \simeq (\square X{:}\iota\alpha.\iota\beta) \times T$$

Some comments, some proofs.

212. The exception can perhaps be understood by looking at the set $A \simeq \alpha$ as the extent of α. The sum $\Sigma X{:}K.A$ is then the set $\{X{:}K/\ \alpha(X)\}$ of *all* the witnesses of $\alpha(X)$, while $\exists X{:}K.\alpha$ just says that there is *a* witness. – For a proof that these two types are not isomorphic one should consider a model (e.g. in Hyland-Pitts 1989).

221. The isomorphisms are:

$$X{:}\iota\alpha \Rightarrow \delta\langle X,*\rangle : \iota(\iota\alpha \times T)$$
$$Y{:}\iota(\iota\alpha \times T) \Rightarrow \delta\nu(\tau Y,\ (X,*).\tau X) : \iota\alpha.$$

We check one of two identities that must be proved:

$$\delta\langle\delta\nu(\tau Y,(X,*).\tau X),\ *\rangle \stackrel{\eta}{=} \delta\nu\Big(\tau Y,\ (X,*).\big\langle\delta\nu(\langle X,*\rangle,(X,*).\tau X),\ *\big\rangle\Big) \stackrel{\beta}{=}$$
$$\delta\nu\big(\tau Y,\ (X,*).\langle\delta\tau X,*\rangle\big) = \delta\tau Y = Y.$$

231. $x : \alpha \Rightarrow \delta^* x : \iota\alpha \times T,$

$$z{:}\ \iota\alpha \times T \Rightarrow \nu\big(z,\ (X,*).\tau Z\big){:}\ \alpha.$$

One identity:

$$\delta^*\nu\big(z,\ (X,*).\tau X\big) \stackrel{\eta}{=}$$
$$\nu\Big(z,\ (X,*).\delta^*\nu\big(\langle X,*\rangle,\ (X',*).\tau X'\big)\Big) \stackrel{\beta}{=}$$
$$\nu\big(z,\ (X,*).\delta^*(\tau X)\big) = \nu\big(z,\ (X,*).\langle X,*\rangle\big) = z$$

232. Everything follows from previous results, plus:

$$\Sigma x{:}\alpha.\beta \simeq \exists X{:}\iota\alpha.\beta(\tau X) \text{ and}$$
$$\iota(\Pi x{:}\alpha.\beta) \simeq \Pi X{:}\iota\alpha.\iota\big(\beta(\tau X)\big).$$

The second isomorphism is obtained using 231 and

$$\iota(\Pi x{:}(\iota\alpha\times T).\gamma(x)) \simeq \Pi X{:}\iota\alpha.\iota\big(\gamma(\langle X,*\rangle)\big)$$

And this last iso is definable in the theory of predicates:

$$Z : \iota(\Pi x{:}(\iota\alpha\times T).\gamma(x)) \Rightarrow \lambda X.\delta\big((\tau Z)\langle X,*\rangle\big) : \Pi X{:}\iota\alpha.\iota\big(\gamma(\langle X,*\rangle)\big)$$
$$Y{:}\Pi X{:}\iota\alpha.\iota\big(\gamma(\langle X,*\rangle)\big) \Rightarrow \delta\lambda x.\nu\big(x,\ (X,*).\tau(YX)\big){:}\ \iota(\Pi x{:}(\iota\alpha\times T).\gamma(x))$$

As for the first of the above isomorphisms, we have

$$\Sigma x{:}\alpha.\beta \simeq \iota(\Sigma x{:}\alpha.\beta)\times T \simeq (\Sigma X{:}\iota\alpha.\iota\beta)\times T \stackrel{\#}{\simeq} \exists X{:}\iota\alpha.(\iota\beta\times T) \simeq$$
$$\simeq \exists X{:}\iota\alpha.\beta.$$

The step (#) is a special case of $\exists Z{:}(\Sigma X{:}A.B).\gamma \simeq \exists X{:}A.\exists Y{:}B.\gamma$.

References

Cartmell, J.

(1986) Generalized algebraic theories and contextual categories, *Ann. Pure Appl. Logic* 32, 209-243

Church, A.

(1940) A Formulation of the Simple Theory of Types, *J. Symbolic Logic*, 5(1), pp. 56-68

Coquand, Th.

(1990) Metamathematical Investigations of a Calculus of Constructions, *Logic and Computer Science* (Academic Press)

Coquand, Th., Huet, G.

(1986) Constructions: A higher order proof system of mechanizing mathematics, *EUROCAL 85, Linz* , Lecture notes in Computer Science 203 (Springer, Berlin)

(1988) The Calculus of Constructions, *Information and Computation* 76, 95-120

Ehrhard, T.

(1989) Dictoses, *Category theory in computer science,* Lecture Notes in Computer Science 389 (Springer, Berlin), 213-223

Girard, J.-Y.

(1972) Une extension de l'interpretation de Gödel à l'analyse, et son application à l'élimination des coupures dans l'analyse et la théorie des types, *Proceedings of the Second Scandinavian Logic Symposium* (North-Holland, Amsterdam) 63-92

Harper, R., Mitchell, J.C., Moggi, E.

(1990) Higher-Order Modules and the Phase Distinction, to appear in the *Proceedings of the 17th POPL ACM Conference*

Hyland, J.M.E., Pitts, A.M.

(1989) The theory of constructions: categorical semantics and topos-theoretic models, *Categories in Computer Science and Logic (Proc. Boulder 1987)*, Contemporary Math. (Amer. Math. Soc., Providence RI)

Lambek, J., Scott, P.J.

(1986) *Introduction to higher order categorical logic*, Cambridge studies in advanced mathematics 7 (Cambridge University Press, Cambridge)

Meseguer, J.

(1989) Relating Models of Polymorphism, *Conference Record of the XVI ACM POPL Symposium*, 228-241

Moggi, E.

(1990) A category-theoretic account of program modules, Manuscript

Pavlović, D.

(1990) *Predicates and Fibrations: From Type Theoretical to Category Theoretical Presentation of Constructive Logic*, Thesis (State University Utrecht)

Seely, R.A.G.

(1987) Categorical semantics for higher order polymorphic lambda calculus, *J. Symbolic Logic* 52(4), 969-989

Troelstra, A.S., Dalen, D. van

(1988) *Constructivism in Mathematics. An Introduction*, Studies in Logic and Foundations of Mathematics 121, 123 (North-Holland, Amsterdam)

Relating Models of Impredicative Type Theories

Bart Jacobs, Dep. Comp. Sci., Nijmegen, NL.

Eugenio Moggi, Dip. Informatica, Genova, I.

Thomas Streicher, Fak. Math. & Inform., Passau, D.

Abstract

The object of study of this paper is the categorical semantics of three impredicative type theories, viz. Higher Order λ-calculus $F\omega$, the Calculus of Constructions and Higher Order ML. The latter is particularly interesting because it is a two-level type theory with type dependency at both levels. Having described appropriate categorical structures for these calculi, we establish translations back and forth between all of them. Most of the research in the paper concerns the theory of fibrations and comprehension categories.

1. Introduction

In recent years there has been a considerable amount of work in using category theory to analyse and model type theories. Most of this work has taken inspiration from Categorical Logic (especially Hyperdoctrines), and has developed in two directions: the modelling of Higher order Lambda Calculus $F\omega$ using indexed categories (see Seely [1987]), and the modelling of Martin-Löf Type Theory *TT* (see Cartmell [1987], Seely [1984] and Taylor [1987]) and the Calculus of Constructions *CC* (see Ehrhard [1988], Hyland & Pitts [1989] and Streicher [1989]). More recently new type theories have been proposed, *HML* (see Moggi [1991]) and the Theory of Predicates (see Pavlović [1990]), exhibiting a two-level structure (like in $F\omega$), where the first level (constructors of $F\omega$) is *independent* from the second level (terms of $F\omega$), and type dependency (like in *TT*) but now at both levels. *HML* is an example of such a type theory.

This paper introduces comprehension categories (see also Jacobs [1990]), which generalise D-categories (see Ehrhard [1988]) and classes of display maps (see Hyland & Pitts [1989]), and explains how they can be used (together with fibrations) to give a categorical treatment of type theories. Comprehension categories are unnecessarily general for describing type dependency (see Pitts [1989], Moggi [1991] for similar remarks on D-categories). In practice one uses comprehension categories which are either *cartesian* (corresponding to Cartmell's categories with attributes) or *full* (corresponding to Taylor's relative slice categories). However, most of the relevant definitions and results make sense for comprehension categories in general. Our investigation is divided in two steps:

- general results on fibrations and comprehension categories; this is "category theory over a base *category*".

- the *fibred* version of such results where the base category has been replaced by a fibration. This is "category theory over a fibration". The choice of fibrations and comprehension categories as conceptual framework for type theories makes this passage from the first to the second step particularly straightforward.

We apply some of the above results to clarify the relation between categorical models of $F\omega$, *HML* and *CC*. In comparison with other approaches to type theories ours is particularly suitable to describe type theories with two levels (or more):

- Seely [1987] considers only a two-level type theory without type dependency;
- Hyland & Pitts [1989] considers type theories with type dependency, but only one level (since in *CC* constructors and terms are interdependent);
- Moggi [1991] models two-level type theories without using explicitly comprehension categories over a fibration, but then the less natural concept of *independence* (of a comprehension category from another comprehension category over the same category) has to be used;
- Pavlović [1990] models two level-type theories using classes of display maps (over a fibration). However, the use of fibrations and comprehension categories provides in our opinion great conceptual clarity.

We do realise that the use of fibred category theory makes the paper rather technical. However, whenever possible we try to give underlying type theoretical intuitions. At first reading it might be of help to concentrate on sections 2,6 and 7 and take a brief look only at the first parts of the more technical sections 3,4 and 5.

2. Summary of relations

In this section we give an informal *type-theoretic* formulation of the results in section 7 on the relations among $F\omega$ (Higher order Lambda Calculus), *HML* (Higher order ML) and *CC* (Calculus of Constructions). Type-theoretically $F\omega$, *CC* and *HML* are described by a set of rules for deriving well-formation and equality judgements. The informal description below involves only well-formation judgements and we will use the following notational conventions: k for kinds, u for constructors, v for constructor variables, τ for types, e for terms, x for term variables, Δ for constructor contexts "$v_1 : k_1, \ldots, v_m : k_m$", Γ for term contexts "$x_1 : \tau_1, \ldots, x_n : \tau_n$" and Φ for mixed contexts (allowed only in *CC*).

- $F\omega$ Higher order Lambda Calculus

well-formation judgements			
constructors		*terms*	
$\Delta \vdash_{F\omega}$	*context*	$\Delta;\Gamma \vdash_{F\omega}$	*context*
$\vdash_{F\omega} k$	*kind*	$\Delta \vdash_{F\omega} \tau$	*type*
$\Delta \vdash_{F\omega} u : k$	*constructor*	$\Delta;\Gamma \vdash_{F\omega} e : \tau$	*term*

context extension rules

$$\frac{\Delta \vdash_{F\omega} \quad \vdash_{F\omega} k}{\Delta, v : k \vdash_{F\omega}} \qquad \frac{\Delta;\Gamma \vdash_{F\omega} \quad \Delta \vdash_{F\omega} \tau}{\Delta;\Gamma, x : \tau \vdash_{F\omega}}$$

The judgements for $F\omega$ tell us that:

- kinds do not depend on variables;
- constructors and types may depend only on constructor variables;
- terms may depend both on constructor and on term variables.

Therefore the rules for kinds and constructors can be given independently from those for types and terms, and there are no dependent kinds or types.

- *HML* Higher order ML

well-formation judgements			
constructors		*terms*	
$\Delta \vdash_{HML}$	*context*	$\Delta;\Gamma \vdash_{HML}$	*context*
$\Delta \vdash_{HML} k$	*kind*	$\Delta;\Gamma \vdash_{HML} \tau$	*type*
$\Delta \vdash_{HML} u : k$	*constructor*	$\Delta;\Gamma \vdash_{HML} e : \tau$	*term*

context extension rules

$$\frac{\Delta \vdash_{HML} k}{\Delta, v : k \vdash_{HML}} \qquad \frac{\Delta;\Gamma \vdash_{HML} \tau}{\Delta;\Gamma, x : \tau \vdash_{HML}}$$

The judgements for *HML* tell us that:

- kinds and constructors may depend only on constructor variables;
- types and terms may depend both on constructor and on term variables.

As for $F\omega$ the rules for kinds and constructors can be given independently from those for types and terms, but there may be dependent kinds and types.

- *CC* Calculus of Constructions

well-formation judgements			
constructors		*terms*	
$\Phi \vdash_{CC}$	*context*	$\Phi \vdash_{CC}$	*context*
$\vdash_{CC} k$	*kind*	$\Phi \vdash_{CC} \tau$	*type*
$\Phi \vdash_{CC} u : k$	*constructor*	$\Phi \vdash_{CC} e : \tau$	*term*

context extension rules

$$\frac{\Phi \vdash_{CC} k}{\Phi, v : k \vdash_{CC}} \qquad \frac{\Phi \vdash_{CC} \tau}{\Phi, x : \tau \vdash_{CC}}$$

The judgements for *CC* are all interdependent, and there may be all sorts of dependency between kinds and types.

The relations established in section 7 are summarized by the following picture, where an arrow $S_1 \rightarrow S_2$ is a mapping from S_1-models/-theories to S_2-models/-theories. Alternatively, an arrow $S_1 \rightarrow S_2$ can be viewed as a translation from S_2 to S_1.

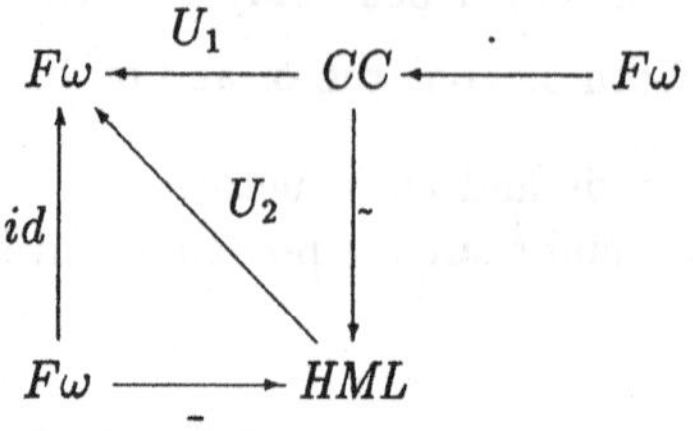

We describe an arrow $S_1 \rightarrow S_2$ as a function f from S_1-theories to S_2-theories, by giving the judgements in $f(T)$ in terms of the judgements in T, where T is an S_1-theory. We consider only few *key* judgements.

- $\bar{\ }: F\omega \rightarrow HML$ is essentially an inclusion, i.e.
  - $\Delta \vdash_{HML} u : k$ iff $\Delta \vdash_{F\omega} u : k$
  - $\Delta; \Gamma \vdash_{HML} e : \tau$ iff $\Delta; \Gamma \vdash_{F\omega} e : \tau$

  so k does not depend on Δ and τ does not depend on Γ.

- $U_2 : HML \rightarrow F\omega$ is removal of dependency, i.e.
  - $u_1 : k_1, \ldots, u_m : k_m \vdash_{F\omega}$ iff $\emptyset \vdash_{HML} k_i$ for $1 \leq i \leq m$
  - $\Delta \vdash_{F\omega} u : k$ iff $\Delta \vdash_{F\omega}$, $\emptyset \vdash_{HML} k$ and $\Delta \vdash_{HML} u : k$
  - $\Delta; x_1 : \tau_1, \ldots, x_n : \tau_n \vdash_{F\omega}$ iff $\Delta \vdash_{F\omega}$ and $\Delta; \emptyset \vdash_{HML} \tau_i$ for $1 \leq i \leq n$
  - $\Delta; \Gamma \vdash_{F\omega} e : \tau$ iff $\Delta; \Gamma \vdash_{F\omega}$, $\Delta; \emptyset \vdash_{HML} \tau$ and $\Delta; \Gamma \vdash_{HML} e : \tau$

- $\tilde{\ }: CC \rightarrow HML$ is restriction to split contexts, i.e.
  - $\Delta \vdash_{HML} u : k$ iff $\Delta \vdash_{CC} u : k$
  - $\Delta; \Gamma \vdash_{HML} e : \tau$ iff $\Delta, \Gamma \vdash_{CC} e : \tau$

  The actual construction in section 7 is a bit more complicated than this: there the context Γ is reduced to a context of length one using unit and strong sums.

- $\cdot : F\omega \rightarrow CC$ is the most complex and has to rely on a translation $\_^*$.
  - $\Phi \vdash_{CC}$ iff $\Phi^* \equiv \Delta; \Gamma$ and $\Delta; \Gamma \vdash_{F\omega}$
  - $\Phi \vdash_{CC} u : k$ iff $\Phi^* \equiv \Delta; \Gamma$, $k^* \equiv [v : k', \tau']$, $u^* \equiv [u', e']$ and $\Delta; \Gamma \vdash_{F\omega}$, $\Delta, v : k' \vdash_{F\omega} \tau'$, $\Delta \vdash_{F\omega} u' : k'$, $\Delta; \Gamma \vdash_{F\omega} e' : [u'/v]\tau'$
  - $\Phi \vdash_{CC} e : \tau$ iff $\Phi^* \equiv \Delta; \Gamma$, $\tau^* \equiv \tau'$, $e^* \equiv e'$ and $\Delta; \Gamma \vdash_{F\omega} e' : \tau'$

  the key clause in the definition of $\_^*$ (used in defining context extension) is $(\Phi, v : k)^* \equiv \Delta, v : k'; \Gamma, x_v : \tau'$, where $\Phi^* \equiv \Delta; \Gamma$ and $k^* \equiv [v : k', \tau']$.

3. Fibrations and Comprehension Categories.

Before starting with the precise mathematical exposition in 3.5 below, we describe the type-theoretic view on the notions defined in this section.

3.1. FIBRATIONS. A fibration $p : \mathbf{D} \to \mathbf{B}$ corresponds to a type theory with two levels, the second depending on the first, but with no other dependency. The judgements interpretable in such structure (and their interpretation) are:

judgement	interpretation
$x : B \vdash_p$	object B in the *base* $\mathbf{B}$
$x : B \vdash_p e(x) : B'$	morphism in the base
$x : B; y : D(x) \vdash_p$	object of *total category* $\mathbf{D}$
	";" indicates the separation between the two levels
$x : B \vdash_p D(x)$	object in the *fibre* $\mathbf{D}_B$ over B
$x : B; y : D(x) \vdash_p [e(x), e'(x,y)] : [x' : B'; D'(x')]$	morphism in the total category
$x : B; y : D(x) \vdash_p e(x,y) : D'(x)$	morphism in the fibre over B

3.2. COMPREHENSION CATEGORIES. A comprehension category

$$\mathbf{D} \underset{p}{\overset{\mathcal{P}_0}{\rightrightarrows}} \mathbf{B} \quad (\Downarrow \mathcal{P})$$

corresponds to a type theory with one level and type dependency (i.e. with two levels, but where the second can be *reflected* into the first). Such a diagram corresponds to a functor from $\mathbf{D}$ to $\mathbf{B}^{\to}$, the "arrow category" of $\mathbf{B}$. The kind of judgements interpretable in such structure are:

judgement	interpretation
$\Delta \vdash_{\mathcal{P}}$	object B in the base
$\Delta \vdash_{\mathcal{P}} D$	object D in the fibre over B
$\Delta \vdash_{\mathcal{P}} \phi : \Gamma$	morphism in the base
$\Delta \vdash_{\mathcal{P}} e : D$	*section* of $\mathcal{P}D$

$$\frac{\Delta \vdash_{\mathcal{P}} D}{\Delta, x : D \vdash_{\mathcal{P}}}$$

is the rule for context extension, and it corresponds to apply $\mathcal{P}_0$ to D. The arrow $\mathcal{P}D$ is the projection morphism $\Delta, x : D \to \Delta$.

3.3. CHANGE-OF-BASE. If $\mathcal{P} : \mathbf{D} \to \mathbf{B}^{\to}$ and $q : \mathbf{E} \to \mathbf{B}$, then the comprehension category $q^*(\mathcal{P})$ over $\mathbf{E}$ is such that $\Delta; z : E \vdash_{q^*(\mathcal{P})} D$ iff $\Delta \vdash_q E$ and $\Delta \vdash_{\mathcal{P}} D$, while context extension is given by

$$\frac{\Delta; z : E \vdash_{q^*(\mathcal{P})} D}{\Delta, y : D; z : E \vdash_{q^*(\mathcal{P})}}$$

3.4. JUXTAPOSITION. If $\mathcal{P} : \mathbf{D} \to \mathbf{B}^{\to}$ and $\mathcal{Q} : \mathbf{E} \to \mathbf{B}^{\to}$, then the comprehension category $\mathcal{P} \cdot \mathcal{Q}$ over $\mathbf{B}$ is such that $\Delta \vdash_{\mathcal{P}\cdot\mathcal{Q}} [y : D, E]$ iff $\Delta \vdash_{\mathcal{P}} D$ and $\Delta, y : D \vdash_{\mathcal{Q}} E$, while context extension is given by

$$\frac{\Delta \vdash_{\mathcal{P}\cdot\mathcal{Q}} [y : D, E]}{\Delta, y : D, z : E \vdash_{\mathcal{P}\cdot\mathcal{Q}}}$$

3.5. FIBRED CATEGORY THEORY. Suppose we have a functor $p : \mathbf{E} \to \mathbf{B}$. An object $E \in \mathbf{E}$ (resp. a morphism f in $\mathbf{E}$) is said to be *above* $A \in \mathbf{B}$ (resp. u in $\mathbf{B}$) if $pE = A$ (resp. $pf = u$). A morphism above an identity is called *vertical.* Every object $A \in \mathbf{B}$ thus determines a so-called "fibre" category $\mathbf{E}_A$ consisting of objects above A and vertical morphisms. One often calls $\mathbf{B}$ the *base* category and $\mathbf{E}$ the *total* category.

A morphism $f : D \to E$ in $\mathbf{E}$ is called *cartesian* over a morphism u in $\mathbf{B}$ if f is above u and every $f' : D' \to E$ with $pf' = u \circ v$ in $\mathbf{B}$, uniquely determines a $\phi : D' \to D$ above v with $f \circ \phi = f'$. The functor $p : \mathbf{E} \to \mathbf{B}$ is called a *fibration* (sometimes a *fibred category* or a *category over* $\mathbf{B}$) if for every $E \in \mathbf{E}$ and $u : A \to pE$ in $\mathbf{B}$, there is a cartesian morphism with codomain E above u. Dually, $f : D \to E$ is *cocartesian* over u if every $f' : D \to E'$ with $pf' = v \circ u$, uniquely determines a $\phi : E \to E'$ above v with $\phi \circ f = f'$. And: p is a *cofibration* if every morphism $pE \to A$ in $\mathbf{B}$ has a "cocartesian lifting" with domain E. In case $\mathbf{B}$ is a category with pullbacks, the functor $cod : \mathbf{B}^{\to} \to \mathbf{B}$ forms an example of a fibration; it is at the same time a cofibration.

If $f : D \to E$ and $f' : D' \to E$ are both cartesian over u, then $f \cong f'$ in $\mathbf{E}/E$ by a vertical isomorphism. Hence given $u : A \to B$ in $\mathbf{B}$ and E above B, it makes sense to *choose* a cartesian lifting of u with codomain E; we often write $\overline{u}(E) : u^*(E) \to E$ for such a choice. Making similar choices for every $E \in \mathbf{E}_B$ determines a functor $u^* : \mathbf{E}_B \to \mathbf{E}_A$, called *inverse image*, *reindexing* or *substitution* functor. Such functors u^* are determined (by choice) up to vertical natural isomorphism. In general, one only has vertical natural isomorphisms $(u \circ v)^* \cong v^* \circ u^*$ and $id^* \cong Id$, as for pullbacks in case of $cod : \mathbf{B}^{\to} \to \mathbf{B}$. A fibration is *split* if it is given together with a choice of inverse images for which these isomorphisms are identities. Often, we suppose that a fibration comes equipped with a *cleavage*, i.e. an arbitrary choice of inverse images.

A morphism between fibrations p and q is given by a commuting square as below, in which the functor H preserves cartesian morphisms, i.e. f is p-cartesian implies that Hf is q-cartesian (such a functor is called *cartesian*).

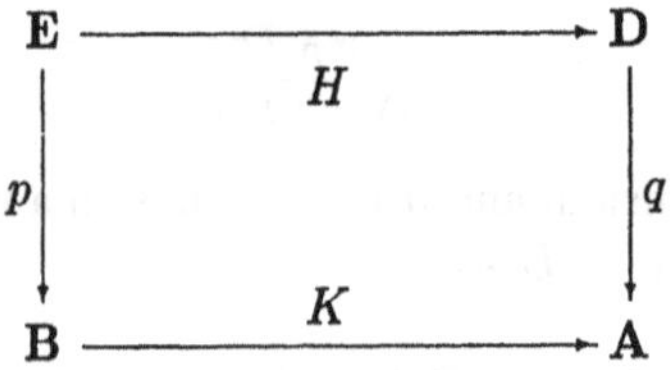

This determines a (very large) "category" *Fib*. Given a fibration $q\colon \mathbf{D} \to \mathbf{A}$ and an arbitrary functor $K\colon \mathbf{B} \to \mathbf{A}$ one can form the pullback

$$\begin{array}{ccc} \mathbf{B} \underset{K,q}{\times} \mathbf{D} & \longrightarrow & \mathbf{D} \\ {\scriptstyle K^*(q)}\downarrow & & \downarrow{\scriptstyle q} \\ \mathbf{B} & \xrightarrow{\quad K \quad} & \mathbf{A} \end{array}$$

and verify that $K^*(q)$ is a fibration again. Consequently, the "functor" $Fib \longrightarrow \mathbf{Cat}$, sending a fibration to its base, is a fibration itself. Usually, one writes $Fib(\mathbf{B})$ for the "fibre" category of fibrations with base $\mathbf{B}$. The above construction is called *change-of-base* (for fibrations). $Fib(\mathbf{B})$ is in fact a 2-category with *vertical* natural transformations as 2-cells. Also *Fib* is a 2-category.

The "fibred" way of doing category theory over a base category was started by A. Grothendieck and further developed notably by J. Bénabou. For example, a *fibred* adjunction between fibrations $p\colon \mathbf{E} \to \mathbf{B}$ and $q\colon \mathbf{D} \to \mathbf{B}$ is given by a pair of cartesian functors $F\colon \mathbf{E} \to \mathbf{D}$ and $G\colon \mathbf{D} \to \mathbf{E}$ forming an adjunction $F \dashv G$ with vertical units and counits. These data determine adjunctions between the fibre categories, which are preserved under reindexing. Similarly, one says that a fibration has *fibred* cartesian products, exponents etc. if such a structure exists in the fibre categories and is preserved under reindexing, see Jacobs [1990] for a (more precise) description in terms of fibred adjunctions. Thus, fibred terminal objects for a fibration $p\colon \mathbf{E} \to \mathbf{B}$ may be described by a functor $1\colon \mathbf{B} \to \mathbf{E}$ such that $p \circ 1 = Id$ with the property that $1A$ is terminal in $\mathbf{E}_A$ and for $u\colon A \to B$ in $\mathbf{B}$ one has $u^*(1B) \cong 1A$.

There is one further notion that should be explained at this point. A fibration $p\colon \mathbf{E} \to \mathbf{B}$ is said to have a *generic object* if there is an object $T \in \mathbf{E}$ such that for every $E \in \mathbf{E}$ one can find a cartesian arrow $E \to T$

3.6. An elementary construction. Suppose a fibration $p\colon \mathbf{E} \to \mathbf{B}$ is given which has fibred finite products. A new fibration $\overline{p}\colon \overline{\mathbf{E}} \to \mathbf{E}$ is constructed in the following way. The category $\overline{\mathbf{E}}$ has pairs $E, E' \in \mathbf{E}$ with $pE = pE'$ as objects; morphisms $(f,g)\colon (E,E') \to (D,D')$ in $\overline{\mathbf{E}}$ are given by arrows $f\colon E \to D$ and $g\colon E \times E' \to D'$ in $\mathbf{E}$ with $pf = pg$. The first projection $\overline{p}\colon \overline{\mathbf{E}} \to \mathbf{E}$ is then a fibration. One easily verifies that $\overline{p}$ has fibred finite products again and that it has a generic object in case p has one. Moreover that there is a change-of-base situation,

$$\begin{array}{ccc} \mathbf{E} & \xrightarrow{\quad H \quad} & \overline{\mathbf{E}} \\ {\scriptstyle p}\downarrow & & \downarrow{\scriptstyle \overline{p}} \\ \mathbf{B} & \xrightarrow{\quad 1 \quad} & \mathbf{E} \end{array}$$

in which both 1 (for terminals) and H are full and faithful functors.

3.7. Definition. (Jacobs [1990]) A *comprehension category* is a functor $\mathcal{P}\colon \mathbf{E} \to \mathbf{B}^{\to}$ satisfying

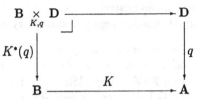

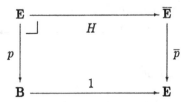

(i) $cod \circ \mathcal{P} : \mathbf{E} \to \mathbf{B}$ is a fibration;

(ii) f cartesian in $\mathbf{E} \Rightarrow \mathcal{P}f$ is a pullback in $\mathbf{B}$.

This $\mathcal{P}$ is called a *full* comprehension category in case $\mathcal{P}$ is a full and faithful functor, and it is called a *cartesian* comprehension category in case every morphisms in $\mathbf{E}$ is cartesian.

3.8. REMARKS. (i) For a comprehension category $\mathcal{P} : \mathbf{E} \to \mathbf{B}^{\to}$ we use the following standard notation: $p = cod \circ \mathcal{P}$ and $\mathcal{P}_0 = dom \circ \mathcal{P}$. The object part of $\mathcal{P}$ then forms a natural transformation $\mathcal{P} : \mathcal{P}_0 \overset{.}{\longrightarrow} p$. (Similarly, for e.g. $\mathcal{Q} : \mathbf{D} \to \mathbf{A}^{\to}$, we write $q = cod \circ \mathcal{Q}$ and $\mathcal{Q}_0 = dom \circ \mathcal{Q}$.) The components $\mathcal{P}E$ are often called *projections* and reindexing functors of the form $\mathcal{P}E^*$ are called *weakening* functors.

(ii) If $\mathcal{P} : \mathbf{E} \to \mathbf{B}^{\to}$ is a comprehension category, one has that every object $E \in \mathbf{E}$ determines a pullback functor $\mathcal{P}E^{\#} : \mathbf{B}/pE \to \mathbf{B}/\mathcal{P}_0E$ by $u \mapsto \mathcal{P}_0(\overline{u}(E))$.

(iii) In Ehrhard [1988] a *D-category* is defined as a fibration $p : \mathbf{E} \to \mathbf{B}$ provided with a terminal object functor $1 : \mathbf{B} \to \mathbf{E}$, which has a right adjoint $\mathcal{P}_0 : \mathbf{E} \to \mathbf{B}$. The ensuing functor $\mathcal{P} : \mathbf{E} \to \mathbf{B}^{\to}$ given by $E \mapsto p(\varepsilon_E)$ then forms a comprehension category (where $\varepsilon : 1\mathcal{P}_0 \overset{.}{\longrightarrow} Id$ is counit). Two things are worth noticing.

(a) This functor $\mathcal{P}$ preserves the terminal objects, i.e. for $A \in \mathbf{B}$, the map $\mathcal{P}1A$ is an isomorphism (i.e. terminal in $\mathbf{B}/A$). Since 1 is a full and faithful functor, the unit $\eta : Id \overset{.}{\longrightarrow} \mathcal{P}_0 1$ is an iso. But $\mathcal{P}1 \circ \eta = p\varepsilon 1 \circ \eta = p\varepsilon 1 \circ p1\eta = p(\varepsilon 1 \circ 1\eta) = id$, which makes $\mathcal{P}1$ an iso as well.

(b) For $E \in \mathbf{E}$ and $u : A \to pE$ in $\mathbf{B}$ one has

$$\mathbf{B}/pE\,(u,\ \mathcal{P}E) \;\cong\; \mathbf{E}_A\,(1A,\ u^*(E)),$$

which can be verified by playing a bit with the adjunction $1 \dashv \mathcal{P}_0$. We understand D-categories as forming a suitable concept of "comprehension categories with a unit". Therefore, we'll say that a comprehension category has (or admits) a unit if it is a D-category. This renaming gives more uniformity.

The essential point about a comprehension category $\mathcal{P} : \mathbf{E} \to \mathbf{B}^{\to}$ is that it determines a class $\{\mathcal{P}E \mid E \in \mathbf{E}\}$ of "display" maps in $\mathbf{B}$, which behave well in a certain sense. The abstract formulation of comprehension categories has technical and methodological advantages. This section concludes with two constructions on comprehension categories; the first one is from Jacobs [1991] and the second one from Moggi [1991].

3.9. DEFINITION (Change-of-base for comprehension categories along fibrations). Given a comprehension category $\mathcal{P} : \mathbf{E} \to \mathbf{B}^{\to}$ and a fibration $q : \mathbf{D} \to \mathbf{B}$, a new comprehension category $q^*(\mathcal{P})$ with base category $\mathbf{D}$ can be constructed as follows. First form the fibration $q^*(p)$ by change-of-base

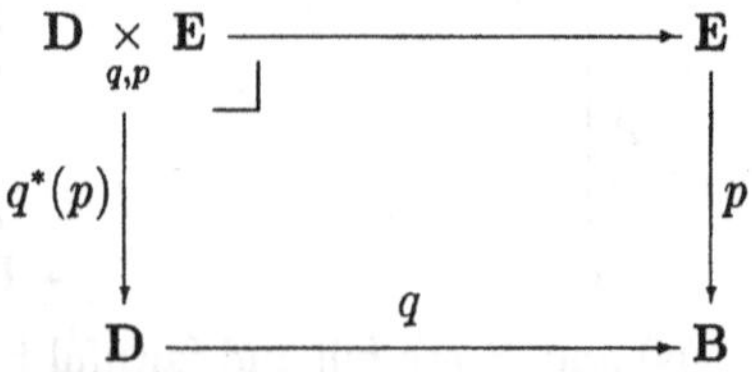

and then choose $q^*(\mathcal{P}) : \mathbf{D} \underset{q,p}{\times} \mathbf{E} \to \mathbf{D}^{\to}$ by $(D, E) \mapsto \overline{\mathcal{P}E}(D) : \mathcal{P}E^*(D) \to D$. On arrows $(f, g) : (D, E) \to (D', E')$ where $qf = pg$ one defines $q^*(\mathcal{P})(f, g) = (f, h)$, in which

$h:\mathcal{P}E^*(D) \to \mathcal{P}E'^*(D')$ is the unique arrow above $\mathcal{P}_0 g$ satisfying $\overline{\mathcal{P}E'}(D') \circ h = f \circ \overline{\mathcal{P}E}(D)$.

3.10. Definition (Juxtaposition of comprehension categories).
Starting from two comprehension categories $\mathbf{E} \xrightarrow{\mathcal{P}} \mathbf{B}^{\to} \xleftarrow{\mathcal{Q}} \mathbf{D}$ one constructs another comprehension category $\mathcal{Q} \cdot \mathcal{P}$ with base category $\mathbf{B}$, by first performing change-of-base

$$\begin{array}{ccc} \mathbf{D} \underset{\mathcal{Q},p}{\times} \mathbf{E} & \longrightarrow & \mathbf{E} \\ {\scriptstyle \mathcal{Q}_0^*(p)} \downarrow & \lrcorner & \downarrow {\scriptstyle p} \\ \mathbf{D} & \xrightarrow{\mathcal{Q}_0} & \mathbf{B} \end{array}$$

and then defining $\mathcal{Q} \cdot \mathcal{P} : \mathbf{D} \underset{\mathcal{Q}_0,p}{\times} \mathbf{E} \to \mathbf{B}^{\to}$ by $(D, E) \mapsto \mathcal{Q}D \circ \mathcal{P}E$ and $(f, g) \mapsto (qf, \mathcal{P}_0 g)$. One has $cod \circ \mathcal{Q} \cdot \mathcal{P} = q \circ \mathcal{Q}_0^*(p)$.

3.11. Lemma. (i) *$\mathcal{P}$ is full $\Rightarrow$ $q^*(\mathcal{P})$ is full;*
$\mathcal{P}$ has a unit $\Rightarrow$ $q^*(\mathcal{P})$ has a unit.

(ii) *$\mathcal{P}, \mathcal{Q}$ have units $\Rightarrow$ $\mathcal{Q} \cdot \mathcal{P}$ has a unit. Moreover, there is a full and faithful functor $\mathcal{I} : cod \circ \mathcal{P} \to cod \circ \mathcal{Q} \cdot \mathcal{P}$ preserving the terminal object and satisfying $\mathcal{Q} \cdot \mathcal{P} \circ \mathcal{I} \cong \mathcal{P}$.* □

4. Fibred Products and Sums.

Comprehension categories will be used in two different, but related ways: in this section to define appropriate fibred notions of product and sum; in the sixth section to provide categories for type dependency.

A comprehension category determines a class of "projection" morphisms. Quantification along such projections is described in the next definition by adjoints to the corresponding "weakening" functors. We first mention that a fibration $p : \mathbf{E} \to \mathbf{B}$ determines a category $Cart(\mathbf{E})$ with all objects from $\mathbf{E}$, but only the cartesian arrows. By restriction one obtains a fibration $|p| : Cart(\mathbf{E}) \to \mathbf{B}$. Similarly a comprehension category $\mathcal{P} : \mathbf{E} \to \mathbf{B}^{\to}$ determines two functors $|p|$, $|\mathcal{P}_0| : Cart(\mathbf{E}) \to \mathbf{B}$ and a natural transformation between them. Hence one obtains a cartesian comprehension category $|\mathcal{P}| : Cart(\mathbf{E}) \to \mathbf{B}^{\to}$.

4.1. Definition. Let $q : \mathbf{D} \to \mathbf{B}$ be a fibration and $\mathcal{P} : \mathbf{E} \to \mathbf{B}^{\to}$ a comprehension category. By change-of-base of q along the above functors $|p|$ and $|\mathcal{P}_0|$ one obtains two fibrations $|p|^*(q)$ and $|\mathcal{P}_0|^*(q)$. There is a cartesian functor $\langle\mathcal{P}\rangle : |p|^*(q) \to |\mathcal{P}_0|^*(q)$ described by $(E, D) \mapsto (E, \mathcal{P}E^*(D))$. We say that q *admits $\mathcal{P}$-products* (resp. *$\mathcal{P}$-sums*) if this functor $\langle\mathcal{P}\rangle$ has a fibred right (resp. left) adjoint.

4.2. Remarks. (i) The definition of the cartesian functor $\langle\mathcal{P}\rangle$ is based on proposition 3 in Ehrhard's [1988]. In fact this whole definition is inspired by his approach.

(ii) The above fibred definition has a more practical equivalent: let q and $\mathcal{P}$ be as above; then q admits $\mathcal{P}$-products (resp. $\mathcal{P}$-sums) iff both

- for every $E \in \mathbf{E}$, every weakening functor $\mathcal{P}E^* : \mathbf{D}_{pE} \to \mathbf{D}_{\mathcal{P}_0 E}$ has a right adjoint Π_E (resp. a left adjoint Σ_E).

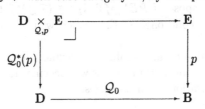

- the "Beck-Chevalley" condition holds, i.e. for every cartesian morphism $f : E \to E'$ in $\mathbf{E}$ one has that the canonical natural transformation

$$(pf)^*\Pi_{E'} \xrightarrow{\cdot} \Pi_E(\mathcal{P}_0 f)^* \qquad \text{(resp. } \Sigma_E(\mathcal{P}_0 f)^* \xrightarrow{\cdot} (pf)^*\Sigma_{E'}\text{)}$$

  is an isomorphism.

The first map is the transpose of $\mathcal{P}E^*\,(pf)^*\,\Pi_{E'} \cong (\mathcal{P}_0 f)^*\,\mathcal{P}E'^*\,\Pi_{E'} \xrightarrow{(\mathcal{P}_0 f)^*(\varepsilon)} (\mathcal{P}_0 f)^*$; similarly one obtains the second one. In the sequel, we'll use products and sums in this "fibrewise" form.

(iii) A more economical approach would be to define $\mathcal{P}$-quantification for full cartesian comprehension categories $\mathcal{P}$ only. However, as will become clear later, the extra generality we have in the above definition has advantages.

4.3. LEMMA. *Suppose q admits $\mathcal{P}$-sums as defined above. For every $E \in \mathbf{E}$ and $D \in \mathbf{D}$ with $qD = \mathcal{P}_0 E$, one has that the morphism $in_{E,D} = \overline{\mathcal{P}E}(\Sigma_E.D) \circ \eta_D : D \to \mathcal{P}E^*(\Sigma_E.D) \to \Sigma_E.D$ is cocartesian.* □

In type theory one finds so-called "weak" and "strong" sums, see section 6. The above definition covers the weak case. For the strong one q must be (part of) a comprehension category.

4.4. DEFINITION. Given two comprehension categories $\mathbf{E} \xrightarrow{\mathcal{P}} \mathbf{B}^{\to} \xleftarrow{\mathcal{Q}} \mathbf{D}$, we say that

(i) $\mathcal{Q}$ has $\mathcal{P}$-products/sums in case $q = cod \circ \mathcal{Q}$ has $\mathcal{P}$-products/sums.

(ii) $\mathcal{Q}$ has *strong* $\mathcal{P}$-sums in case $\mathcal{Q}$ has $\mathcal{P}$-sums in such a way that every morphism $\mathcal{Q}_0(in_{E,D})$ in $\mathbf{B}$ (cf. the previous lemma) is orthogonal to the class $\{\mathcal{Q}D' \mid D' \in \mathbf{D}\}$. The latter means that for every $D' \in \mathbf{D}$ and u, v forming a commuting square,

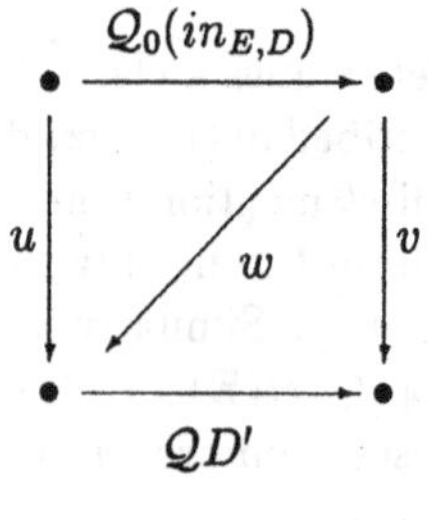

there is a unique w satisfying $\mathcal{Q}D' \circ w = v$ and $w \circ \mathcal{Q}_0(in_{E,D}) = u$.

We say that a comprehension category $\mathcal{Q}$ admits products/(strong) sums if it admits $\mathcal{Q}$-products/(strong) sums. One easily verifies that $\mathcal{Q}$ has strong sums iff the above morphism $\mathcal{Q}_0(in_{E,D})$ is an isomorphism. The latter formulation is used in Jacobs [1990] to define strong sums for comprehension categories.

The notion to be introduced next is of great importance — see e.g. the subsequent examples. It provides a suitable "unit" or "building-block" to describe more complicated categories later.

4.5. DEFINITION. A *closed* comprehension category (abbr. CCompC) is a full comprehension category with unit, products and strong sums; moreover, the base category is required to have a terminal object.

4.6. EXAMPLES. (i) Let $\mathbf{B}$ be a category with finite limits. The identity functor on $\mathbf{B}^{\to}$ then forms a full comprehension category with unit and strong sums. Moreover,

$$id_{\mathbf{B}^{\to}} \text{ is a CCompC} \quad \Leftrightarrow \quad \mathbf{B} \text{ is a LCCC.}$$

(ii) Let $\mathbf{B}$ be a category with finite products. The functor $\mathbf{B} \to \mathbf{1}$ (the terminal category) then forms a fibration with finite products. Hence the construction from 3.6 yields a fibration $\overline{\mathbf{B}} \to \mathbf{B}$. The functor $Cons_{\mathbf{B}} : \overline{\mathbf{B}} \to \mathbf{B}^{\to}$ given by $(A, A') \mapsto [\pi : A \times A' \to A]$ forms a full comprehension category with unit and strong sums. Moreover,

$$Cons_{\mathbf{B}} \text{ is a CCompC} \quad \Leftrightarrow \quad \mathbf{B} \text{ is a CCC.}$$

The rest of this section will be devoted to technical results about products and sums and closed comprehension categories.

4.7. LEMMA. *Let $\mathcal{P} : \mathbf{E} \to \mathbf{B}^{\to}$ be a comprehension category and $q : \mathbf{D} \to \mathbf{B}$ a fibration.*

(i) *A fibration r admits $\mathcal{P}$-products/sums $\Rightarrow$ $q^*(r)$ admits $q^*(\mathcal{P})$-products/ sums.*

(ii) *A comprehension category $\mathcal{R}$ admits strong $\mathcal{P}$-sums $\Rightarrow$ $q^*(\mathcal{R})$ admits strong $q^*(\mathcal{P})$-sums.*

Proof. Suppose the fibration r has $\mathbf{C}$ as total category, i.e. one has $r : \mathbf{C} \to \mathbf{B}$.

(i) Assume $\Sigma_E \dashv \mathcal{P}E^*$ in $\mathbf{C}$; we have to construct $\exists_{(D,E)} \dashv (q^*(\mathcal{P})(D,E))^*$. This is done by defining $\exists_{(D,E)} : (\mathbf{D} \underset{q,r}{\times} \mathbf{C})_{\mathcal{P}E^*(D)} \to (\mathbf{D} \underset{q,r}{\times} \mathbf{C})_D$ as $(\mathcal{P}E^*(D), C) \mapsto (D, \Sigma_E.C)$. Products are handled similarly.

(ii) Notice that one has $in = in_{(D,E),(\mathcal{P}E^*(D),C)} = (\overline{\mathcal{P}E}(D), in_{E,C}) : (\mathcal{P}E^*(D), C) \to \exists_{(D,E)}.(\mathcal{P}E^*(D), C)$ and that $q^*(\mathcal{P})_0(in)$ is by definition above $\mathcal{R}_0(in_{E,C})$. Orthogonilty can then be lifted. □

4.8. LEMMA. *Let $\mathbf{E} \xrightarrow{\mathcal{P}} \mathbf{B}^{\to} \xleftarrow{\mathcal{Q}} \mathbf{D}$ be comprehension categories.*

(i) *A fibration r admits both $\mathcal{P}$- and $\mathcal{Q}$-products/sums $\Rightarrow$ r admits $\mathcal{Q}{\cdot}\mathcal{P}$-products/sums.*

(ii) *A comprehension category $\mathcal{R}$ admits both strong $\mathcal{P}$- and strong $\mathcal{Q}$-sums $\Rightarrow$ $\mathcal{R}$ admits strong $\mathcal{Q} \cdot \mathcal{P}$-sums.*

Proof. (i) By composition of adjoints.

(ii) By successive application of orthogonality. □

4.9. LEMMA. *Consider two comprehension categories $\mathbf{E} \xrightarrow{\mathcal{P}} \mathbf{B}^{\to} \xleftarrow{\mathcal{Q}} \mathbf{D}$.*

(i) *$\mathcal{Q}$ has strong sums $\Rightarrow$ $\mathcal{Q} \cdot \mathcal{P}$ has strong $\mathcal{Q}$-sums.*

(ii) *In case $\mathcal{P}$ has a unit,*
$\mathcal{P}$ and $\mathcal{Q}$ both have $\mathcal{Q}$-products $\Rightarrow$ $\mathcal{Q} \cdot \mathcal{P}$ has $\mathcal{Q}$-products.

Proof. (i) Let $\Sigma_D \dashv \mathcal{Q}D^*$ in $\mathbf{D}$ be given; we have to construct $\exists_D \dashv \mathcal{Q}D^*$ in $\mathbf{D} \underset{\mathcal{Q}_0,p}{\times} \mathbf{E}$. This is done by taking for $(D_1, E_1) \in (\mathbf{D} \underset{\mathcal{Q}_0,p}{\times} \mathbf{E})$ above $\mathcal{Q}_0 D$,

$$\exists_D.(D_1, E_1) = (\Sigma_D.D_1,\ \mathcal{Q}_0(in_{D,D_1})^{-1*}(E_1)).$$

(ii) We may assume adjunctions $\mathcal{Q}D^* \dashv \Pi_D$ in $\mathbf{D}$ and $\mathcal{Q}D^* \dashv \forall_D$ in $\mathbf{E}$. One takes

$$\forall_D.(D_1, E_1) = (D',\ \forall_{\mathcal{Q}D'^*(D)}.\ \phi^*\ \mathcal{Q}_0(\varepsilon_{D_1})^*(E_1)\,),$$

where $D' = \Pi_D.D_1$ and $\varepsilon_{D_1} : \mathcal{Q}D^*(\Pi_D.D_1) \to D_1$ is unit and ϕ is an obvious mediating isomorphism in $\mathbf{B}$. In order to obtain the required adjunction, one has to use that $\mathcal{P}$ preserves products, see the proof of 4.11 (ii) below. □

4.10. Lemma. *Let $q:\mathbf{D} \to \mathbf{B}$ be a fibration and $\mathcal{P}:\mathbf{E} \to \mathbf{B}^{\to}$ be a comprehension category.*

(i) *If there is a fibred reflection $r \to q$ (i.e. a fibration $r:\mathbf{C} \to \mathbf{B}$ and a full and faithful cartesian functor $\mathbf{C} \to \mathbf{D}$ which has a fibred left adjoint), then*

$$q \text{ has } \mathcal{P}\text{-products/sums} \quad \Rightarrow \quad r \text{ has } \mathcal{P}\text{-products/sums.}$$

Moreover, the functor $\mathbf{C} \to \mathbf{D}$ preserves the products.

(ii) *In case $\mathcal{P}$ is a full comprehension category with unit and sums and q has a fibred terminal object, which is preserved by a full and faithful cartesian functor $G:\mathbf{D} \to \mathbf{E}$, then*

$$G \text{ has a fibred left adjoint} \quad \Leftrightarrow \quad q \text{ has } \mathcal{P}\text{-sums.}$$

Proof. (i) Standard.

(ii) ($\Rightarrow$) By (i).

($\Leftarrow$) Define $F:\mathbf{E} \to \mathbf{D}$ by $E \mapsto \Sigma_E(\top\mathcal{P}_0E)$, where $\top:\mathbf{B} \to \mathbf{D}$ describes the terminal object for q. By 4.3, F extends to a functor, which is cartesian by Beck-chevalley. □

4.11. Lemma. *Let $\mathcal{P}:\mathbf{E} \to \mathbf{B}^{\to}$ be a CCompC. Then*

(i) *$p = cod \circ \mathcal{P}:\mathbf{E} \to \mathbf{B}$ is a fibred CCC;*

(ii) *Considered as a functor, $\mathcal{P}$ preserves units, sums and products.*

Proof. (i) One takes $E \times E' = \Sigma_E.\mathcal{P}E^*(E')$ and $E \Rightarrow E' = \Pi_E.\mathcal{P}E^*(E')$. In fact, strongness of the sums is not needed to obtain this, see Jacobs [1990], 5.2.3.

(ii) Units are preserved by remark 3.8 (iii) (a) and sums by strongness: $\mathcal{P}(\Sigma_E.E') \cong \mathcal{P}E \circ \mathcal{P}E' = \Sigma_{\mathcal{P}E}.\mathcal{P}E'$ in $\mathbf{B}/pE$. As to products we obtain for $u:A \to pE$ in $\mathbf{B}$,

$$\begin{aligned}
\mathbf{B}/pE\,(u,\ \mathcal{P}(\Pi_E.E')\,) &\cong \mathbf{E}_A\,(1A,\ u^*(\Pi_E.E')\,) \\
&\cong \mathbf{E}_A\,(1A,\ \Pi_{u^*(E)}.(\mathcal{P}E^{\#}(u))^*(E')\,), \text{ by Beck-Chevalley} \\
&\cong \mathbf{E}_{\mathcal{P}_0u^*(E)}\,(\,(\mathcal{P}u^*(E))^*(1A),\ (\mathcal{P}E^{\#}(u))^*(E')\,) \\
&\cong \mathbf{E}_{\mathcal{P}_0u^*(E)}\,(\,1\mathcal{P}u^*(E),\ (\mathcal{P}E^{\#}(u))^*(E')\,) \\
&\cong \mathbf{B}/\mathcal{P}_0E\,(\mathcal{P}E^{\#}(u),\ \mathcal{P}E')
\end{aligned}$$

in which the pullback functor $\mathcal{P}E^{\#}$ comes from 3.8 (ii). The first and last step hold by 3.8 (iii) (b) □

4.12. Proposition. *$\mathcal{P}$ is a CCompC $\Rightarrow$ $q^*(\mathcal{P})$ is a CCompC.*

Proof. Let $\mathcal{P}:\mathbf{E} \to \mathbf{B}^{\to}$ be a full comprehension category with unit, products and strong sums. $q^*(\mathcal{P})$ is again full and has a unit by 3.11 (ii); it admits products and strong sums by 4.7. □

5. Category Theory over a Fibration

In the first section it was explained how a fibration forms a category fibred over a base category. Now we go one step up and consider categories over a fibration. This is not as bad as it may seem, since it turns out that one can reduce matters to the previous level. The following lemma lies at the basis of all this.

5.1. LEMMA. *Let* $p: \mathbf{E} \to \mathbf{B}$ *and* $r: \mathbf{B} \to \mathbf{A}$ *be fibrations.*

(i) *The functor* $rp: \mathbf{E} \to \mathbf{A}$ *is a fibration, with*

$$f \text{ is } rp\text{-cartesian} \quad \Leftrightarrow \quad f \text{ is } p\text{-cartesian and } pf \text{ is } r\text{-cartesian.}$$

(ii) *The functor p is cartesian from rp to r.*

(iii) *If* $q: \mathbf{D} \to \mathbf{B}$ *is another fibration, then*

$$F: p \to q \text{ in } Fib(\mathbf{B}) \quad \Rightarrow \quad F: rp \to rq \text{ in } Fib(\mathbf{A}). \; \square$$

5.2. A FIBRATION AS A BASIS. Suppose a cartesian functor p is given as in the following diagram.

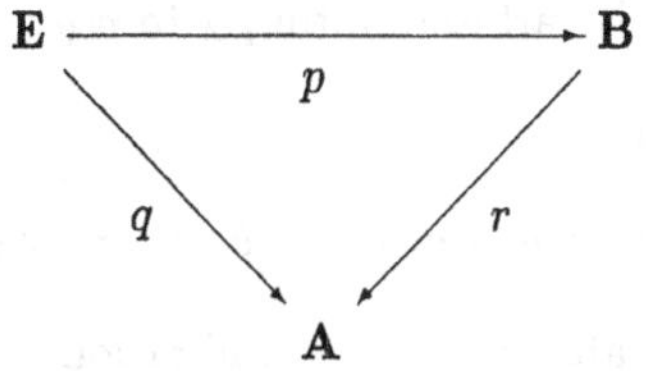

Every object $A \in \mathbf{A}$ determines a "fibrewise" functor $p|_A: \mathbf{E}_A \to \mathbf{B}_A$ by restriction. One calls p a *fibration over r* if all these fibrewise functors are fibrations and reindexing functors preserve the relevant cartesian structure (similarly to the definition of e.g. fibred cartesian products). More explicitly, p is a fibration over r if both

- for very $A \in \mathbf{A}$, $p|_A$ is a fibration;
- for every $u: A \to A'$ in $\mathbf{A}$ and every r-reindexing functor $u^*: \mathbf{B}_{A'} \to \mathbf{B}_A$, there is a q-reindexing functor $u^\#: \mathbf{E}_{A'} \to \mathbf{E}_A$ forming a morphism of fibrations:

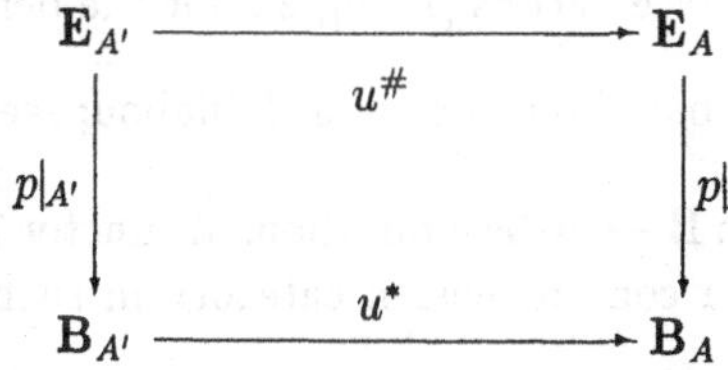

This rather complicated notion is equivalent to a more simple one; namely

$$p \text{ is a fibration over } r \;\Leftrightarrow\; p \text{ is a fibration itself.}$$

To verify the implication ($\Leftarrow$), notice that $p|_A$ can be obtained from p by change-of-base. This yields that f in $\mathbf{E}_A$ is $p|_A$-cartesian iff f is p-cartesian. The rest is not difficult. As to the implication ($\Rightarrow$), observe that if p is a fibration over r, then f in $\mathbf{E}$ is p-cartesian iff f can be written as $g \circ \alpha$ where g is q-cartesian and α is $p|_A$-cartesian (with $A = q(dom\ f)$).

Next, suppose we have a diagram,

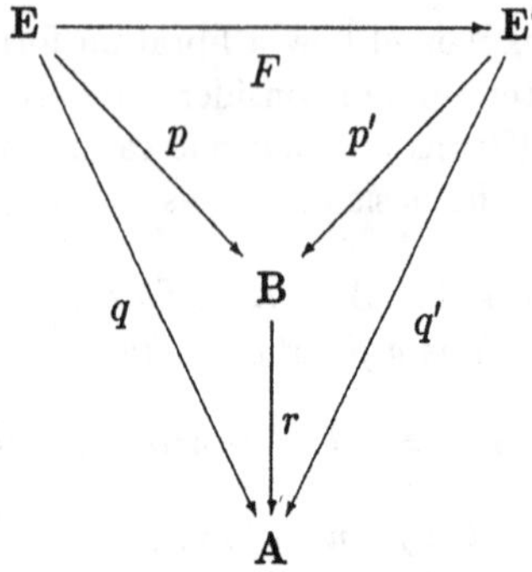

in which r, q, q', p and p' are fibrations with $q = rp$, $q' = rp'$ and F is a cartesian functor from q to q'. One calls F a *cartesian functor from p to p' over r* if both

- $p' \circ F = p$.
- for every $A \in \mathbf{A}$, $F|_A$ is cartesian form $p|_A$ to $p|_{A'}$;

As before, one can show that

$$F \text{ is cartesian } p \to p' \text{ over } r \;\Leftrightarrow\; F \text{ is cartesian } p \to p' \text{ in } Fib(\mathbf{B}).$$

In this way, one obtains a category $Fib(r)$ of fibrations and cartesian functors over r. As shown, one has $Fib(r) = Fib(\mathbf{B})$. It is left to the reader to formulate what natural transformations over r are and that the previous identification also concerns the 2-structure. Hence adjunctions over $r : \mathbf{B} \to \mathbf{A}$ are adjunctions over $\mathbf{B}$ (i.e. in the 2-category $Fib(\mathbf{B})$). In order to get an even better picture, the reader may want to verify that for $F : p \to p'$ in $Fib(\mathbf{B})$ as above and $G : p' \to p$ one has that $F \dashv G$ is an adjunction over r iff both

- for every $A \in \mathbf{A}$, there is fibred adjunction $F|_A \dashv G|_A$ in $Fib(\mathbf{B}_A)$;
- for every morphism $A \to A'$ in $\mathbf{A}$ there is a morphism $F|_{A'} \dashv G|_{A'} \longrightarrow F|_A \dashv G|_A$ of fibred adjunctions (see Jacobs [1990], 2.2 for the definition).

The above exposition is based on work of J. Bénabou; see also Pavlović [1990].

5.3. Definition. Let $r : \mathbf{B} \to \mathbf{A}$ be a fibration. A functor $\mathcal{P} : \mathbf{E} \to \mathbf{B}^{\to}$ is a *comprehension category over r* if $\mathcal{P}$ is a comprehension category in such a way that the functor $\mathcal{P}_0 = dom \circ \mathcal{P}$ is cartesian in

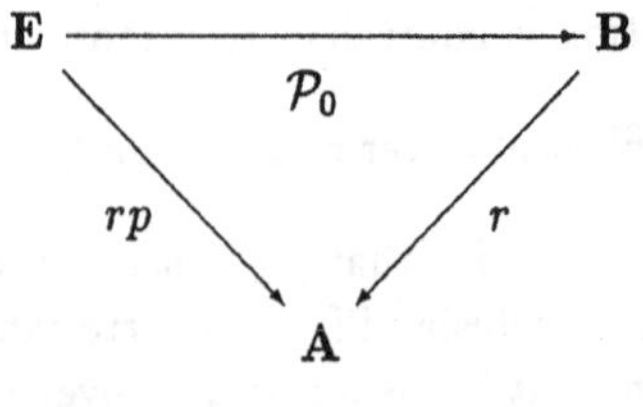

and $\mathcal{P}$ has r-vertical components.

5.4. REMARKS. (i) It takes a bit of effort, but one can show that given fibrations $\mathbf{E} \xrightarrow{p} \mathbf{B} \xrightarrow{r} \mathbf{A}$ and a 2-cell $\mathcal{P}: \mathcal{P}_0 \stackrel{\cdot}{\longrightarrow} p : rp \to r$ in $Fib(\mathbf{A})$, one has that $\mathcal{P}$ is a comprehension category over r iff for every $A \in \mathbf{A}$ one has a comprehension category $\mathcal{P}|_A : \mathbf{E}_A \to \mathbf{B}_A^{\to}$ and reindexing functors form suitable maps between these. Moreover, $\mathcal{P}$ is a full comprehension category iff all $\mathcal{P}|_A$'s are full, see Jacobs [1991] for more details.

(ii) We mention two examples.

(a) Let $\mathbf{B}$ be a category with pullbacks. The (obvious) functor $cod^{\to} : \mathbf{B}^{\to\to} \to \mathbf{B}^{\to}$ forms a fibration over $cod : \mathbf{B}^{\to} \to \mathbf{B}$. One obtains a full comprehension category $\mathbf{B}^{\to\to} \longrightarrow \mathbf{B}^{\to\to}$ over cod by $[\xrightarrow{v}\xrightarrow{u}] \mapsto [(id, v) : u \circ v \to u]$ in $\mathbf{B}^{\to}$.

(b) Let $p : \mathbf{E} \to \mathbf{B}$ be a fibration with finite products as in 3.6. One defines functor $\overline{\mathcal{P}} : \overline{\mathbf{E}} \to \mathbf{E}^{\to}$ by $(E, E') \mapsto \pi : E \times E' \to E$, which forms a full comprehension category over p. This generalizes $Cons_{\mathbf{B}} : \overline{\mathbf{B}} \to \mathbf{B}^{\to}$ from 4.6.

5.5. DEFINITION. Let $p : \mathbf{E} \to \mathbf{B}$ and $r : \mathbf{B} \to \mathbf{A}$ be fibrations; p forms a *comprehension category with unit over* r if there is

- a terminal object functor $1 : \mathbf{B} \to \mathbf{E}$ for p in $Fib(\mathbf{B})$;
- a fibred right adjoint $\mathcal{P}_0$ of $1 : r \to rp$ in $Fib(\mathbf{A})$.

(Ordinary comprehension categories with unit are described in remark 3.8 (iii).)

5.6. DEFINITION. A *closed comprehension category over a fibration* r is a full comprehension category with unit over r which admits products and strong sums (as a comprehension category in itself, not over r); moreover, r is required to have a fibred terminal object.

5.7. EXAMPLES. (i) If $\mathbf{B}$ is a LCCC, one obtains a CCompC over $cod : \mathbf{B}^{\to} \to \mathbf{B}$, see remark 5.4 (ii) (a). This comes from the fact that the slice categories $\mathbf{B}/A$ are LCCC's again using the same ingredients.

(ii) The other example mentioned in remark 5.4 (ii) is also of interest; it gives rise to a generalization of the equivalence obtained in 4.6 (ii). For a fibration $p : \mathbf{E} \to \mathbf{B}$ with finite products one has

$$\overline{\mathcal{P}} : \overline{\mathbf{E}} \to \mathbf{E}^{\to} \text{ is a CCompC over } p \quad \Leftrightarrow \quad p \text{ is a fibred CCC.}$$

In the next construction (from Jacobs [1991]), a generalization of $\overline{p}$ from 3.6 is obtained by using strong sums instead of cartesian products. In fact, the first example above is obtained in this way from $id_{\mathbf{B}^{\to}}$.

5.8. PROPOSITION. *Let $\mathcal{P} : \mathbf{E} \to \mathbf{B}^{\to}$ be a closed comprehension category. By change-of-base, we form the fibration $\tilde{p} : \tilde{\mathbf{E}} \to \mathbf{E}$.*

$$\begin{array}{ccc}
\tilde{\mathbf{E}} = \mathbf{E} \underset{\mathcal{P}_0, p}{\times} \mathbf{E} & \longrightarrow & \mathbf{E} \\
\Big\downarrow \tilde{p} & & \Big\downarrow p \\
\mathbf{E} & \xrightarrow{\mathcal{P}_0} & \mathbf{B}
\end{array}$$

Then

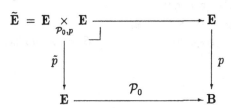

(i) $\tilde{p}: \tilde{\mathbf{E}} \to \mathbf{E}$ *forms part of a CCompC* $\tilde{\mathcal{P}}: \tilde{\mathbf{E}} \to \mathbf{E}^{\to}$ *over* p;

(ii) *there is a "pseudo" change-of-base situation (in which* 1 *is terminal object functor),*

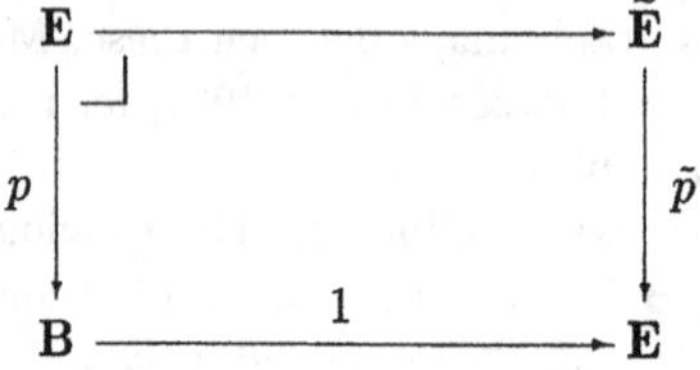

By "pseudo" we mean that the pullback of $\tilde{p}$ along 1 yields a fibration which is equivalent instead of isomorphic to p (in the fibred sense).

Proof. (i) One defines $\tilde{\mathcal{P}}: \tilde{\mathbf{E}} \to \mathbf{E}^{\to}$ by $(E, E') \mapsto$ [the projection $\Sigma_E.E' \to E$]; it is the unique map f with $\mathcal{P}_0 f = \mathcal{P}E' \circ \mathcal{P}_0(in_{E,E'})^{-1}$, using the morphism described in lemma 4.3 and the fact that $\mathcal{P}$ is full. The cocartesianess obtained in this lemma is needed to define $\tilde{\mathcal{P}}$ on morphisms. The rest is laborious but straightforward.

(ii) Easy. □

The constructions $\overline{p}$ and $\tilde{p}$ provide two ways to obtain closed comprehension categories over a fibration. The next two lemmas state that quantification for the base fibration p can be lifted to $\overline{p}$ and $\tilde{p}$. One gets strongness of the lifted sums for free. Proofs are omitted, but may be found in Jacobs [1991].

5.9. LEMMA. *Let* $p: \mathbf{E} \to \mathbf{B}$ *be a fibred CCC and* $\mathcal{Q}: \mathbf{D} \to \mathbf{B}^{\to}$ *a comprehension category. Then*

p *admits* $\mathcal{Q}$ *products and sums* $\Rightarrow$ $\overline{\mathcal{P}}$ *admits* $p^*(\mathcal{Q})$ *products and strong sums.* □

5.10. LEMMA. *Let* $\mathcal{P}: \mathbf{E} \to \mathbf{B}^{\to}$ *be a closed comprehension category and* $\mathcal{Q}: \mathbf{D} \to \mathbf{B}^{\to}$ *an arbitrary comprehension category. Then*

p *admits* $\mathcal{Q}$*-products and sums* $\Rightarrow$ $\tilde{\mathcal{P}}$ *admits* $p^*(\mathcal{Q})$*-products and strong sums.* □

In order to obtain the result of the next lemma, Moggi [1991] used the auxiliary notion of *independence*. This can be avoided with comprehension categories *over a fibration.*

5.11. LEMMA. *Let* $\mathcal{P}: \mathbf{E} \to \mathbf{B}^{\to}$ *be a comprehension category over* $r: \mathbf{B} \to \mathbf{A}$ *and let* $\mathcal{Q}: \mathbf{D} \to \mathbf{A}^{\to}$ *be an ordinary comprehension category. We form* $\mathcal{S} = r^*(\mathcal{Q}) \cdot \mathcal{P}$ *with base category* $\mathbf{B}$*, see 3.9 and 3.10.*

(i) $\mathcal{P}, \mathcal{Q}$ *are full* $\Rightarrow$ $\mathcal{S}$ *is full.*

(ii) $\mathcal{P}, \mathcal{Q}$ *have a unit* $\Rightarrow$ $\mathcal{S}$ *has a unit.*

(iii) $\mathcal{P}$ *has products (resp. strong sums)* $\Rightarrow$ $\mathcal{S}$ *has* $\mathcal{P}$*-products (resp. strong sums).* □

6. Models

In this section we describe categorical versions of the typed λ-calculi $F\omega$, the Calculus of Constructions(CC) and HML. The first system (due to J.-Y. Girard and J. Reynolds) and the second one (due to Th. Coquand and G. Huet) are considered to be well-known. The latter one comes from Moggi [1991] — see also Pavlović [1990] for a comparable system — and is essentially $F\omega$ extended with the possibility of types depending on types and propositions on propositions.

The next definition is based on Seely [1987].

6.1. Definition. A *PL-category* is a fibred CCC $p: \mathbf{E} \to \mathbf{B}$ with a CCC $\mathbf{B}$ as basis, admitting a generic object and $Cons_{\mathbf{B}}$-products and sums.

For the Calculus of Construction, we give a fibred version of a notion from Hyland & Pitts [1989]. Later, we briefly mention a weaker version.

6.2. Definition. A *CC-category* is described by the following data.

(i) A CCompC $\mathcal{Q}: \mathbf{D} \to \mathbf{B}^{\to}$.

(ii) A fibration $p: \mathbf{E} \to \mathbf{B}$ together with a fibred terminal object and a full and faithful cartesian functor $\mathcal{I}: \mathbf{E} \to \mathbf{D}$ which preserves this terminal. Further, we require that the comprehension category (with unit) $\mathcal{P} = \mathcal{Q}\mathcal{I}: \mathbf{E} \to \mathbf{B}^{\to}$ admits strong $\mathcal{Q}$-sums. In that case also $\mathcal{P}$ is a CCompC, see 4.10.

(iii) An object $\Omega \in \mathbf{D}$ such that $q\Omega \in \mathbf{B}$ is terminal and there is a generic object for p in $\mathbf{E}$ above $\mathcal{Q}_0\Omega \in \mathbf{B}$.

Summarizing all this in a picture, we have

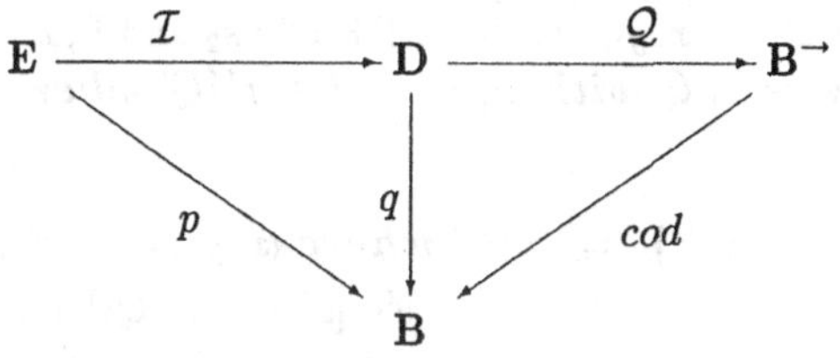

As a simple example of such a category, one can take $\mathbf{B}$ to be a topos, $\mathbf{D} = \mathbf{B}^{\to}$ and $\mathcal{Q} = id_{\mathbf{B}^{\to}}$. For $\mathbf{E}$, we take the full subcategory $Sub(\mathbf{B}) \hookrightarrow \mathbf{B}^{\to}$ of monic arrows.

In this figure, the objects of $\mathbf{E}$ are to be understood as types and the objects of $\mathbf{D}$ as kinds. For $s_1, s_2 \in$ {type, kind}, one has that all "(s_1, s_2)-sums" are strong. Let's explain weak and strong sums type-theoretically: there is no difference in the formation and introduction rules.

$$\frac{\Gamma \vdash A : s_1 \qquad \Gamma, x : A \vdash B : s_2}{\Gamma \vdash \Sigma x{:}\, A.B : s_2}\,(s_1, s_2) \qquad\qquad \frac{\Gamma \vdash M : A \qquad \Gamma \vdash N : [M/x]B}{\Gamma \vdash \langle M, N\rangle : \Sigma x{:}\, A.B.}$$

The *weak* elimination rule is given by

$$\frac{\Gamma \vdash P : \Sigma x{:}\, A.B \qquad \Gamma \vdash C : s_2 \qquad \Gamma, x : A, y : B \vdash Q : C}{\Gamma \vdash Q \;\boldsymbol{where}\; \langle x, y\rangle := P : C.}$$

In the *strong* elimination rule, the type C may contain an extra variable $w : \Sigma x{:}\, A.B$.

$$\frac{\Gamma \vdash P : \Sigma x{:}\, A.B \qquad \Gamma, w : \Sigma x{:}\, A.B \vdash C : s_2 \qquad \Gamma, x : A, y : B \vdash Q : [\langle x, y\rangle/w]C}{\Gamma \vdash Q \ \textbf{\textit{where}}\ \langle x, y\rangle := P : [P/w]C,}$$

both with conversions

$$\begin{array}{rcl} Q\ \textbf{\textit{where}}\ \langle x, y\rangle := \langle M, N\rangle & = & [M, N/x, y]Q \\ [\langle x, y\rangle/w]Q\ \textbf{\textit{where}}\ \langle x, y\rangle := P & = & [P/w]Q. \end{array}$$

6.3. Proposition. (i) *The elimination and conversion rules for strong (s, s)-sums can equivalently be described by the following rules with explicit projections.*

$$\frac{\Gamma \vdash P : \Sigma x{:}\, A.B}{\Gamma \vdash \pi P : A \qquad \Gamma \vdash \pi' P : [\pi P/x]B}$$

$$\frac{\Gamma \vdash M : A \qquad \Gamma \vdash N : [M/x]B}{\Gamma \vdash \pi\langle M, N\rangle = M : A \qquad \Gamma \vdash \pi'\langle M, N\rangle = N : [M/x]B}$$

$$\frac{\Gamma \vdash P : \Sigma x{:}\, A.B}{\Gamma \vdash \langle \pi P, \pi' P\rangle = P : \Sigma x{:}\, A.B}$$

(ii) *weak (s_1, s_2)-sums $+$ strong (s_2, s_2)-sums $\Rightarrow$ strong (s_1, s_2)-sums.*

Proof. (i) Standard.

(ii) Let's use "$\exists$" for the (s_1, s_2)-sums and "Σ" for the (strong) (s_2, s_2)-sums. Assume types $\Gamma \vdash A : s_1$ and $\Gamma, x : A \vdash B : s_2$ are given together with terms $\Gamma \vdash P : \exists x{:}\, A.B$ and $\Gamma, x : A, y : B \vdash Q : [\langle x, y\rangle/w]C$, where $\Gamma, w : \exists x{:}\, A.B \vdash C : s_2$. Write $C' \equiv \Sigma w{:}\, \exists x{:}\, A.B.\ C$ and $Q' \equiv \langle\langle x, y\rangle, Q\rangle$. Then $\Gamma \vdash C' : s_2$ and $\Gamma, x : A, y : B \vdash Q' : C'$. Hence one can take as new term $Q\ \textbf{\textit{with}}\ \langle x, y\rangle := P \equiv \pi'\{Q'\ \textbf{\textit{where}}\ \langle x, y\rangle := P\}$, which is of type $[P/w]C$, since

$$\begin{array}{rcl} \pi\{Q'\ \textbf{\textit{where}}\ \langle x, y\rangle := P\} & = & \pi\{Q'\ \textbf{\textit{where}}\ \langle x, y\rangle := \langle x', y'\rangle\}\ \textbf{\textit{where}}\ \langle x', y'\rangle := P \\ & = & \pi\{\langle\langle x', y'\rangle, [x', y'/x, y]Q\rangle\}\ \textbf{\textit{where}}\ \langle x', y'\rangle := P \\ & = & \langle x', y'\rangle\ \textbf{\textit{where}}\ \langle x', y'\rangle := P \\ & = & P.\ \square \end{array}$$

Based on this result (and on the fact that we use type-kind inclusion via the above functor $\mathcal{I}$), we conclude that a variant of *CC*-categories with weak $(s,$ type$)$-sums can only have at the same time weak (type, type) and weak (kind, type) sums. Hence one can define a *weak* *CC*-category similarly to definition 6.2, except that $\mathcal{P}$ should only have *weak* $\mathcal{Q}$-sums — instead of the strong ones used in 6.2 (ii).

The next definition introduces a new notion. The subsequent term model example may help to convey the underlying ideas. The fact that in the calculus *HML* one can separate kind- and type-contexts plays an important structural role.

6.4. Definition. A *HML-category* is given by the following data.

(i) A CCompC $\mathcal{Q}{:}\, \mathbf{D} \to \mathbf{A}^{\to}$.

(ii) A CCompC $\mathcal{P}: \mathbf{E} \to \mathbf{B}^{\to}$ over a fibration $r: \mathbf{B} \to \mathbf{A}$ (provided with terminal object functor $1: \mathbf{A} \to \mathbf{B}$); moreover, we require that $\mathcal{P}$ admits $r^*(\mathcal{Q})$-products and strong sums.

(iii) An object $\Omega \in \mathbf{D}$ such that $q\Omega \in \mathbf{A}$ is terminal; further, the fibration p' obtained by change-of-base as below should have a generic object above $\mathcal{Q}_0\Omega \in \mathbf{A}$.

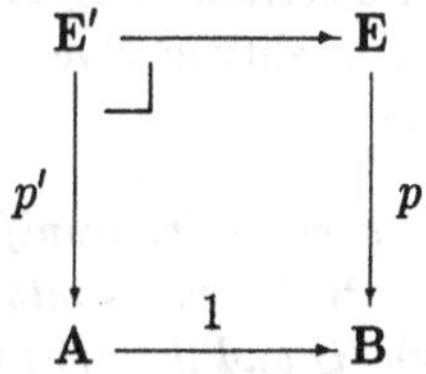

Summarising the constituents of a *HML*-category in a figure, we obtain

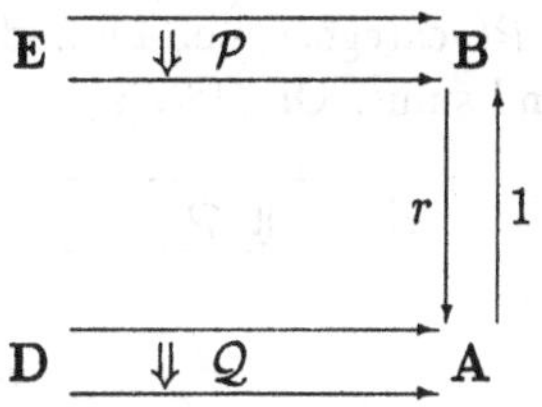

The upper comprehension category $\mathcal{P}$ concerns the run-time part of the system; the one below concerns the compile-time part.

6.5. TERM MODEL EXAMPLE. The generic object in the definition of a *HML*-category is not as in the original system in Moggi [1991]. To obtain it in the term model we must add $\Delta \vdash \tau : \Omega \Leftrightarrow \Delta; \emptyset \vdash \tau$. We define categories as in the above figure.

A obj. $[\Delta]$, where Δ is a "constructor context" $\langle v_1 : k_1, \ldots, v_m : k_m \rangle$; the brackets $[-]$ denote that we take the equivalence class (wrt. conversion).

mor. $([u_1], \ldots, [u_n]) : [\Delta] \to [\langle v_1 : k_1, \ldots, v_n : k_n \rangle]$ consist of equivalence classes of constructors $\Delta \vdash u_i : [u_1, \ldots, u_{i-1}/v_1, \ldots, v_{i-1}]k_i$.

D obj. $[\Delta \vdash k]$.

mor. $([\vec{u}], [u']) : [\Delta \vdash k] \to [\Delta' \vdash k'] \Leftrightarrow [\vec{u}] : [\Delta] \to [\Delta']$ in **A** and $\Delta, w : k \vdash u' : [\vec{u}/\vec{v}]k'$.

B obj. $[\Delta; \Gamma]$, where Γ is a "term context" $\langle x_1 : \tau_1, \ldots, x_n : \tau_n \rangle$ with $\Delta;\ \langle x_1 : \tau_1,\ \ldots,\ x_{i-1} : \tau_{i-1} \rangle \vdash \tau_i : \Omega$.

mor. $([\vec{u}]; [\vec{e}]) : [\Delta; \Gamma] \to [\Delta'; \langle x_1 : \tau_1, \ldots, x_m : \tau_m \rangle] \Leftrightarrow [\vec{u}] : [\Delta] \to [\Delta']$ in **A** and $\Delta; \Gamma \vdash e_j : [e_1, \ldots, e_{j-1}/x_1, \ldots, x_{j-1}][\vec{u}/\vec{v}]\tau_j$.

E obj. $[\Delta; \Gamma \vdash \sigma]$.

mor. $([\vec{u}]; [\vec{e}], [e']) : [\Delta; \Gamma \vdash \sigma] \to [\Delta'; \Gamma' \vdash \sigma'] \Leftrightarrow ([\vec{u}]; [\vec{e}]) : [\Delta; \Gamma] \to [\Delta'; \Gamma']$ in **B** and $\Delta; \Gamma, y : \sigma \vdash e' : [\vec{u}/\vec{v}, \vec{e}/\vec{x}]\sigma'$.

In this way the setting is built. The presence of the required additional structure is easily verified.

7. Relating models

The relations that we are about to establish between the categorical versions of the calculi $F\omega$, CC and HML described in the previous section, are of the following kind: given a model of calculus 1, we can perform certain categorical constructions and obtain a model of calculus 2. There will be no functoriality involved, since we did not describe appropriate morphisms between such categories.

7.1. THEOREM. (i) *Every PL-category can be transformed into a HML-category.*

(ii) *Every HML-category can be transformed into a PL-category.*

(iii) *The output of first applying (i) and then (ii) yields a result which is isomorphic to the input.*

Proof. (i) Let $p: \mathbf{E} \to \mathbf{B}$ be a *PL*-category, i.e. a fibred CCC on a CCC $\mathbf{B}$, with a generic object and $Cons_{\mathbf{B}}$-products and sums. One forms

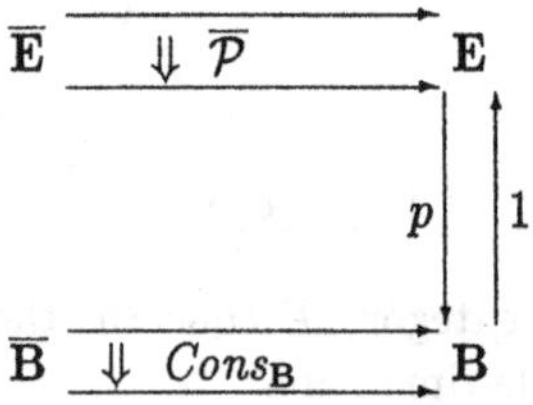

This structure forms an *HML*-category since

- $Cons_{\mathbf{B}}$ is a CCompC, see example 4.6 (ii).
- $\overline{\mathcal{P}}: \overline{\mathbf{E}} \to \mathbf{E}^{\to}$ is a CCompC over p, see example 5.7 (ii); moreover, it has $p^*(Cons_{\mathbf{B}})$-products and strong sums by lemma 5.9.
- The generic object for p also works here, by the change-of-base situation described in 3.6.

(ii) Suppose an *HML*-category as in the figure after definition 6.4 is given. We form the fibration p'' by change-of-base

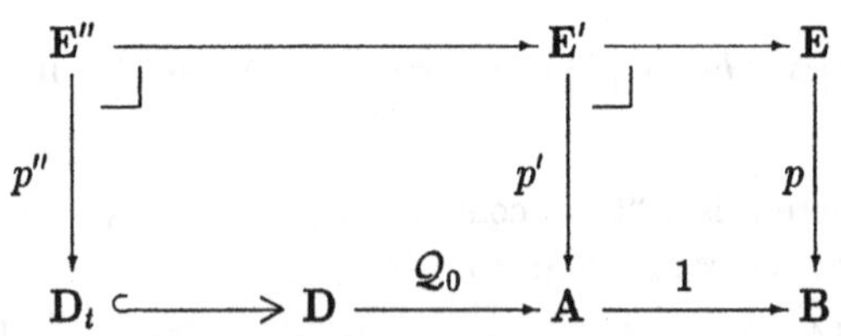

where $t \in \mathbf{A}$ is terminal object. Then

- $\mathbf{D}_t$ is CCC, since $q = cod \circ \mathcal{Q}$ is a fibred CCC, see 4.11 (i).
- p'' is a fibred CCC, since fibred CCC's are preserved by change-of-base.
- The generic object T for p' above $\mathcal{Q}_0\Omega \in \mathbf{A}$ where $\Omega \in \mathbf{D}_t$ yields a generic object for p'': for every $E \in \mathbf{E}$ and $D \in \mathbf{D}_t$ with $pE = 1\mathcal{Q}_0 D$, there is a morphism $u: \mathcal{Q}_0 D \to \mathcal{Q}_0\Omega$ in $\mathbf{A}$ with $u^*(T) \cong E$ in $\mathbf{E}'$. Since $\mathcal{Q}$ is a *full* comprehension category there is a (unique) $f: D \to \Omega$ in $\mathbf{D}_t$ with $\mathcal{Q}_0 f = u$. But then we are done.

- p'' admits $Cons_{\mathbf{D}_t}$-products and sums: the essential point to verify is that p' admits $\mathcal{Q}$-products and sums; then one easily obtains that p'' admits $\tilde{\mathcal{Q}}|_t$-products and sums — where $\tilde{\mathcal{Q}}|_t$ denotes the restriction of $\tilde{\mathcal{Q}}$ from definition 5.8 to the fibre above the terminal object t, see also remark 5.4 (i). As a special case we obtain that p'' admits $Cons_{\mathbf{D}_t}$-products and sums, since the projection $D \times D' \to D$ in $\mathbf{D}_t$ is $\tilde{\mathcal{Q}}|_t(D, \mathcal{Q}D^*(D')) : \Sigma_D.\mathcal{Q}D^*(D') \to D$.

(iii) By the change-of-base situation $p \to \overline{p}$ from 3.6 and the fact that $\overline{\mathbf{B}}_t \cong \mathbf{B}$. □

7.2. Theorem. (i) *Every PL-category can be transformed into a CC-category.*
(ii) *Every CC-category can be transformed into a PL-category.*

Proof. (i) Let $p : \mathbf{E} \to \mathbf{B}$ be a *PL*-category. We form

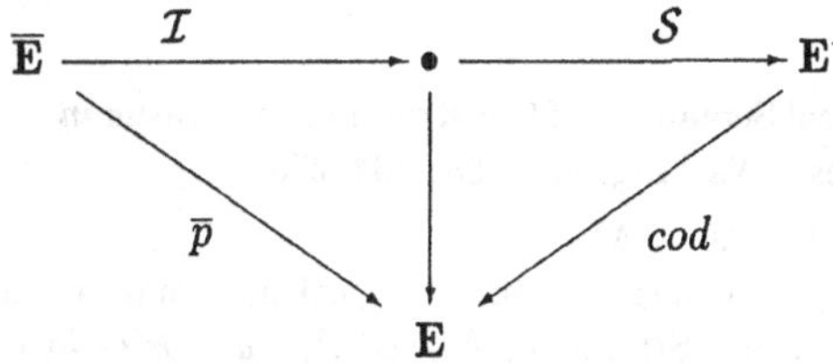

where $\mathcal{S} = p^*(Cons_{\mathbf{B}}) \cdot \overline{\mathcal{P}}$. It is a CCompC by lemmas 3.11, 4.8, 4.9 and 5.11. The first lemma 3.11 yields the functor $\mathcal{I} : \overline{p} \to cod \circ \mathcal{S}$; $\mathcal{SI} \cong \overline{\mathcal{P}}$ has strong $\mathcal{S}$-sums by 5.9, 5.7 (ii) and 4.8 (ii). Hence we are done.

(ii) Assume a *CC*-category as described in definition 6.2 is given. We obtain a *PL*-category $p'' : \mathbf{E}'' \to \mathbf{D}_t$, much in the same way as in (ii) of the previous proof:

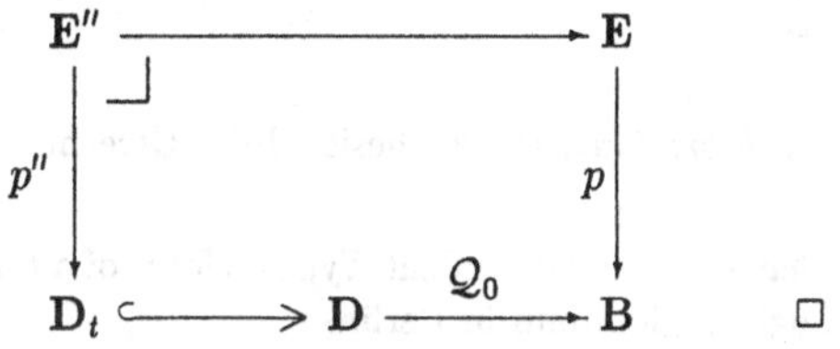

□

With these two theorems one can transform *HML*-categories into *CC*-categories via *PL*-categories and vice-versa. There is however a direct "canonical" way to go from *CC* to *HML*. Whether one can do similar things the other way round is not clear.

7.3. Theorem. (i) *Every CC-category can be transformed directly into a HML-category.*
(ii) *Doing CC $\longrightarrow$ HML $\longrightarrow$ PL and CC $\longrightarrow$ PL yields equivalent results.*

Proof. (i) Assume we have a *CC*-category as in definition 6.2. One forms

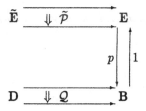

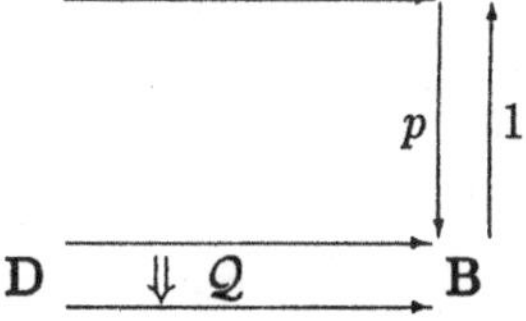

where $\mathcal{P} = \mathcal{QI}$ is a CCompC, see 6.2 (ii). Hence $\widetilde{\mathcal{P}}$ is a CCompC over p by 5.8, admitting $p^*(\mathcal{Q})$-products and strong sums by lemma 5.10. The generic object of the CC-category also works here, because of the "pseudo" change-of-base situation $\widetilde{p} \rightarrow p$ from 5.8.

(ii) Again by the "pseudo" change-of-base situation $\widetilde{p} \rightarrow p$. □

References

CARTMELL, J.

[1978] *Generalised Algebraic Theories and Contextual Categories*, Ph.D. thesis, Univ. Oxford.

COQUAND, T. AND HUET G.

[1988] The Calculus of Constructions, in: *Information and Computation* (1988), vol. 73, num. 2/3.

EHRHARD, TH.

[1988] A Categorical Semantics of Constructions, in: *Logic in Computer Science* (Computer Society Press, Washington, 1988) 264–273.

HYLAND, J.M.E. and PITTS, A.M.

[1989] The Theory of Constructions: categorical semantics and topos-theoretic Models, in: GRAY, J. and SCEDROV, A., (eds.), *Categories in Computer Science and Logic* (Contemp. Math. 92, AMS, Providence, 1989).

JACOBS, B.P.F.

[1990] Comprehension Categories and the Semantics of Type Dependency, manuscript.

[1991] *Categorical Type Theory*, Ph.D. thesis, Univ. Nijmegen.

MOGGI, E.

[1991] A Category Theoretic Account of Program Modules, *Math. Struct. in Comp. Sc.* (1991), vol. 1.

PAVLOVIĆ, D.

[1990] *Predicates and Fibrations*, Ph.D. thesis, Univ. Utrecht.

PITTS, A.M.

[1989] Categorical Semantics of Dependent Types, Notes of a talk given at SRI Menlo Park and at the Logic Colloquium in Berlin.

SEELY, R.A.G.

[1984] Locally cartesian closed Categories and Type Theory, *Math. Proc. Camb. Phil. Soc.* **95** 33 – 48.

[1987] Categorical Semantics for higher order Polymorphic Lambda Calculus, *J. Symb. Log.* **52** 969 – 989.

STREICHER, TH.

[1989] *Correctness and Completeness of a Categorical Semantics of the Calculus of Constructions*, Ph.D. thesis, Univ. Passau.

TAYLOR, P.

[1987] *Recursive Domains, indexed Categories and Polymorphism*, Ph.D. thesis, Univ. Cambridge.

Two results on set-theoretic polymorphism

Wesley Phoa
School of Mathematics and Statistics
University of Sydney
NSW 2006, Australia

Abstract

Moggi and Hyland showed how to model various polymorphic λ-calculi inside the effective topos and other realizability toposes; types are modelled by the so-called *modest sets*, which form an internal category **Mod** in the topos that is, in a certain sense, complete. Polymorphic types are modelled as products indexed by the object of modest sets. The same idea lets us model polymorphism in reflective subcategories of the category of modest sets—for example, the categories of *synthetic domains* studied by various authors.

This paper presents two alternative interpretations of polymorphic types. The first is the **groupoid interpretation**. Unlike the Moggi-Hyland interpretation, it is stable under equivalence; but it is also very easy to define, and makes sense for any small 'complete' category in a topos.

The second is the **uniformized interpretation** applicable to reflective subcategories of **Mod**. It clarifies the way in which they can be regarded as PER models, and has applications to the interpretation of subtyping and bounded quantification.

Introduction

Data types are sets; that's the obvious way to think of them. Unfortunately, this view is inconsistent with set theory: for cardinality reasons, recursive data types cannot always exist; and similarly Reynolds showed that polymorphic types can have no set-theoretic interpretation either: [22].

The traditional solution has been to abandon sets and use more complicated mathematical structures: Scott domains, metric spaces... These approaches have been very successful—within limits—but some intuition has been lost. What's more, there are still often technical obstacles preventing us from obtaining really satisfactory results—for example, in constructing models for polymorphism, or nondeterminism.

More recently an alternative approach has emerged, proposed (it seems) by Dana Scott. Namely: keep the view that data types are simply sets, but use, instead of classical set theory, *constructive* set theory. Then the proofs that certain phenomena cannot occur cease to be valid; though we still have to work to show that those phenomena *do* occur. But once we've done this, the resulting models should be easy to understand and use.

In fact, it turns out that an appropriate version of constructive set theory to use is *topos theory* (see [13]). A topos is a category which can be regarded as a model of constructive set theory; associated to each topos is a formal language—its 'internal language'—that resembles the language of set theory. This allows us to reason about it as if its objects were sets, with elements, in the usual way—so long as we reason constructively. (That is, we simply pretend that the topos is 'the' category of sets.) But the fact that we are working with a topos also makes available many useful category-theoretic tools, and this can be very convenient.

It was Moggi and Hyland who first showed that set-theoretic models of polymorphism do in fact exist in certain toposes. They found one inside $\mathcal{E}\!f\!f$, the effective topos.

In this paper—which is partly expository—we will study two refinements of the Moggi-Hyland interpretation of polymorphism. One, the *groupoid interpretation*, deals with the problem of stability under equivalence; it gives us a model that we can use when we are working with equivalent representations of the 'same' category. The other, the *uniformized interpretation*, is more special: it can be thought of as a way of generating submodels of PER models.

Contents of the paper Section 1 contains a very brief review of how small complete categories can be used to model polymorphism; we assume some familiarity with toposes and with polymorphic type theories. The problem of invariance under equivalence is introduced.

In section 2 we propose a very simple-minded solution to the problem of invariance under equivalence: for any small complete category $\mathbb{C}$ we construct the **groupoid interpretation** $(\Pi\alpha.T\alpha)_{\mathbb{C}}$ of a polymorphic type $\Pi\alpha.T\alpha$ relative to $\mathbb{C}$, and show that it is indeed stable under any equivalence $\mathbb{D} \to \mathbb{C}$ (and indeed under a weaker condition, which is useful).

Section 3 introduces the effective topos $\mathcal{E}\!f\!f$ and modest sets; the category **Mod** of modest sets in $\mathcal{E}\!f\!f$ was the first example of a small complete category in a topos. We present a result of Rosolini (discovered since this paper was first submitted) on the parametricity of the groupoid interpretation in this case.

In section 4 we show that, given any complete full subcategory $\mathbb{C}$ of **Mod**, there is an equivalent category $\bar{\mathbb{C}}$ whose object-of-objects $\bar{\mathbb{C}}_0$ is a constant set. This allows us to define a **uniformized interpretation** of a polymorphic type $\Pi\alpha.T\alpha$ in $\mathbb{C}$, which is an indexed product that can in fact be calculated as an intersection of PERs. It depends, however, on the choice of a particular reflection functor. The result has a number of applications.

In fact, all the 'new' results presented in this paper are essentially extremely simple.

But if one insists on talking about the external PER model, instead of the category of modest sets in $\mathcal{E}ff$, then setting them up is a much more unwieldy process.

Finally, the appendix contains a discussion of the senses in which categories in $\mathcal{E}ff$ are complete, or reflective, and so on; it is mainly intended to point out that the informal-looking reasoning in the rest of the paper is indeed valid.

Note 1. In this paper we only produce models of ML-style 'shallow' polymorphism. The uniformized interpretation can actually be extended to the full theory of constructions; but the groupoid interpretation can, at the moment, only be extended to systems like System F (the second order typed λ-calculus) and System $F\omega$.

Note 2. We will work throughout in the internal language of a topos, but informally; that is, we will pretend that the topos is the category of sets, while ensuring that all our reasoning is constructive. Strictly speaking, we should write everything out in the formal internal language of the topos; but this would be very tedious. The arguments can actually be cast in a purely category-theoretic form, but that would make them hard to follow and take up too much room.

Note 3. Nothing about this paper is special to $\mathcal{E}ff$; everything applies equally well to other realizability toposes, such as those arising from various models of the λ-calculus.

Acknowledgements This paper owes a great deal to the insights of Edmund Robinson; in fact, I believe that the result of section 4 was observed first by him. I also learnt a great deal from Peter Freyd, who might not approve of this paper; from my former supervisor Martin Hyland; and especially from conversations and correspondence with Pino Rosolini. He contributed the parametricity theorem sketched in section 3.

Fu Yuxi, Edmund Robinson, Pino Rosolini and an anonymous referee all helped me improve this paper. Both the original version and the final draft were prepared with the assistance of the Australian Research Council.

1 Small complete categories and polymorphism

Let $\mathcal{E}$ be a topos: we think of it as a model of set theory, and regard its objects as sets. When we use $\mathcal{E}$ in giving the semantics of a language, we interpret the types in the language using objects of $\mathcal{E}$: 'sets'.

This set-theoretic approach will then carry over to the interpretation of type-forming operations. For example, if X and Y are types in the language and $X, Y \in \mathcal{E}$ are the objects we use to interpret them, then the interpretation of the arrow type $\mathsf{X} \to \mathsf{Y}$ is just the exponential Y^X; that is, the ordinary set-theoretic function space in $\mathcal{E}$.

Usually $\mathcal{E}$ will have more objects than necessary for interpreting a language; we generally pick out some full subcategory $\mathcal{C}$ of $\mathcal{E}$ and regard only the objects of $\mathcal{C}$ as data types. For instance, $\mathcal{C}$ could be some category of 'predomains'—see [16]. Actually, we need a *subfibration* rather than just a subcategory; we gloss over this point here, but cf. the appendix.

It is often the case that there's an *internal*—i.e. 'small'—category $\mathbb{C}$ in $\mathcal{E}$ corresponding to $\mathcal{C}$: $\mathcal{C}$ is the 'externalization' of $\mathbb{C}$ in a sense which won't be made precise here (but see [10], [11]). In particular, the collection of all types (all objects of $\mathcal{C}$) can itself be regarded as an object of $\mathcal{E}$, namely the object-of-objects $\mathbb{C}_0$ of $\mathbb{C}$. (Note that $\mathbb{C}_0$ is rarely an object of $\mathcal{C}$, though.)

It was Moggi who saw that if $\mathbb{C}$ is *complete* then we can use it to interpret polymorphic types. Note that in *Sets* there are no nontrivial small complete categories (this was proved long ago by Freyd; it's related to Reynolds' result in [22] that there are no set-theoretic models of polymorphism). To find some, we need to change our model of set theory; that is, to look inside a different topos.

Actually, it's surprisingly hard to give a good definition of completeness for an internal category in a topos (see [21]). There is no room to treat this issue in detail here, though a brief discussion appears in the appendix. For the moment, we take 'completeness' to mean that for any internal category $\mathbb{I}$ in $\mathcal{E}$, there is a functor $\varprojlim: \mathbb{C}^{\mathbb{I}} \to \mathbb{C}$ taking an $\mathbb{I}$-diagram in $\mathbb{C}$ to its limit (here $\mathbb{C}^{\mathbb{I}}$ is the interal functor category).

The Moggi-Hyland model The interpretation of polymorphic types in a small complete category $\mathbb{C}$ suggested by Moggi goes as follows: If T is some type-constructor, modelled as a function $T: \mathbb{C}_0 \to \mathbb{C}_0$, then the polymorphic type

$$\Pi\alpha.\mathrm{T}\alpha$$

can be interpreted as the $\mathbb{C}_0$-indexed product

$$\prod_{X \in \mathbb{C}_0} TX.$$

Call this the **Moggi-Hyland interpretation** of $\Pi\alpha.\mathrm{T}\alpha$ relative to **Mod**; it is of course equipped with a projection map into TX, where X is any modest set.

Let's pause to see why this is actually an interpretation. The terms of a type X (with, say, a single free variable y : Y) are interpreted as arrows $Y \to X$ in $\mathbb{C}$. Now, if we have a uniform way of constructing a term of type TX, for any X, then in the model this corresponds to an (internally defined) family of maps $Y \to TX$, one for each X; but then this gives rise to a uniquely determined may $Y \to \prod_{X \in \mathbb{C}_0} TX$. Conversely, given a term of type $\Pi\alpha.\mathrm{T}\alpha$ and a particular type X, we want to instantiate the term to obtain a term of type TX; but if we have a map $Y \to \prod_{X \in \mathbb{C}_0} TX$, we can simply compose it with the projection $\prod_{X \in \mathbb{C}_0} TX \to TX$ to obtain a map $Y \to TX$. Of course, some verifications are still required.

A problem with the interpretation Note that the Moggi-Hyland interpretation does not take into account the fact that T is generally 'natural'; simply taking the product throws away all information about naturality. An odd consequence, raised by Freyd, Phil Scott and others and also by Robinson, is that if $\mathbb{C}$ and $\mathbb{D}$ are equivalent small complete categories then the Moggi-Hyland interpretations of $\Pi\alpha.\mathsf{T}\alpha$ relative to $\mathbb{C}$ and $\mathbb{D}$ may be quite different—simply because their objects-of-objects $\mathbb{C}_0$ and $\mathbb{D}_0$ may be completely different objects. That is,

- *The Moggi-Hyland interpretation is not stable under equivalence*

The reason is exactly that naturality data is discarded. But in fact it's not straightforward to fix this. The problem is that T often does not give rise to a functor; for example we allow $\mathsf{T}\alpha = \alpha \to \alpha$. This is related to the notorious fact that it is very difficult to express the sense in which $X, Y \mapsto Y^X$ is natural.

Several ways of dealing with this problem have been proposed: approaches involving dinaturality (see [2], [5]) or Freyd's theory of structors (see [6]). An odd feature of some of these results is that they are special to PER models. A more satisfying approach may emerge from unpublished work of Carboni, Kelly and Wadler.

The approach proposed in the next section is considerably simpler. It skates over some of the issues tackled head-on by the papers mentioned previously; but it does provide a very simple solution to the problem of invariance under equivalence.

2 The groupoid interpretation

We mentioned that type constructors $T: \mathbb{C}_0 \to \mathbb{C}_0$ were often not functors. Typically, though, we have a bifunctor

$$B: \mathbb{C}^{op} \times \mathbb{C} \to \mathbb{C}$$

(for example, $X, Y \mapsto (Y^X)^{(X^Y)}$) and on objects, T acts as follows:

$$TX = BXX.$$

This, and what follows, should look quite familiar; cf. [20].

Let $\mathbb{C}^{\cong}$ be the non-full subcategory of $\mathbb{C}$ with the same objects but only the isomorphisms as arrows. We have a diagonal functor

$$\Delta: \mathbb{C}^{\cong} \to \mathbb{C}^{op} \times \mathbb{C}$$

and hence a functor

$$T = B\Delta: \mathbb{C}^{\cong} \to \mathbb{C}$$

taking $X \mapsto BXX$.

Define the object

$$(\Pi\alpha.\mathsf{T}\alpha)_{\mathbb{C}}$$

to be the limit of T. We will call this the **groupoid interpretation** of $\Pi\alpha.\mathsf{T}\alpha$ in $\mathbb{C}$. Informally, one thinks of it as a product over all the isomorphism classes.

Why does this give us a sound interpretation? The main point is that if we have a family of maps $Y \to TX$ arising as the interpretation of a term of the polymorphic λ-calculus of type $\mathsf{T}\alpha$ (with free type variable α), then because these maps arose in a 'uniform' way, they must commute with all the isomorphisms $TX \cong TX'$ arising from isomorphisms $X \cong X'$; hence they form a cone over the functor $T: \mathbb{C}^{\cong} \to \mathbb{C}$ and we get a map $Y \to (\Pi\alpha.\mathsf{T}\alpha)_{\mathbb{C}}$. A full proof would involve a lengthy induction, but this fact is essentially obvious.

Stability under equivalence Now suppose that $F: \mathbb{D} \to \mathbb{C}$ is an equivalence. Then we have

$$GB(F^{op} \times F): \mathbb{D}^{op} \times \mathbb{D} \to \mathbb{D}$$

where G is the adjoint of F. Hence we get a functor

$$T': \mathbb{D}^{\cong} \to \mathbb{D}$$

and we can define, as before,

$$(\Pi\alpha.\mathsf{T}\alpha)_{\mathbb{D}} = \varprojlim T'.$$

Saying that the interpretation of $\Pi\alpha.\mathsf{T}\alpha$ is invariant under the equivalence F amounts to showing that

$$F(\Pi\alpha.\mathsf{T}\alpha)_{\mathbb{D}} \cong (\Pi\alpha.\mathsf{T}\alpha)_{\mathbb{C}}.$$

Now, F restricts to an equivalence $F: \mathbb{D}^{\cong} \to \mathbb{C}^{\cong}$, so we have a functor $TF: \mathbb{D}^{\cong} \to \mathbb{C}$; and it suffices just to prove that

$$\varprojlim TF \cong \varprojlim T.$$

This follows from

- *Let $\mathbb{C}$ be a (small) complete category, $\mathbb{I}$ and $\mathbb{J}$ small categories and $F: \mathbb{I} \to \mathbb{J}$ an equivalence with adjoint $G: \mathbb{J} \to \mathbb{I}$. Then for any diagram $D: \mathbb{J} \to \mathbb{C}$, the canonical map*

  $$\varprojlim DF \to \varprojlim D$$

  is an isomorphism.

Proof. It suffices to prove that F is *initial* (see [14, p.213]); that is,

$$\forall J \in \mathbb{J}_0. \exists I \in \mathbb{I}_0. \quad \text{there is a map } FI \to J$$

and

$$\forall J \in \mathbb{J}_0. \forall I, I' \in \mathbb{I}_0. \forall f: FI \to J, f': FI' \to J.$$

there exists a path $I \to\leftarrow \cdots \to\leftarrow I'$ in $\mathbb{I}$ such that we can apply F and fill in arrows to make a commutative diagram

$$\begin{array}{ccccccccc} FI & \longrightarrow & \bullet & \longleftarrow & \cdots & \longrightarrow & \bullet & \longleftarrow & FI' \\ \downarrow & & \downarrow & & & & \downarrow & & \downarrow \\ J & = & J & = & \cdots & = & J & = & J \end{array}$$

But this is immediate: given $J \in \mathbb{J}_0$ we have the counit $FGJ \to J$; and given any $I \in \mathbb{I}_0$ with $FI \to J$ we have a unique factorization

$$\begin{array}{ccc} FI & \longrightarrow & FGJ \\ & \searrow & \downarrow \\ & & J \end{array}$$

■

Alternative interpretations We could perhaps obtain different interpretations by using the same idea, but with categories other than $\mathbb{C}^{\cong}$: for example, the category of retractions in $\mathbb{C}$, or even the category $\mathbb{C}^{\leftrightarrows}$ whose objects are the same as the objects of $\mathbb{C}$ but where a morphism from X to Y consists of a pair of morphisms $X \to Y$, $Y \to X$ in $\mathbb{C}$.

Modelling stronger theories We have just shown how to model ML-style polymorphism. In (for example) System F, polymorphic types that look like

$$\Pi\alpha.\alpha \to (\Pi\beta.\beta \to \alpha)$$

can occur: see [8]. These can be handled inductively. The point is that although

$$X \mapsto (\Pi\beta.\beta \to X)_{\mathsf{C}}$$

no longer comes from a functor $\mathbb{C}^{op} \times \mathbb{C} \to \mathbb{C}$, it does define a functor $\mathbb{C}^{\cong} \to \mathbb{C}$; so we can do the same thing as before.

System $F\omega$ also presents no problems: here we are allowed to quantify, not just over the collection C of all types, but also over $\mathsf{C} \to \mathsf{C}$ and so on. But this just corresponds to taking limits in $\mathbb{C}$ indexed by the functor groupoid

$$(\mathbb{C}^{\cong})^{(\mathbb{C}^{\cong})}$$

and it's certainly the case that if $\mathbb{C}$ and $\mathbb{D}$ are equivalent then so are the corresponding endofunctor categories. So the groupoid interpretation of polymorphic types in $F\omega$ is also stable under equivalence.

The calculus of constructions and its variants present a bigger problem. How can we regard a general kind as a groupoid? I have no answer to this question at the moment.

3 Modest sets and parametricity

From now on we will work in a particular topos: Hyland's effective topos $\mathcal{E}\!f\!f$. However, what follows also applies to other realizability toposes.

$\mathcal{E}\!f\!f$ can be thought of as the 'computable universe', or the world of recursive mathematics. In it, a statement ϕ is valid precisely when there is a recursive function producing evidence for ϕ. Such a universe seems to be a very natural setting for a semantics, and a logic, of computations (as suggested by Scott, Rosolini and many others...); in fact, proofs of mathematical statements actually give rise to recursive functions—programs—via the realizability interpretation.

A warning, though—the objects and arrows of $\mathcal{E}\!f\!f$ cannot be thought of as 'recursive sets' and 'recursive maps'; this is an oversimplification. For example, $\mathcal{E}\!f\!f$ contains the ordinary category *Sets* of sets as a full reflective subcategory; it also contains the realizability toposes associated to the various Turing degrees: [15].

But it is often possible to regard an object X of $\mathcal{E}\!f\!f$ as a set equipped with a certain amount of intensional information. For example, the exponential $\mathbb{N}^{\mathbb{N}}$ is the set of total recursive functions together with, for each recursive function, the collection of Gödel numbers for it.

Taking this point of view, a map $X \to Y$ in $\mathcal{E}\!f\!f$ must necessarily come equipped with a recursive procedure that produces this information. (I should emphasize, though, that from the point of view of the internal logic, all objects look like unadorned sets; it is only in the metatheory that they are seen to be more than this.) A proper explanation of this requires a precise definition of $\mathcal{E}\!f\!f$: see [9].

The objects in the image of the embedding $\mathit{Sets} \hookrightarrow \mathcal{E}\!f\!f$ can then be thought of as given by sets with no extra intensional information; we will refer to them as **constant sets.** In fact, up to equivalence, *Sets* is actually the category of 'double-negation sheaves' in $\mathcal{E}\!f\!f$ associated to the Gödel double-negation translation of classical logic into intuitionistic logic. The double negation operator is in fact a very important tool in understanding $\mathcal{E}\!f\!f$: for example, the category *Sep* of $\neg\neg$-separated objects is equivalent to Moggi's category of ω-sets (giving an immediate proof that that category is locally cartesian closed); and a $\neg\neg$-closed subobject of an object X can generally be regarded as a 'pure subset' of X, not requiring any extra intensional information to specify. (For example, the $\neg\neg$-closed subobjects of the constant sets just correspond to their ordinary subsets in *Sets*.)

[For a proper account of these things, see [9]. A more accessible account of some of these matters appears in [25], and a summary in [18].]

In contrast to the constant sets, $\mathcal{E}\!f\!f$ contains objects which are wholly determined by the intensional information supplied with them; these are the *modest sets.* They were first studied by Hyland, but it was Dana Scott who realized their importance in computer science.

Modest sets [10],[11] $\mathcal{E}\!f\!f$ contains a certain small (i.e. internal) category **Mod**; the object-of-objects of **Mod** is

$$\mathbf{Mod}_0 = \{R \subseteq \mathbb{N} \times \mathbb{N} \colon R \text{ is a PER and } \forall n, m.\, \neg\neg(nRm) \to (nRm)\}$$

The condition that a partial equivalence relation R be $\neg\neg$-closed just ensures that R comes from a subset of $\mathbb{N} \times \mathbb{N}$ in the usual (external) sense; that is, it is a PER on $\mathbb{N}$ in the usual sense. (This is not quite precise, since $\mathbb{N}$ is not itself a constant set; but this is the intuition, anyway.) In fact, $\mathbf{Mod}_0$ is a constant set and can be identified with the usual set—in the ordinary sense—of PERs on $\mathbb{N}$ in *Sets*.

Mod actually corresponds to a full subcategory $\mathcal{M}od$ of $\mathcal{E}\!f\!f$; that is, there is a category $\mathcal{M}od$ that is the externalization of **Mod**. The objects of $\mathcal{M}od$ are exactly the quotients of $\mathbb{N}$ by $\neg\neg$-closed partial equivalence relations; they can be thought of as (nice) partitions of (nice) subsets of $\mathbb{N}$: see [25]. Let me emphasize again that, in a precise treatment, all this must be stated in terms of fibrations; and then certain complications arise: [11]. But it is safe to ignore those matters for the moment.

So **Mod** is actually the counterpart in $\mathcal{E}\!f\!f$ of the usual category of PERs on $\mathbb{N}$. But the difference is that, via the internal logic of $\mathcal{E}\!f\!f$, its objects can be regarded as *sets*—and indeed are called **modest sets**—because it is a *full* subcategory of $\mathcal{E}\!f\!f$. A map between PERs has to be defined as something that is tracked by a recursive function; but when we speak of a map between modest sets, we mean any set-theoretic function (and it is a theorem, not an assumption, that any such map is tracked by a recursive function; this is not part of the structure).

Modest sets and polymorphism The remarkable fact about **Mod** is that it is *complete.* This was noticed by Moggi and proved by Hyland: see [10] and also [11]. They used this fact to produce the first set-theoretic models of polymorphism along the lines described at the beginning of section 1; however, one could equally well use the groupoid interpretation to obtain polymorphic types in **Mod**. And, as Rosolini noticed, it is in some respects nicer.

Under some circumstances there is a more concrete description of the Moggi-Hyland polymorphic types. Recall that a modest set X can be identified with a ($\neg\neg$-closed) partial equivalence relation R on $\mathbb{N}$. If $K \subset \mathbf{Mod}_0$ is a set of PERs which is actually a constant set (this amounts to saying that K is a $\neg\neg$-closed subobject of $\mathbf{Mod}_0$), then in fact

$$\prod_{X \in K} TX \cong \bigcap_{R \in K} TR.$$

That is, "products are calculated as intersections". In particular, this applies when $K = \mathbf{Mod}_0$; so the Moggi-Hyland interpretation of $\Pi\alpha.\mathrm{T}\alpha$ here agrees with the interpretation in the usual, external, PER model. It also applies in other interesting cases, for example the category of predomains described in [7] (and this turns out to be of some practical interest: [24]). In general we can identify such a $\mathbb{C}$ with a submodel of the PER model.

I should emphasize, though, that the Moggi-Hyland interpretation relative to a small complete category $\mathbb{C}$ in $\mathcal{E}\!f\!f$ still makes sense when $\mathbb{C}_0$ is not a constant set—as in, for example,[16]; the only difference is that we don't know that it can be calculated as an intersection (but see section 4). It is still a sound interpretation of the rules of the polymorphic λ-calculus.

Polymorphic parametricity Under more restricted circumstances one can use the description of products as intersections to prove 'polymorphic parametricity' results for the Moggi-Hyland interpretation. Such theorems are very useful when reasoning about programs whose type is polymorphic.

Roughly speaking, we say that an interpretation $|\Pi\alpha.\mathsf{T}\alpha|$ of a polymorphic type $\Pi\alpha.\mathsf{T}\alpha$ in a category of PERs (or modest sets) is *parametric* if it "contains only what it is forced to"; for example, an interpretation of

$$\Pi\alpha.((\alpha \to \alpha) \times \alpha) \to \alpha$$

is parametric if it contains only the Church numerals.

More generally if $\mathsf{T}\alpha$ is of the form

$$((\alpha^{n_1} \to \alpha) \times \cdots \times (\alpha^{n_k} \to \alpha)) \to \alpha,$$

where the left-hand side looks like the signature of an algebraic data type, then $|\Pi\alpha.\mathsf{T}\alpha|$ is parametric if it is an initial algebra for that signature. (A thing of type $\Pi\alpha.\mathsf{T}\alpha$ is a procedure that, given a type X and an algebra structure of the given signature on X, produces a thing of type X; that is, the elements of $|\Pi\alpha.\mathsf{T}\alpha|$ should just be the *terms* of that signature.) Results like this do in fact hold in $\mathcal{E}\!f\!f$ for the Moggi-Hyland interpretation (this amounts to saying that they hold in the classical PER model): see [12], [23].

Note that we might hope to prove parametricity results more easily for the groupoid interpretation than for the Moggi-Hyland interpretation. The reason is that the invariant interpretation is 'smaller'. More precisely, there is a canonical map

$$\tau: (\Pi\alpha.\mathsf{T}\alpha)_{\mathbb{C}} \to \prod_{X \in \mathbb{C}_0} TX$$

from the invariant interpretation into the Moggi-Hyland interpretation, arising from the family of projections $(\Pi\alpha.\mathsf{T}\alpha)_{\mathbb{C}} \to X$; and it's easy to see that τ is mono.

Theorem 3.1 (Rosolini) *The groupoid interpretation in* **Mod** *is parametric.*

The proof of this theorem relies on results in the sequence of papers beginning with [2]. We only give a sketch here, due to Rosolini.

The point is that we only need to show that elements of $(\Pi X.\mathsf{T}X)_{\mathbb{C}}$ are dinatural transformations $1 \to B$. But such an element is dinatural for isomorphisms by definition; and any internal functor **Mod** $\to$ **Mod** (is isomorphic to one which) preserves maps coded by the identity recursive function (see [24]), so it is dinatural with respect to such maps too. So all we need to prove is

Lemma 3.2 (Rosolini) *Any map between modest sets factors as a composite of isomorphisms and maps tracked by the identity.*

Proof. There are several proofs; one is as follows. We will talk in terms of PERs instead of modest sets (quotients of PERs).

Let R and S be $\neg\neg$-closed PERs on $\mathbb{N}$, and suppose we have a map $f: R \to S$. Recall that the product PER $R \times S$ is defined by

$$\langle n_1, n_2 \rangle \,(R \times S)\, \langle m_1, m_2 \rangle \qquad \longleftrightarrow \qquad n_1 R m_1 \wedge n_2 S m_2$$

where $\langle \cdot, \cdot \rangle$ is a recursive pairing function.

The graph of f is a sub-PER $G \subset R \times S$, and clearly $R \cong G$. Enlarge G into a larger PER H by setting

$$\langle n_1, n_2 \rangle \, H \, \langle m_1, m_2 \rangle \qquad \longleftrightarrow \qquad f(n_1) S f(n_2).$$

(A little care is needed to make this constructive.)

H can be thought of as the image of f; we have a map $G \twoheadrightarrow H$ tracked by the identity recursive function. But now H is isomorphic to a sub-PER K of S; and $K \hookrightarrow S$ is tracked by the identity too. ∎

In the next section we'll see how to extend the theorem to full reflective subcategories of **Mod**; this provides the first proof of parametricity for interpretations of polymorphism in a general such subcategory.

4 Uniformized interpretations

Let $\mathbb{C}$ be a full reflective subcategory of **Mod**, and let F: **Mod** $\to \mathbb{C}$ be a reflection functor. If $\mathbb{C}_0$ is a constant set, we mentioned in the previous section that the Moggi-Hyland interpretation of a polymorphic type $\Pi\alpha.T\alpha$ relative to $\mathbb{C}$ could be calculated as a $\mathbb{C}_0$-indexed intersection of PERs. In general, however, this does not hold.

But consider the category $\bar{\mathbb{C}}$ defined as follows:

$$\bar{\mathbb{C}}_0 = \mathbf{Mod}_0$$

and if $X, Y \in \bar{\mathbb{C}}_0$,

$$\bar{\mathbb{C}}(X, Y) = \mathbb{C}(FX, FY).$$

There is a full embedding $G: \mathbb{C} \to \bar{\mathbb{C}}$ and a functor $F: \bar{\mathbb{C}} \to \mathbb{C}$ (which acts as the reflection functor F on objects and as the identity on arrows); and this defines an equivalence between $\mathbb{C}$ and $\bar{\mathbb{C}}$.

Let T be a type constructor on $\mathbb{C}$ that respects isomorphisms (as noted before, all of them generally do). Then we get a type constructor on $\bar{\mathbb{C}}$, namely GTF. The **uniformized Moggi-Hyland interpretation** of $\Pi\alpha.T\alpha$ is the product

$$F\Big(\prod_{X \in \bar{\mathbb{C}}_0} GTFX \Big)$$

indexed by $\bar{\mathbb{C}}_0 = \mathbf{Mod}_0$. That is, we compute the Moggi-Hyland interpretation in $\bar{\mathbb{C}}$ and transport it to $\mathbb{C}$ via the equivalence.

Note that this interpretation is sound; for suppose we have an object $Y \in \mathbb{C}$. Then a map $Y \to F(\prod_{X \in \bar{\mathbb{C}}_0} GTFX)$ in $\mathbb{C}$ corresponds, under the equivalence, to a map $GY \to \prod_{X \in \bar{\mathbb{C}}_0} GTFX$ in $\bar{\mathbb{C}}$; composing with the projections gives us a family of maps $GY \to GTFX$ and hence, using the equivalence again, a family $Y \to FGTFX \cong TFX \cong TX$ in $\mathbb{C}$ (remember that $X \cong FX$).

Conversely, suppose we have a family of maps $Y \to TX$, one for each $X \in \mathbb{C}$. Then for any $X' \in \mathbf{Mod}_0$ we have a map $Y \to TFX' \cong FGTFX'$ in $\mathbb{C}$; hence a map $GY \to GTFX'$ in $\bar{\mathbb{C}}$. But these uniquely determine a map $Y \to \prod_{X' \in \bar{\mathbb{C}}_0} GTFX'$, and thus a map $GY \to F(\prod_{X \in \bar{\mathbb{C}}_0} GTFX)$ back in $\mathbb{C}$.

The uniformized interpretation as an intersection The point here is that

$$
\begin{aligned}
F(\textstyle\prod_{X \in \bar{\mathbb{C}}_0} GTFX) &\cong \textstyle\prod_{X \in \bar{\mathbb{C}}_0} FGTFX \\
&\cong \textstyle\prod_{X \in \bar{\mathbb{C}}_0} TFX \\
&\cong \textstyle\prod_{X \in \mathbf{Mod}_0} TFX
\end{aligned}
$$

which can be calculated in **Mod**; and there, since it is indexed by the constant set $\mathbf{Mod}_0$, it can be calculated as the intersection of PERs

$$\bigcap_{R \in \mathbf{Mod}_0} TFR.$$

One way of interpreting this is that it shows how $\mathbb{C}$ gives rise to a 'submodel' of the PER model (i.e. another model where types are PERs and polymorphic types are computed as intersections, instead of more general products or limits). It is not a submodel in the ordinary sense: rather than restricting the collection of PERs, we keep all the PERs but change the maps.

I should emphasize that this interpretation depends on the particular reflection functor F; so if we want to use in in calculations, we need to have a reasonably explicit description of F. This is not always the case (e.g. [16]).

Application to bounded quantification PER models have been used in the past to model typed λ-calculi with subtyping and bounded quantification. It is reasonably straightforward to do this using the full PER model—or, which comes to almost the same thing, the whole category **Mod** of modest sets—but extending the idea to interesting submodels of the PER model (cf. [1]) or subcategories of **Mod** (such as categories of 'synthetic domains') has proved more problematic, though it can sometimes be done: [24].

The problem is that the subtyping relation is interpreted as inclusion of PERs; and this only interacts well with type quantification if polymorphic types can be computed as intersections. This will be the case if we use the Moggi-Hyland interpretation, so long as the relevant products are indexed by constant sets. So we want $\mathbb{C}_0$ to be a constant set, and we want powerkinds to be constant sets; but in general this will not be the case.

However, given any full reflective subcategory $\mathbb{C}$ of **Mod**, the idea of replacing $\mathbb{C}$ by an equivalent category whose object-of-objects is a constant set (namely $\mathbf{Mod}_0$) suggests a solution to this problem. Details will appear in [18]. In any case, this shows us that we can use the methods of [24] to model (say) Quest using (say) the complete Σ-spaces of [16].

A parametric interpretation in $\mathbb{C}$ We can combine the two interpretations by defining the **uniformized groupoid interpretation** computed by applying F to a certain limit in $\bar{\mathbb{C}}$ indexed by $\bar{\mathbb{C}}^{\cong}$. As in the end of the last section, one can show that this interpretation is *parametric*; the main point is that the object-of-objects of $\bar{\mathbb{C}}$ is $\mathbf{Mod}_0$, a constant set, so the results of [24] can be used.

The closure of a category Let me mention a red herring. Given a full subcategory $\mathbb{C}$ of **Mod** with $\mathbb{C}_0 \subset \mathbf{Mod}_0$, one can consider the full subcategory $\mathbb{C}_{\neg\neg}$ of **Mod** whose object-of-objects is $\neg\neg\mathbb{C}_0$ (the double-negation closure of $\mathbb{C}_0$ in $\mathbf{Mod}_0$); intuitively, an object of $\mathbb{C}_{\neg\neg}$ is a PER in $\mathbb{C}$, but without the intensional information telling us that the PER is actually in $\mathbb{C}$. Then $\mathbb{C}$ is still a full reflective subcategory of $\mathbb{C}_{\neg\neg}$; but there seems to be no reason why the two categories should be equivalent, without extra assumptions.

For example, it suffices that the reflection $F: \mathbf{Mod} \to \mathbb{C}$ satisfies $FX = X$ for $X \in \mathbb{C}$: equality, not isomorphism. But this is not usually the case unless $\mathbb{C}_0$ was a constant set in the first place, and R was calculated as an intersection: for example, this was the situation in [18].

Appendix: The truth about completeness

The statement that **Mod** is complete is not precise, nor is it true in the strongest possible sense. There are many ways to formulate the notion of completeness, not all equivalent: see [11]. An accessible account of the ways in which a small category can be complete, and the versions of polymorphic types λ-calculus that these completeness properties suffice to model, may be found in [21]. Here we only give a very brief summary.

Completeness Let $\mathbb{I}$ be an internal category in $\mathcal{E}\!f\!f$; then we could ask that there be a functor

$$\varprojlim : \mathbf{Mod}^{\mathbb{I}} \to \mathbf{Mod}.$$

Mod is in fact complete in this sense.

However, there is a weaker sense in which **Mod** could have all $\mathbb{I}$-indexed limits: namely,

$$\forall D: \mathbb{I} \to \mathbf{Mod}.\, \exists X \in \mathbf{Mod}.\, \text{“there is a limit cone over } D \text{ with vertex } X\text{”}$$

could be valid. Of course, **Mod** has this property too. (The reason that this doesn't imply the first property is because the axiom of choice fails in $\mathcal{E}\!f\!f$.)

But there is another subtlety. If we want to interpret iterated polymorphic types like

$$\Pi\alpha.\alpha \to (\Pi\beta.\alpha \to \beta),$$

or theories of dependent types, then completeness properties like this are not enough. One needs to talk about *families* of diagrams.

Let K be an object of $\mathcal{E}\!f\!f$. One thinks of an object of the topos $\mathcal{E}\!f\!f/K$ as being a K-indexed family of objects of $\mathcal{E}\!f\!f$; so we can talk about K-indexed families of diagrams by working in $\mathcal{E}\!f\!f/K$. That is, one has a K-indexed family of categories—viz. a category $\mathbb{I}$ in $\mathcal{E}\!f\!f/K$—and a K-indexed family of diagrams—viz. a functor $\mathbb{I} \to K^*\mathbf{Mod}$, where $K^*\mathbf{Mod}$ denotes the constant family of categories all equal to **Mod**.

Then we could either ask that there be a functor

$$\varprojlim: (K^*\mathbf{Mod})^{\mathbf{I}} \to K^*\mathbf{Mod}$$

between the two categories in $\mathcal{E}\!f\!f/K$, or simply that

$$\forall D: \mathbb{I} \to K^*\mathbf{Mod}.\, \exists X \in K\mathbf{Mod}.\, \text{“there is a limit cone over } D \text{ with vertex } X\text{”}$$

be valid in $\mathcal{E}\!f\!f/K$(i.e. "given a family of diagrams, there exists a family of objects in **Mod** and a family of limit cones..."). Again, the second property is weaker since the axiom of choice is not internally valid in $\mathcal{E}\!f\!f/K$.

Mod has all K-indexed families of $\mathbb{I}$-limits in the second sense but not the first. However, if $K \in \mathcal{S}ep$, the category of $\neg\neg$-separated objects ("ω-sets" as they are known elsewhere) mentioned before, then **Mod** actually has $\mathbb{I}$-limits in the first, stronger sense. Details may be found in [11]; the basic idea is that **Mod** is 'complete relative to $\mathcal{S}ep$' and that its completeness properties relative to $\mathcal{E}\!f\!f$ are just those that follow from this.

Throughout this paper we always have $K \in \mathcal{S}ep$, as all the categories considered actually lie in $\mathcal{S}ep$ (basically because they are subcategories of **Mod**, which is an internal category in $\mathcal{S}ep$). So it was in fact safe to ignore the different notions of completeness, and to write $\varprojlim D$, for example, because the functor $\varprojlim$ does exist. In general, when working with modest sets one expects the situation to be similarly nice; everything happens in $\mathcal{S}ep$.

Equivalence [11] There are also different ways in which two categories $\mathbb{C}$ and $\mathbb{D}$ in $\mathcal{E}\!f\!f$ can be equivalent. We might have a pair of functors

$$F: \mathbb{C} \to \mathbb{D},\ G: \mathbb{D} \to \mathbb{C}$$

with $FG \cong id_{\mathbb{D}}$, $GF \cong id_{\mathbb{C}}$; or we might simply have a functor

$$F\colon \mathbb{C} \to \mathbb{D}$$

that is full, faithful and essentially surjective on objects (one can express this in the internal language of the topos). Again, these are not equivalent because of the failure of the axiom of choice.

The proofs in section 2 assume that we have the former, stronger kind of equivalence. In practice we often have simply the functor F (i.e. we are in the latter situation) and we must construct the functor G using the adjoint functor theorem (cf. [17]); this amounts to finding a map that takes each $X \in \mathbb{D}$ to an initial object of $(X \downarrow F)$. But the categories we use are all subcategories of **Mod** and hence live inside *Sep*, and so this can be done; we omit the details.

In any case there is a 'weak' form of the main result in section 2 that only requires the second form of equivalence, and only asserts that for each diagram D, a limit of D and a limit of FD must be isomorphic (i.e. it does not refer to the existence of any functors $\varprojlim$). This result has essentially the same proof.

Reflectivity [17] Similar remarks apply: that is, in practice the subcategories $\mathbb{C}$ we deal with (subcategories of **Mod**) are reflective in the strong sense that there is actually a reflection functor $F\colon \mathbf{Mod} \to \mathbb{C}$. In fact, in contrast to the remarks above, this actually seems to be necessary for our purposes: for example, for the construction in section 4.

By the way, note that while we can formulate in the internal language the way in which a subcategory $\mathbb{C}$ of **Mod** is reflective, there's no really satisfactory way of expressing the fact that **Mod** is itself reflective in $\mathcal{E}\!f\!f$ except in the metatheory (e.g. by using the language of indexed categories, or fibrations, as in [11]). This is because the internal language can't talk about 'large' categories relative to $\mathcal{E}\!f\!f$—such as $\mathcal{E}\!f\!f$ itself. In practice, when such a problem comes up there is generally an *ad hoc* solution; but the general question of how to define an internal language adequate for doing category theory over a general base topos "as if it were *Sets*" remains open. And only by settling it can we cast the 'set-theoretic' explanation of polymorphism in its most convincing form.

References

[1] R. Amadio: Domains in a realizability framework, LIENS-TR 19-90.

[2] E. S. Bainbridge, P. J. Freyd, A. Scedrov, P. J. Scott: Functorial polymorphism: preliminary report, Proceedings of the Programming Institute on Logical Foundations of Functional Programmin (Austin, 1987).

[3] M. Beeson: *Foundations of Constructive Mathematics*, Springer, 1985.

[4] P. J. Freyd: POLYNAT in Per, in *Categories in Computer Science and Logic (Proc. Boulder 1987)*, *Contemp. Math.* **92** (1989).

[5] P. J. Freyd: Functorial polymorphism, to appear.

[6] P. J. Freyd: Structural polymorphism, to appear.

[7] P. J. Freyd, P. Mulry, G. Rosolini, D.S. Scott: Extensional PERs, in: Proc. of 5th Annual Symposium on Logic in Computer Science, 1990.

[8] J.-Y. Girard, Y. Lafont, P. Taylor: *Proofs and Types*, Cambridge University Press, 1989.

[9] J.M.E. Hyland: The effective topos, in: *The L.E.J. Brouwer Centenary Symposium* (ed. A. S. Troelstra, D. van Dalen), North-Holland, 1982.

[10] J.M.E. Hyland: A small complete category, *Annals of Pure and Applied Logic* 40 (1988) 135-165.

[11] J.M.E. Hyland, E.P. Robinson, G. Rosolini: The discrete objects in the effective topos, *Proc. Lond. Math. Soc.* (3) 60 (1990) 1–36.

[12] J. M. E. Hyland, E. P. Robinson, G. Rosolini: Algebraic types in PER models.

[13] P. T. Johnstone: *Topos Theory*, Academic Press, 1977.

[14] S. Mac Lane: *Categories for the Working Mathematician*, Springer-Verlag, 1971.

[15] W. K.-S. Phoa: Relative computability in the effective topos, *Math. Proc. Camb. Phil. Soc.* **106** (1989), 419–422.

[16] W. K.-S. Phoa: Effective domains and intrinsic structure, in: Proc. of 5th Annual Symposium on Logic in Computer Science, 1990.

[17] W. K.-S. Phoa: Reflectivity for categories of synthetic domains, submitted, 1990.

[18] W. K.-S. Phoa: From term models to domains, to appear in the proceedings of Theoretical Aspects of Computer Software (Sendai, 1990).

[19] W. K.-S. Phoa: Using fibrations to understand subtypes, in preparation.

[20] G. D. Plotkin, M. B. Smyth: The category-theoretic solution of recursive domain equations, *SIAM Journal of Computing* vol. 11 no. 4, 1982.

[21] E. Robinson: How complete is PER? in: Proc. of 4th Annual Symposium on Logic in Computer Science, 1989.

[22] J. C. Reynolds: Polymorphism is not set-theoretic, Symposium on Semantics of Data Types (ed. Kahn, McQueen, Plotkin), Springer LNCS 173, 1984.

[23] G. Rosolini: Uniformity, dinaturality and 2.

[24] G. Rosolini: An Exper model for Quest, to appear in the proceedings of MFPS91.

[25] G. Rosolini: About modest sets, to appear.

[26] P. Taylor: The fixed point property in synthetic domain theory, to appear in the proceedings of the 6th Annual Symposium on Logic in Computer Science (Amsterdam, 1991).

An algebra of graphs and graph rewriting*

Andrea Corradini **Ugo Montanari**

Università di Pisa
Dipartimento di Informatica
Corso Italia 40
56125 Pisa - ITALY

Abstract. In this paper we propose an axiomatization of 'partially abstract graphs', i.e., of suitable classes of monomorphisms in a category of graphs, which may be interpreted as graphs having both a concrete part and an abstract part (defined up to isomorphism). Morphisms between pa-graphs are pushout squares. We show that the basic notions of the algebraic theory of graph grammars [Eh79] (instantiated to a suitable category of graphs) can be rephrased in a natural way using partially abstract graphs. The terms of the algebra we propose are built over a small set of operators, including parallel composition, substitution application, and restriction. By equipping the algebra of terms with a categorical structure (arrows are equivalence classes of monadic contexts), we show that there is a full and faithful embedding (with a right adjoint) of the category of partially abstract graphs into the category of (well-formed) terms. This embedding is exploited to show that rewriting (in the sense of term rewriting systems) over this algebra models faithfully the direct derivations of graphs, described by a double pushout construction along the guidelines of [Eh79]. In particular, we show that also graph productions having non-discrete gluing graphs can be represented as term rewrite rules without loss of information, unlike a similar approach proposed in [BC87].

1 Introduction

The "theory of graph grammars" basically studies a variety of formalisms which extend the theory of formal languages in order to deal with structures more general than strings, like graphs and maps. A graph grammar allows one to describe finitely a (possibly infinite) collection of graphs, i.e., those graphs which can be obtained from an initial graph through repeated application of graph productions. The form of graph productions, and the rule stating how a production can be applied to a graph and what the resulting graph is, depend on the specific formalism.

The development of this theory, originated in the late 60's, is well motivated by many fruitful applications in different areas of computer science: among them we recall data bases, software specification, incremental compilers, pattern recognition, and term rewriting (see [CER79, ENR83, ENRR87, EKR91], the proceedings of the international workshops on graph grammars).

Among the various approaches to graph grammars, which differ for the kind of graphs they deal with and for the rules stating when and how a production can be applied to a certain graph, one of the most

* Research partially supported by the GRAGRA Basic Research Esprit Working Group n. 3299.

successful has been the *algebraic* (or *Berlin*) *approach* [EPS73, Eh79]. Thanks to the clean categorical description used, this approach has been applied in the last decade to various categories of structures (including various kinds of graphs [Eh79, PEM87, HK87, HKP88, CMREL91], relational structures [EKMRW81], software modules [PP90], pictures [Ha91], etc.). Recently, moreover, the algebraic theory of graph grammars has been generalized to arbitrary underlying categories [EHKP91a/b, Ke91].

The kernel of the algebraic theory of graph grammars includes, besides the basic definitions of production, direct derivation, derivation, etc., also some Church-Rosser results based on a notion of independence among productions. Instantiating the general definitions to a category of graphs, a production p = $(L \overset{l}{\leftarrowtail} K \overset{r}{\rightarrowtail} R)$ is a pair of graph monomorphisms having as common source a graph K, the 'gluing' graph, indicating which edges and nodes have to be preserved by the application of the production. Production p can be applied to a graph G yielding H (written $G \Rightarrow_p H$) if there is an 'occurrence' (i.e., a graph morphism) $g: L \rightarrow G$, and H is obtained as the result of a suitable 'double pushout' construction.

Given a graph to be rewritten, two production applications are 'parallel independent' if their redexes are either disjoint, or their intersection is preserved by both productions. In this case the two productions can be applied to the graph in any order (and in parallel as well) producing isomorphic graphs, because the application of one of them does not invalidate the applicability of the other. This result provides a reasonable notion of equivalence among the finite derivations producing the same graph: two derivations are equivalent if one can be obtained from the other by 'switching' consecutive independent productions any number of times. This equivalence can be exploited as the basis of a truly concurrent semantics for graph grammars, which provides an explicit representation of the causal dependencies among the production applications in a derivation [MR91].

As far as the generative power of graph rewriting is concerned, it is sound to consider just productions having *discrete* gluing graphs [EPS73, Eh79]. However, the possibility of considering productions with non-discrete gluing graphs as well is fundamental for the Church-Rosser properties (see [Eh79], pp. 28-37), and thus for a truly concurrent semantics of graph grammars.

A complete axiomatization of expressions denoting finite graphs has been presented by Bauderon and Courcelle in [BC87], where it is shown that rewriting (in the sense of term rewriting systems) in the algebra of graph expressions is as powerful as the rewriting formalism which results by the application of the algebraic approach to a suitable category of graphs. However, an intrinsic limit of this proposal is that term rewrite rules over the algebra of graphs are able to model just graph productions with discrete gluing graphs. Therefore, the result of equivalence between the two rewriting formalisms just holds with respect to the generative power, but not for the truly concurrent aspects.

In this paper we propose a new axiomatization of graphs and graph rewriting, called the algebra of *partially abstract graphs*, which allows a faithful representation as term rewrite rules of graph productions having non-discrete gluing graphs as well. As in [BC87], we are able to show that term rewriting in the algebra of partially abstract graphs provides a sound and complete representation (w.r.t. the generative power) of graph rewriting, but the more faithful representation of graph productions should allow us to capture in our framework also many results about parallelism and concurrency presented in the theory of graph grammars. Although this topic is not explored here, in the concluding section we sketch the possible developments of this work. In this paper we focus instead on the presentation of the axiomatization, and on the study of the relationship between the terms of the algebra and suitable graphs.

For *graphs* we mean *directed hypergraphs*, i.e., graphs where each edge can be connected to a finite list of nodes, rather than to exactly two nodes. The terms of our algebra include atoms of the form 0 (representing the empty graph), *a* (meaning '*a* is a node'), and $p(a_1, \ldots, a_n)$ (representing a single edge p connected to the

A node $a \in N$ of a graph $G = \langle N, E, c \rangle$ is *isolated* if there is no edge $e \in E$ such that a occurs in $c(e)$. The set of isolated nodes of a graph G will be denoted by IN_G. The category having graphs as objects and graph morphisms as arrows is called **Graph**. The full subcategory of **Graph** containing just finite graphs will be denoted by **FGraph**. ♦

It is worth stressing that all the definitions and results presented in this paper could be restated for other categories of graphs as well (for example, for colored graphs) with some minor modifications.

2.1 Basic definitions about graph rewriting

A graph production, analogously to a string production, describes how to replace the occurrence of a subgraph L of a graph G with another graph R. While in the case of strings the embedding of the right hand side of a production inside the string to be rewritten is uniquely determined, this is not true in the more general case of graphs. Thus a third graph K is needed to give the connection points of R into G. In fact, all the items of K are preserved by the application of the production.

2.2 Definition *(graph productions)*

A *graph production* p is written as $p = (L \stackrel{l}{\leftarrowtail} K \stackrel{r}{\rightarrowtail} R)$, where L, R, and K are graphs and l and r are injective graph morphisms. L, K, and R are called the *left-hand side* (*lhs*), the *gluing graph*, and the *right-hand side* (*rhs*) of p, respectively. ♦

The application of a graph production p to a graph G is modeled by a double-pushout construction.

2.3 Definition *(direct derivation)*

Given a graph G, a graph production $p = (L \stackrel{l}{\leftarrowtail} K \stackrel{r}{\rightarrowtail} R)$, and an *occurrence* of the lhs of p in G, i.e., a morphism $g: L \to G$, a *direct derivation* from G to H exists iff the two pushouts depicted in the following diagram can be constructed. In this case D is called the *context* graph, and we write $(p, g): G \Rightarrow H$ or $G \Rightarrow_{(p,g)} H$. We also write $G \Rightarrow_p H$ if $G \Rightarrow_{(p,g)} H$ for some occurrence g.

$$\begin{array}{ccccc} L & \xleftarrow{\;l\;} & K & \xrightarrow{\;r\;} & R \\ {\scriptstyle g}\downarrow & & \downarrow{\scriptstyle k} & & \downarrow{\scriptstyle h} \\ G & \xleftarrow{\;d\;} & D & \xrightarrow{\;b\;} & H \end{array} \quad ♦$$

From a pragmatical point of view, to apply a production p to a graph G we first need to find an occurrence of its lhs L in G, i.e., a morphism g from L to G. Next, to model the deletion of such occurrence of L in G, we construct the 'pushout complement' of g and l, i.e., we have to find a context graph D (with morphism $k: K \to D$ and $d: D \to G$) such that the resulting square is a pushout. Finally, we have to embed the rhs R in D: such an embedding is expressed by a second pushout. It must be noticed that this construction can fail, because although all pushouts exist in category **Graph**, the pushout complement of two arrows of **Graph** does not always exist. Moreover, the existence of a pushout complement object does not imply its uniqueness, because it is not characterized by a universal property. The next result gives sufficient conditions for the existence of the pushout complement of two arrows in **Graph** (see [Eh79] and the references therein).

2.4 Proposition *(existence of pushout complements in* **Graph***)*

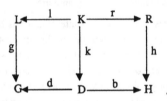

Given two graph morphisms $l: K \rightarrowtail L$ (injective) and $g: L \to G$, their pushout complement (D, k, d) as in

list of nodes $a_1 \cdot \ldots \cdot a_n$; p is an 'edge variable' while each a_i is a 'node variable'); moreover, they are closed under 'parallel composition' (interpreted as union), substitutions (which can be non-injective), and under an operator of restriction, which 'hides' the concrete identity of a variable. The axioms describe in an intuitive way the interplay of the various operations. Indeed, the algebra we propose is very simple, and it would be not surprising to discover that it is a fragment of some richer logical system. We do not explore this possibility, focusing instead on the direct presentation of the axiomatization and on the correspondence with suitable graphs.

If restricted to the sub-algebra of 'concrete' terms (i.e., terms without restriction operators), our axiomatization is proved to be complete for finite graphs. Furthermore, we show that there is a full and faithful embedding (having a right adjoint) from the category of finite graphs to the category having concrete terms as objects and concrete contexts (i.e., concrete terms with a 'hole') as arrows.

On the other hand, 'well-formed' terms (i.e., terms possibly containing restrictions, but subjected to a suitable syntactic constraint) are shown to denote faithfully 'partially abstract graphs', i.e., equivalence classes of injective morphisms having the same source graph (the concrete part) and isomorphic target graphs (the abstract part). Morphisms between partially abstract graphs are pushout squares. Intuitively, a well-formed term denotes a partially abstract graph the concrete part of which is the subgraph composed by all the variables of the term which are not restricted (its 'free' variables), while its abstract part includes all the edge and node variables which appear in the term. The 'abstractness' is given by the fact that restricted variables are defined up to renaming.

The correspondence between partially abstract graphs and well-formed terms is similar to the one for concrete graphs: there is a full and faithful embedding functor with a right adjoint from the category of finite partially abstract graphs to the category having well-formed terms as objects and concrete contexts as arrows. An important consequence of this result is that the application of a concrete context to a well-formed term corresponds to a pushout in the category of graph.

A term rewrite rule over our algebra is a pair of terms having the same set of free variables. Such a rewrite rule can be used to represent faithfully a graph production possibly having a non-discrete gluing graph. Since the application of a concrete context to a term rewrite rule corresponds to a double pushout in the category of graph, term rewriting sequences accurately model graph derivations.

2 The algebraic theory of graph grammars

In this section we first recall the basic notions of the algebraic theory of graph grammars, as summarized in [Eh79]. Next we introduce the category of *partially abstract graphs*, and use them to rephrase the basic notions related to graph rewriting. The kind of graphs we refer to throughout the paper are non-colored, directed hypergraphs.

2.1 Definition *(the category of hypergraphs)*

A *directed (hyper)graph* G is a triple $G = \langle N, E, c \rangle$, where N is a set of *nodes* (or *vertices*), E is a set of *hyper-edges*, and $c: E \to N^*$ is the *connection function.* From now on we will omit the *hyper-* prefix. Edges and nodes of a graph G will be called *items* of G (written $x \in G$).

Let $G = \langle N_G, E_G, c_G \rangle$ and $H = \langle N_H, E_H, c_H \rangle$ be two graphs. A *graph morphism* $h: G \to H$ is a pair of functions $\langle h_N: N_G \to N_H, h_E: E_G \to E_H \rangle$ such that $c_H \circ h_E = h_N^* \circ c_G$ (h_N^* is the extension of h_N to lists). A graph morphism h is *injective* (or a *monomorphism*, written $h: G \rightarrowtail H$) (resp. an *isomorphism*) if both components are injective (resp. isomorphisms).

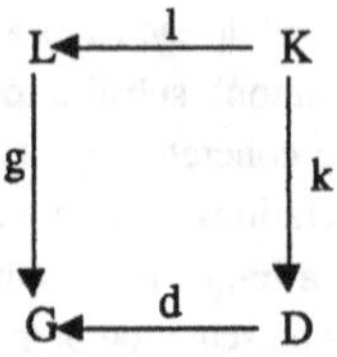

exists and is unique (up to isomorphism) if the following conditions are satisfied:

Identifying condition: all the items of L which are identified by g are in the image of l. Formally,

$$\forall x \in L\,.\,(\exists y \in L\,.\,x \neq y \wedge g(x) = g(y)) \Rightarrow x \in l(K).$$

Dangling condition: all the nodes of L which are connected in G to an edge which is not in the image of g, are in the image of l. Formally,

$$\forall n \in N_L\,.\,(\exists e \in E_G \backslash g_E(E_L)\,.\,g_N(n) \text{ occurs in } c_G(e)) \Rightarrow n \in l_N(N_K).\ \blacklozenge$$

2.5 Definition *(derivation)*

Given two graphs G and H, and a set of graph productions P, a *derivation* from G to H over P, denoted by $G \Rightarrow_P^* H$, is a finite sequence of direct derivations of the form $G \Rightarrow_{(p_1,g_1)} G_1 \Rightarrow_{(p_2,g_2)} \ldots \Rightarrow_{(p_n,g_n)} G_n = H$, where $p_1, \ldots, p_n$ are in P. ♦

2.6 Definition *(graph grammars and graph languages)*

A **graph grammar** $\mathcal{G}$ is a triple $\mathcal{G} = (T, P, S_0)$ where T is a collection of *terminal* graphs, P is a finite set of graph productions, and S_0 is the *initial* graph. The *graph language* generated by graph grammar $\mathcal{G}$, $\mathcal{L}(\mathcal{G})$, is the set of terminal graphs derivable from the initial graph, i.e., $\mathcal{L}(\mathcal{G}) = \{H \in T \mid S_0 \Rightarrow_P^* H\}$. ♦

As shown in [EPS73], if one considers just productions with discrete gluing graphs, the generative power of graph grammars (i.e., the collection of graph languages generated) remains the same. However, the possibility of considering productions having edges in the gluing graph is fundamental for the Church-Rosser properties, which are based on the notion of parallel independence.

2.7 Definition *(parallel independence of direct derivations)*

Given two productions $p = (L \overset{l}{\leftarrow} K \overset{r}{\rightarrow} R)$ and $p' = (L' \overset{l'}{\leftarrow} K' \overset{r'}{\rightarrow} R')$, two direct derivations $(p, g): G \Rightarrow H$ and $(p', g'): G \Rightarrow H'$ are *parallel independent* if the intersection of the images of the occurrences of p and p' in G consists of common gluing items, i.e., $g(L) \cap g'(L') = g \circ l(K) \cap g' \circ l'(K')$. ♦

Intuitively, an edge *e* appearing in the gluing graph K corresponds to a read-only operation on the graph G to be rewritten, which checks for the existence of the image of *e*, but does not affect it. Thus, two direct derivations such that the intersection of their occurrences in G is formed only by gluing items (i.e., items that both direct derivations access in read-only mode) are parallel independent, in the sense that they can be applied in either order to G, or also in parallel. On the other hand, if an edge *e* is removed from the gluing graph K of production p, the resulting production p' would be clearly sound (i.e., $G \Rightarrow_{p'} H$ implies $G \Rightarrow_p H$), but the read-only access of p to *e* would be replaced in p' by two consecutive operations which first delete *e* and then create a new copy of it. Thus production p' would not result independent of other productions which access edge *e* in read-only mode.

2.2 Graph rewriting using partially abstract graphs

When defining an axiomatization of graphs (as in [BC87]), one crucial choice is wether to axiomatize *concrete* graphs (those of Definition 2.1) or *abstract* graphs (i.e., graphs defined up to isomorphism).

Indeed, from one side the derivations of a grammar produce graphs which are defined up to isomorphism (because they are characterized as pushout objects), and this would suggest one to consider just abstract graphs. On the other side, morphisms between abstract graphs can be defined only by using universal constructions, whereas the basic definitions of graph production and of occurrence are not expressible in this way.

The solution we propose is an algebra which axiomatizes *partially abstract graphs*, i.e., graphs where some items are concrete and the others are determined up to isomorphism. In categorical terms, a partially abstract graph with concrete part G is a *super-object* (a notion analogous to that of sub-object) of G, i.e., an equivalence class of monomorphisms with common source G. A morphism between such graphs should be reasonably a pair of arrows (one of which defined up to isomorphism), such that the resulting square commutes. There are many possible choices. Since the basic constructions used in graph rewriting are pushouts, we require that the square induced by a morphism be a pushout. We shall need the following facts about pushouts and colimits.

2.8 Proposition *(properties of pushouts and colimits)*

- In any category **C**, if squares (1) and (2) are pushouts, then the composite diagram (1) + (2) is a pushout. Moreover, if (1) + (2) and (1) are pushouts, then (2) is a pushout as well.

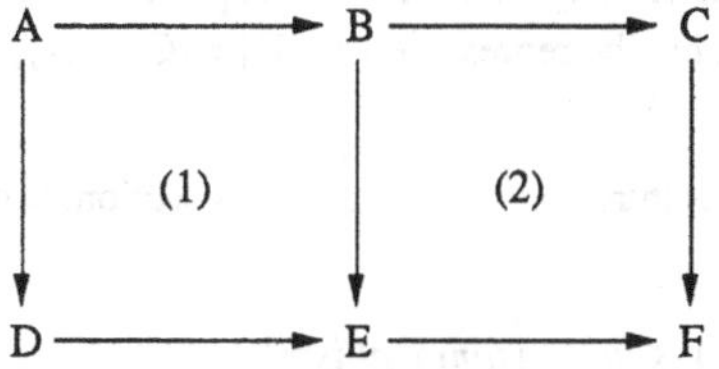

- In category **Graph**, let the following be colimit diagrams. Then if f is injective (resp. if f' and f" are injective) then h (resp. h') is injective as well. Similarly, in any category **C**, if f (resp. f' and f") is an isomorphism, then h (resp. h') is an isomorphism too.

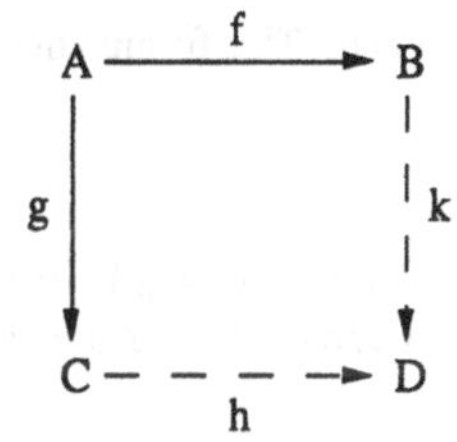

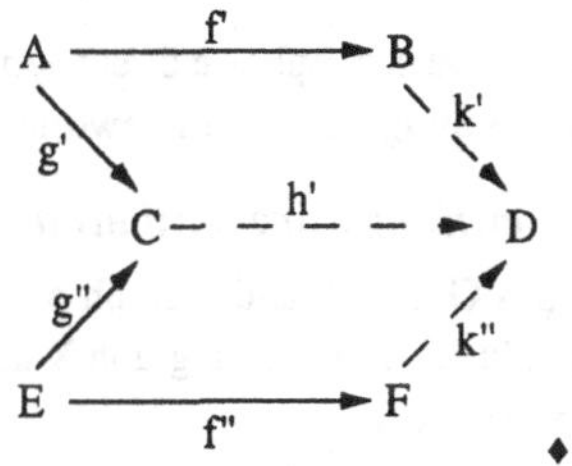

♦

2.9 Definition *(the category of partially abstract graphs)*

A *partially abstract graph X* (shortly *pa-graph*) *with concrete part G* is an equivalence class of monomorphisms of **Graph**, X = [h: G ↣ H], where h: G ↣ H and h': G ↣ H' are equivalent iff there exists an isomorphism f: H ↣ H' such that f ∘ h = h'. The concrete part of a pa-graph X is denoted by $\mathcal{G}$(X), while its abstract part (which is an equivalence class of graphs up to isomorphism) is indicated by $\mathcal{A}$(X).

Let X = [h: G ↣ H], X' = [h': G' ↣ H'] be pa-graphs, and f: G → G' be a graph morphism. Then f is a *morphism of pa-graphs* f: X → X' if there exists a morphism g: H → H' such that the following square is a pushout in **Graph**:

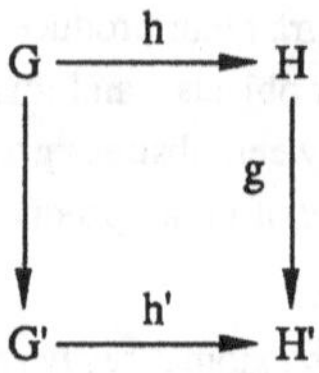

It is easy to check that morphisms are well defined (i.e., they are independent of the choice of the representatives in the equivalence classes), and that pa-graphs and pa-graph morphisms form a category (using Proposition 2.8 to check the correctness of arrow composition).

The category of pa-graphs will be denoted by **PA-Graph**, while **PA-FGraph** will denote its full subcategory including all finite pa-graphs. The function returning the concrete part of a pa-graph, $\mathfrak{G}$, can be turned into a forgetful functor $\mathfrak{G}$: **PA-Graph** → **Graph** by defining $\mathfrak{G}(f) = f$ on arrows. ♦

2.10 Notation *(from graphs to pa-graphs)*

If h is a monomorphism of **Graph**, the unique pa-graph containing h will be denoted by [h]. Both concrete graphs (i.e., objects of **Graph**) and abstract graphs (equivalence classes of concrete graphs up to isomorphism) are special cases of pa-graphs: a concrete graph G is faithfully represented by the pa-graph $[id_G]$, while an abstract graph K can be represented as $[K] \triangleq [\varnothing \rightarrowtail K]$, where $\varnothing$ is the empty graph, the initial object of **Graph**. ♦

The fact that a direct derivation, as introduced in the previous section, is defined up to isomorphism can be formalized as follows.

2.11 Fact *(graph derivation is up to isomorphism)*

Let $G, H \in |\mathbf{Graph}|$ and $p = (L \overset{l}{\leftarrowtail} K \overset{r}{\rightarrowtail} R)$ be a graph production such that $G \Rightarrow_p H$. If $G' \cong G$ and $H' \cong H$ then $G' \Rightarrow_p H'$. Moreover if $p' = (L' \overset{l'}{\leftarrowtail} K' \overset{r'}{\rightarrowtail} R')$ is a production isomorphic to p (i.e., there exist three isomorphisms $g: L \to L'$, $k: K \to K'$, and $f: R \to R'$ such that $g \circ l = l' \circ k$ and $f \circ r = r' \circ k$), then $G \Rightarrow_{p'} H$ as well. ♦

As a consequence, we can rephrase direct derivations in terms of pa-graphs. This formulation will be used to model graph rewriting trough term rewriting in Section 4.

2.12 Fact *(rewriting using partially abstract graphs)*

Given two graphs G and H, and a graph production $p = (L \overset{l}{\leftarrowtail} K \overset{r}{\rightarrowtail} R)$, then $G \Rightarrow_p H$ iff there exists a graph morphism $k: K \to D$ and pa-graph X and Y such that $k: [l] \to X$ and $k: [r] \to Y$ are pa-morphisms, $G \in \mathcal{A}([l])$, and $H \in \mathcal{A}([r])$. ♦

3 An axiomatization of graphs and graph rewriting

The terms of the algebra of partially abstract graphs, introduced in Section 3.1, include a constant 0 (denoting the empty graph), atomic propositions of the form *a* ("*a* is a node variable"), or $p(a_1, \ldots, a_n)$, to be read "p is an edge variable connected to node variables $a_1, \ldots, a_n$"; a binary operator '|' (union); an infinite number of postfixed operators '[Φ]', where Φ is a substitution from nodes to nodes and from edges to edges; and an infinite number of restriction operators \x, where x is a variable.

Terms which do not include restriction operators are called concrete, since we will show, in Section 3.2, that concrete graphs are a model for the axioms when restricted to terms with no restrictions. More precisely, with every concrete term A we will associate a graph $\mathbf{Gr}(A)$, and every graph G will be coded into a term $\mathcal{CT}(G)$ such that $\mathcal{Gr}(\mathcal{CT}(G)) \equiv G$. The correspondence is even stronger: by equipping the terms of the algebra with a categorical structure (using as arrows unary contexts, i.e., terms with a 'hole') we are able to show that there is a full and faithful embedding $\mathcal{CT}$ from category **FGraph** to the category of concrete terms, having $\mathcal{Gr}$ as right adjoint.

This result is extended in Section 3.3 to partially abstract graphs. They are shown to be a model for 'well-formed' terms, i.e., terms possibly containing restriction operators, satisfying a reasonable syntactical constraint. Intuitively, a restriction operator deletes an item from the concrete graph associated with the term, making it not directly accessible. Thus the 'free' variables of a term (i.e., variables which are not restricted) correspond to the concrete part of the associated partially abstract graph. The well-formedness requirement guarantees that if a node is restricted, then all the edges directly connected to it are restricted as well.

Analogously to the concrete case, we show how to associate with every well-formed term A a pa-graph $\mathcal{PaG}(A)$, and *viceversa* with each pa-graph X a well-formed term $\mathcal{WfT}(X)$. The relevant fact is that this correspondence can be extended to arrows as well, i.e., that the application of a context to a pa-graph corresponds to a pushout in the category of graphs. This allows us to turn both functions into functors, showing that $\mathcal{WfT}$ is a full and faithful embedding of the category of finite pa-graphs into the category of well-formed terms, with right adjoint $\mathcal{PaG}$. This correspondence will be exploited in Section 4 to model graph derivations through graph rewriting.

3.1 The algebra of partially abstract graphs

The algebra of pa-graphs $\mathcal{A}$ we introduce now is defined in terms of a fixed pair of disjoint infinite collections $\langle \mathcal{E}, \mathcal{N} \rangle$, called 'edge' variables and 'node' variables respectively. For the results about the correspondence between terms of the algebra and pa-graphs, we will assume that all the edges (resp. nodes) of the graphs we consider belong to the collection $\mathcal{E}$ (resp. $\mathcal{N}$). Since we will consider just finite graphs, this assumption causes no loss of generality.

3.1 Definition *(syntax of the algebra)*

Let $\mathcal{E}$ be a set of *edge variables* (ranged over by e, p, q, …), and $\mathcal{N}$ be a set of *node variables* (ranged over by a, b, …). Each $x \in \mathcal{E} \cup \mathcal{N}$ is called a *variable*. The *algebra of pa-graphs* $\mathcal{A}$ is the algebra having as elements the equivalence classes of terms generated by the following syntax, modulo the minimal equivalence relation induced by the axioms listed below in Definition 3.3:

$$A ::= 0 \mid a \mid p(a_1, \ldots, a_n) \mid A|A \mid A[\Phi] \mid A\backslash x$$

where $a, a_1, \ldots, a_n \in \mathcal{N}$; $p \in \mathcal{E}$; '|' is called *parallel composition*; Φ is a *(finite domain) substitution*, i.e., a function $\Phi: \mathcal{E} \cup \mathcal{N} \rightarrow \mathcal{E} \cup \mathcal{N}$ such that $\Phi(\mathcal{N}) \subseteq \mathcal{N}$, $\Phi(\mathcal{E}) \subseteq \mathcal{E}$, and such that the set of variables for which $x \neq \Phi(x)$ is finite; and x is a variable ($\backslash x$ is called a *restriction operator*). A term of the form 'a' is also called a *node atom*, while $p(a_1, \ldots, a_n)$ is an *edge atom* and '0' is the *empty atom*. ♦

To present the axioms we need the following definitions.

3.2 Definition *(free variables and concrete, open, and closed terms)*

Given a term $A \in \mathcal{A}$, its set of *free variables* $\mathcal{F}(A)$ is inductively defined as $\mathcal{F}(0) = \varnothing$; $\mathcal{F}(a) = \{a\}$; $\mathcal{F}(p(a_1, \ldots, a_n)) = \{p, a_1, \ldots, a_n\}$; $\mathcal{F}(A_1 \mid A_2) = \mathcal{F}(A_1) \cup \mathcal{F}(A_2)$; $\mathcal{F}(A[\Phi]) = \Phi(\mathcal{F}(A)) = \{\Phi(z) \mid z \in \mathcal{F}(A)\}$; and $\mathcal{F}(A\backslash x) = \mathcal{F}(A) - \{x\}$, where '-' denotes set difference (i.e., $\mathcal{F}(A) - \{x\} \triangleq \mathcal{F}(A) \cap \overline{\{x\}}$).

A term A is *closed* iff $\mathcal{F}(A) = \varnothing$. A term is *concrete* if it does not contain any restriction operator. A term is *open* if no variable is really restricted. Formally, all atoms are open; $A_1 \mid A_2$ is open if both A_1 and A_2 are open; $A[\Phi]$ is open if A is open; and $A\backslash x$ is open if A is open and $x \notin \mathcal{F}(A)$. Clearly, all concrete terms are open. We will denote by $\mathbf{C}\mathcal{A}$ the sub-algebra of $\mathcal{A}$ including all *concrete* terms. By Corollary 3.7 below, $\mathbf{C}\mathcal{A}$ can be regarded as the sub-algebra of *open* terms as well. ♦

3.3 Definition *(axioms of the algebra of pa-graphs)*

The terms of algebra $\mathcal{A}$ introduced in Definition 3.1 are subjected to the following conditional axioms.

ACI-|	$(A_1 \mid A_2) \mid A_3 = A_1 \mid (A_2 \mid A_3)$;	$A_1 \mid A_2 = A_2 \mid A_1$;	$A \mid 0 = A$
D-|	$p(a_1, \ldots, a_n) \mid a_i = p(a_1, \ldots, a_n)$,	for $1 \le i \le n$;	$A \mid A = A$, if A is open
COMP-[]	$A[\Phi][\Psi] = A[\Psi \circ \Phi]$		
C-\	$A\backslash x\backslash y = A\backslash y\backslash x$		
EL-\	$A\backslash x = A$, if x is not free in A		
MAP	$p(a_1, \ldots, a_n)[\Phi] = \Phi(p)(\Phi(a_1), \ldots, \Phi(a_n))$;	$a[\Phi] = \Phi(a)$;	$0[\Phi] = 0$
DIS-[]	$(A_1 \mid A_2)[\Phi] = A_1[\Phi] \mid A_2[\Phi]$		
FAC-\	$A_1\backslash x \mid A_2 = (A_1 \mid A_2)\backslash x$,	if x is not free in A_2	
SWAP	$(A\backslash x)[\Phi] = A[\Phi]\backslash\Phi(x)$,	if $\neg\exists\, y \in \mathcal{F}(A\backslash x).\ \Phi(y) = \Phi(x)$	

Two terms A and A' are *equivalent* (written $A \approx A'$) if they are in the least congruence relation (w.r.t. all the operators of the algebra) induced by the above axioms. ♦

All the axioms and the respective side conditions are very intuitive under the interpretation of the operators sketched above. Only the second axiom of D-| (i.e., $A \mid A = A$, if A is open) deserves a comment. As stated in the Introduction, the terms of the algebra are intended to represent pa-graphs, i.e., graphs with a concrete and an abstract part. This axiom formalizes the fact that the 'union' of two pa-graphs is obtained by taking the (set-theoretical) union of their concrete parts, and the disjoint union of the abstract parts.

By Definitions 3.2 and 3.3 it is clear that restriction is a binding operator. The classical α-conversion is derivable from the axioms.

3.4 Proposition *(some derived axioms)*

The following equivalences are derivable from the above axioms:

(1) $A[\Phi] \approx A[\Psi]$ if $\Phi(x) = \Psi(x)$ for each $x \in \mathcal{F}(A)$

(2) $A \approx A[I\delta]$ where $I\delta: \mathcal{E} \cup \mathcal{N} \to \mathcal{E} \cup \mathcal{N}$ is the identity substitution.

(3) $A \approx A[\Phi]$ if $\Phi(x) = x$ for each $x \in \mathcal{F}(A)$

(4) *α-conversion*: $A\backslash x_1\backslash \ldots \backslash x_n \approx A[\Phi]\backslash\Phi(x_1)\backslash\ldots\backslash\Phi(x_n)$ if Φ is a *renaming* (i.e., a bijection $\Phi: \mathcal{E} \cup \mathcal{N} \leftrightarrow \mathcal{E} \cup \mathcal{N}$), and $\Phi(x) = x$ for each $x \in \mathcal{F}(A)$. ♦

The next fact states, *a posteriori*, that the axioms are well defined, because the definitions used in the side conditions are preserved by the equivalence relation.

3.5 Proposition *(definitions are stable under equivalence)*

Let $A \approx A'$. Then $\mathcal{F}(A) = \mathcal{F}(A')$; A is open iff A' is open; and A is closed iff A' is closed.

Proof The statements are easily shown to hold true for lhs and rhs of each axiom. ♦

Using the axioms, the terms of the algebra $\mathcal{A}$ can be reduced into a canonical form which in general is not unique.

3.6 Proposition *(canonical form)*

A term A is in *canonical form* if $A = 0$ or $A = \underline{A}\backslash x_1\backslash\ldots\backslash x_n$, where $\{x_1, \ldots, x_n\} \subseteq \mathcal{F}(\underline{A})$ and $\underline{A}$ is a parallel composition of distinct atoms not including 0, and such that each node atom does not occur in any edge atom. Every term A has a canonical form $\mathrm{can}(A) \approx A$.

Proof outline. Given a term A, we can apply repeatedly axioms SWAP and FAC-\ in order to lift all the restriction operators towards the root of the term. It is always possible to make the side conditions of these two axioms true, possibly by applying first an α-conversion which does not generate new restrictions. In this way we obtain a term equivalent to A of the form $B\backslash y_1\backslash\ldots\backslash y_m$, with B concrete. All the substitution operators can be eliminated from B (yielding an equivalent term $\underline{B}$) by pushing them towards the leafs of the syntactical tree (via DIS-[]), and then by repeated applications of MAP. Finally, the canonical term can(A) can be obtained from $\underline{B}\backslash y_1\backslash\ldots\backslash y_m$ by deleting all $\backslash y_i$ such that y_i is not free in $\underline{B}\backslash y_1\backslash\ldots\backslash y_{i-1}$ (by axiom EL-\); then by removing all the multiple occurrences of atoms and all the node atoms which occur in an edge atom from $\underline{B}$ (using D-|); and finally by erasing any occurrence of 0 (by I-|) if there is at least one non-empty atom. ♦

Clearly, the canonical form of a term is in general not unique. Indeed, if $B\backslash x_1\backslash\ldots\backslash x_n$, is a canonical form of A, an equivalent term in canonical form can be obtained by applying the AC-| axioms to the concrete subterm B, or also by performing an α-conversion of the whole term, i.e., by applying to $B\backslash x_1\backslash\ldots\backslash x_n$ a renaming which is the identity on $\mathcal{F}(B)$, and then by reducing the resulting term in canonical form as in the proof of the last proposition.

The next corollary states that when dealing with equivalence classes of terms, instead of considering all the open terms we can restrict to concrete terms without loss of generality.

3.7 Corollary *(open canonical terms are concrete)*

Every open term in canonical form is concrete. Moreover, A is open iff it is equivalent to a concrete term A'.

Proof Let A be open and in canonical form. If A is not concrete it should be of the form $A = B\backslash x$, with $x \notin \mathcal{F}(B)$ because A is open, contradicting the hypothesis that A is in canonical form. Thus, A is open iff it is equivalent to an open term A' in canonical form (by Propositions 3.5 and 3.6), iff A' is concrete. ♦

Two concrete terms (i.e., elements of $\mathcal{CA}$) in canonical form are equivalent iff they are permutations of the same list of atoms.

3.8 Proposition *(sets of atoms are identical for equivalent concrete canonical terms)*

Let $A = A_1 \mid \ldots \mid A_n$ and $A' = A'_1 \mid \ldots \mid A'_m$ be two equivalent canonical terms in $\mathcal{CA}$, where A_i, A'_j are atoms. Then it can be proved that $A \approx A'$ by using only the axioms AC-|. As a consequence, n = m and the sets $\{A_i\}_{i \le n}$ and $\{A'_j\}_{j \le m}$ are equal.

Proof Let us show that any proof of $A \approx A'$ can be transformed into an equivalent proof which uses just the AC-| axioms. Let $\langle A = B_1, \ldots, B_s = A' \rangle$ be a proof of $A \approx A'$, i.e., a sequence of terms such that $B_i \sim B_{i+1}$ is an axiom instance for each $1 \le i < s$. Since by Proposition 3.5 all B_i are open, we can safely erase all the restriction operators from them (by repeated applications of axiom EL-\). The resulting sequence of concrete terms is still a proof for $A \approx A'$, since the instances of axioms involving restrictions (i.e., C-\, EL-\, FAC-\, SWAP) become identities when the restrictions are eliminated from both sides. Next we can eliminate all the substitutions from the terms of the sequence by applying repeatedly axioms DIS-[], COMP-[], and MAP to each one of them. The resulting sequence of terms is again a proof that $A \approx A'$, since all the instances of axioms DIS-[], COMP-[], and MAP become identities when substitutions are evaluated in both sides. Then we can erase all the occurrences of 0 from the sequence by use of I-|, if there is at least one atom different from 0. Finally we have to delete all multiple occurrences of atoms from all terms, and all node atoms which occur in some edge atom. We can mark all the occurrences of an atom in a term with unique identifiers, starting from the first term of the sequence, and keeping the marking consistent from one term to the next. Since the first and the last terms are in canonical form, all the copies of an atom generated by axiom D-| will eventually be deleted by an application of D-| in the opposite direction. Thus it is possible to choose the marking in such a way that each atom has the same mark in the first and in the last term. Then we can delete from the terms of the sequence all the atoms with a mark not occurring in the first or last term of the sequence. Clearly, in this way all node atoms which occur in edge atoms are eliminated from the sequence, because they do not appear in the first and last term, which are in canonical form. The resulting sequence is equivalent to the initial one, and it is again a proof because the AC-| axioms become trivial when one or more atoms are consistently deleted from both sides. Moreover, all terms of the sequence are in canonical form: then the statement follows by observing that if C and C' ($\neq$ C) are concrete and in canonical form, then C ~ C' can be an instance only of an AC-| axiom. ♦

In the next section we will show that finite graphs are a model for open terms. In order to obtain a similar result for partially abstract graphs, in Section 3.3 we have to restrict to well-formed terms. This is due the fact that not all terms can be modeled faithfully by partially abstract graphs. Intuitively, the well-formedness condition warrants that if a node is abstract, then all the edges connected to that node are abstract too. Since a node is made abstract by a restriction operator, this condition will be expressed as a requirement over the restricted variables of a term.

3.9 Definition *(well-formed terms)*

The set of free variables of a term A is naturally equipped with a dependency relation $\mathfrak{D}(A) \subseteq (\mathcal{F}(A) \cap \mathcal{N}) \times (\mathcal{F}(A) \cap \mathcal{E})$. Intuitively, $\langle a, p \rangle \in \mathfrak{D}(A)$ (read 'a depends on p') if the node variable *a* occurs as argument of the edge variable p. $\mathfrak{D}(A)$ is defined inductively as follows:

- $\mathfrak{D}(0) = \mathfrak{D}(a) = \emptyset$
- $\mathfrak{D}(p(a_1, \ldots, a_n)) = \{\langle a_i, p \rangle \mid 1 \leq i \leq n\}$
- $\mathfrak{D}(A_1 \mid A_2) = \mathfrak{D}(A_1) \cup \mathfrak{D}(A_2)$
- $\mathfrak{D}(A[\Phi]) = \Phi(\mathfrak{D}(A)) = \{\langle \Phi(x), \Phi(y) \rangle \mid \langle x, y \rangle \in \mathfrak{D}(A)\}$
- $\mathfrak{D}(A\backslash x) = \mathfrak{D}(A) - \{\langle z, y \rangle \mid z = x \vee y = x\}$

It is easy to check that if A ≈ A', then $\mathfrak{D}(A) = \mathfrak{D}(A')$. A term A is *truly well-formed* if in any subterm B\x of A, x in $\mathcal{F}(B)$ implies that x does not depend on any variable. More formally,

- All terms of the form $p_n(a_1, \ldots, a_n)$, *a*, or 0 are truly well-formed
- $A_1 \mid A_2$ is truly well-formed if both A_1 and A_2 are truly well-formed
- A[Φ] is truly well-formed if A is truly well-formed
- A\x is truly well-formed if A is truly well-formed and either $x \notin \mathcal{F}(A)$ or $\neg\exists$ y. $\langle x, y \rangle \in \mathfrak{D}(A)$.

Clearly, all open terms are truly well-formed. A term A is *well-formed* if there exists a term A' such that A ≈ A' and A' is truly well-formed. The sub-algebra of $\mathcal{A}$ including all well-formed terms will be denoted by $\mathcal{W}f\mathcal{A}$. ♦

3.2 Open terms and graphs

In this section we first show how to associate with each term A of $\mathcal{A}$ a finite graph **Gr**(A), the concrete graph of A. As expected, parallel composition is interpreted as union, the substitutions are evaluated on their arguments, and the restriction of a variable removes that item from the graph: if the item is a node, also the connected edges have to be eliminated to avoid 'dangling' edges. We also show that the restriction of **Gr** to **C**$\mathcal{A}$, called **CGr**, is injective. Conversely, each graph G will be mapped to a term **CT**(G) in **C**$\mathcal{A}$, such that **CGr**(**CT**(G)) ≅ G. Finally, by defining a suitable categorical structure over the terms, both **CGr** and **CT** can be extended to arrows and the resulting functors form an adjunction: **CT** is the left adjoint, and it is a full and faithful embedding.

3.10 Definition *(from terms to graphs: function $\mathfrak{Gr}$: $\mathcal{A} \to |FGraph|$)*

The finite *concrete graph* associated with a term A, $\mathfrak{Gr}(A) = \langle N_A, E_A, c_A \rangle$, is defined inductively as follows

- $\mathfrak{Gr}(0) = \langle \varnothing, \varnothing, \varnothing \rangle$, the empty graph.
- $\mathfrak{Gr}(a) = \langle \{a\}, \varnothing, \varnothing \rangle$
- $\mathfrak{Gr}(p(a_1, \ldots, a_n)) = \langle \{a_1, \ldots, a_n\}, \{\underline{p(a_1, \ldots, a_n)}\}, c \rangle$, with $c(\underline{p(a_1, \ldots, a_n)}) = a_1 \cdot \ldots \cdot a_n$.
- $\mathfrak{Gr}(A \mid A') = \mathfrak{Gr}(A) \cup \mathfrak{Gr}(A')$, where union is componentwise, possibly non-disjoint.
- $\mathfrak{Gr}(A[\Phi]) = \langle \Phi(N_A), \Phi(E_A), c' \rangle$, where

  $\Phi(N_A) = \{\Phi(a) \mid a \in N_A\}$

  $\Phi(E_A) = \{\underline{q(b_1, \ldots, b_n)} \mid \underline{p(a_1, \ldots, a_n)} \in E_A, \Phi(p) = q, \Phi(a_i) = b_i\}$

  $c'(\underline{q(b_1, \ldots, b_n)}) = b_1 \cdot \ldots \cdot b_n$.
- $\mathfrak{Gr}(A \backslash x) = \langle N_A - \{x\}, E_A - \{\underline{p(a_1, \ldots, a_n)} \mid p = x \text{ or } a_i = x \text{ for some } i\}, c' \rangle$, where c' is the obvious restriction of c_A.

From the definition it is obvious that for each term A, $\mathfrak{Gr}(A)$ is a finite graph. Thus $\mathfrak{Gr}$ is a function $\mathfrak{Gr}$: $\mathcal{A} \to$ **|FGraph|**. Function $\mathfrak{CGr}$ is defined as the restriction of $\mathfrak{Gr}$ to $\mathfrak{C}\mathcal{A}$. ♦

3.11 Proposition *(properties of $\mathfrak{Gr}$)*

Function $\mathfrak{Gr}$ is well-defined w.r.t. equivalence classes (i.e., $A \approx A' \Rightarrow \mathfrak{Gr}(A) = \mathfrak{Gr}(A')$). Moreover, $\mathfrak{CGr}$: $\mathfrak{C}\mathcal{A} \to$ **|FGraph|** is injective.

Proof. The fact that $\mathfrak{Gr}$ is well-defined w.r.t. the equivalence relation comes from the fact that $\mathfrak{Gr}(A) = \mathfrak{Gr}(B)$ for each axiom A = B. This is obvious for the ACID-| axioms, because the parallel composition '|' is interpreted as union, which is associative, commutative and idempotent, and has the empty set as identity; for axioms MAP and COMP-[], because substitutions are interpreted by explicitly evaluating them on all variables; for DIS-[], because the evaluation of a function distributes w.r.t. union, for C-\, because restriction is interpreted as set difference, which satisfies (S - S') - S" = (S - S") - S'; for FAC-\, because set difference distributes w.r.t. union (i.e., $(S \cup S') - T = (S - T) \cup (S' - T)$), and because $\mathfrak{Gr}(A \backslash x) = \mathfrak{Gr}(A)$ if x is not free in A; and for SWAP because for any function f and sets S and $\{x\}$, $f(S - \{x\}) = f(S) - \{f(x)\}$ if $f(s) \neq f(x)\ \forall s \neq x$.

Finally, since $\mathfrak{CGr}$ is well defined w.r.t. the equivalence relation it easily follows that if $A_1 \mid \ldots \mid A_n \mid B_1 \mid \ldots \mid B_m$ is any canonical form of an open term A, where each A_i is an edge atom and each B_j is a node atom, then $E_A = \{\underline{A_i}\}_{i \le n}$ and $IN_A = \{B_j\}_{j \le m}$ (the set of isolated nodes of $\mathfrak{Gr}(A)$). On the other hand, if A and A' are open and $\mathfrak{CGr}(A) \neq \mathfrak{CGr}(A')$, then either $IN_A \neq IN_{A'}$, or $E_A \neq E_{A'}$, since the connection function is completely coded in the edges. Thus $A \not\approx A'$ by Proposition 3.8, i.e., $\mathfrak{CGr}$: $\mathfrak{C}\mathcal{A} \to$ **|FGraph|** is injective. ♦

The translation of finite graphs to concrete terms is very simple.

3.12 Definition *(from graphs to concrete terms: function $\mathfrak{CT}$: |FGraph| $\to \mathfrak{C}\mathcal{A}$)*

The *concrete term representing a graph* $G = \langle N, E, c \rangle$, $\mathfrak{CT}(G)$, is defined as

$$\mathfrak{CT}(G) = a_1 \mid \ldots \mid a_m \mid p_1(a_{11}, \ldots, a_{1k_1}) \mid \ldots \mid p_n(a_{n1}, \ldots, a_{nk_n}) \mid 0$$

where $\{a_i\}_{i \le m}$ is an enumeration of set N, $\{p_j\}_{j \le n}$ is an enumeration of the set E, and $c(p_i) = a_{i1} \cdot \ldots \cdot a_{ik_i}$ for all $1 \le i \le n$. Note that choosing a different enumeration of E or of N one gets an equivalent term. If both E and N are empty then $\mathfrak{CT}(G) = 0$, otherwise the empty atom 0 is absorbed by axiom I-|. ♦

The following fact is a direct consequence of the definitions.

3.13 Fact *($G \cong \mathcal{CGr}(\mathcal{CT}(G))$)*

If G is a graph, then $G \cong \mathcal{CGr}(\mathcal{CT}(G))$. If G is not empty, the isomorphism h: $G \rightarrow \mathcal{CGr}(\mathcal{CT}(G))$ is the identity on nodes, and maps an edge p to the unique edge of the form $\underline{p(a_1, \ldots, a_n)}$. ♦

This correspondence between graphs and terms can be extended to an adjunction between the corresponding categories. We first define a categorical structure over terms. The arrows of the category are contexts, i.e., terms with a 'hole' denoted by a dot (•), under a suitable equivalence relation.

3.14 Definition *(contexts)*

A *context* is a term generated by the following grammar:

$$C ::= \bullet \mid C \,|\, A \mid C[\Phi] \mid C\backslash x$$

where A is a term of the algebra $\mathcal{A}$, Φ is a substitution, and x is a variable (as in Definition 3.1). The set of free variables of a context is determined as in Definition 3.2, adding the clause $\mathcal{F}(\bullet) = \mathcal{N}$ (i.e., the whole set of variables, since any term can replace the placeholder '•'). As for terms, a context is *concrete* if it does not contain any restriction operator.

Contexts are subjected to the same axioms as for terms, treating the operators in their syntax as the corresponding operators in the syntax of terms, and the placeholder • as a new unstructured atom with $\mathcal{N}$ as set of free variables. Two contexts C and C' are *equivalent* (written $C \approx C'$) if they are equivalent as terms, under the mentioned assumptions.

The *application of a context C to a term A*, written C{A}, is the term obtained by substituting the unique occurrence of • in C by A. Formally, $\bullet\{A\} = A$; $(C \,|\, A')\{A\} = C\{A\} \,|\, A'$; $C[\Phi]\{A\} = C\{A\}[\Phi]$); and $C\backslash x\{A\} = C\{A\}\backslash x$. A context C can be *composed* with a context C' by applying it to C' as if it were a term. We denote by $C \circ C'$ the composed context C{C'}.

If C is a concrete context, the *substitution associated with C*, Φ_C, is the substitution inductively defined as follows: $\Phi_\bullet = I\delta$; $\Phi_{C|A} = \Phi_C$; $\Phi_{C[\Psi]} = \Psi \circ \Phi_C$. The substitution Φ_C can also be characterized by using a canonical form of contexts. In fact, every concrete context C can be reduced (using the same techniques as in Proposition 3.6) to a canonical form like $\bullet[\Phi] \,|\, A_1 \,|\, \ldots \,|\, A_n \,|\, B_1 \,|\, \ldots \,|\, B_m$, where every A_i is a distinct edge atom, and every B_j is a distinct node atom not occurring in any A_i. Then $\Phi_C \triangleq \Phi$.

Let $K \subset \mathcal{E} \cup \mathcal{N}$ be a finite set of variables. Two concrete contexts C and C' are *K-equivalent* (written $C \approx_K C'$), iff for each $x \in K$, $\Phi_C(x) = \Phi_{C'}(x)$. This notion of equivalence is quite weak, since it does not take into account the atoms which are put in parallel with the placeholder. As a consequence, $C \approx C' \Rightarrow C \approx_K C'$ for all finite K, but the converse does not hold. ♦

The following fact follows directly from the definitions.

3.15 Fact *(equivalence of contexts is preserved by composition)*

Let A be a term and $C_1, \ldots, C_4$ be concrete contexts. Then if $C_1\{A\} \approx C_2\{A\}$, $C_1 \approx_{\mathcal{F}(A)} C_2$, $C_3\{C_1\{A\}\} \approx C_4\{C_2\{A\}\}$, and $C_3 \approx_{\mathcal{F}(C_1\{A\})} C_4$, then $C_3 \circ C_1 \approx_{\mathcal{F}(A)} C_4 \circ C_2$. Moreover, if C is any context then $\bullet \circ C = C \circ \bullet = C$. ♦

We define now a category having the terms of $\mathcal{A}$ as objects and contexts as arrows.

3.16 Definition *(categories of terms and concrete contexts)*

Category **Term** is defined as follows. The objects of **Term** are terms of the algebra $\mathcal{A}$, i.e., equivalence classes of the terms of Definition 3.1 under the axioms of Definition 3.3. The arrows of **Term** from [A] to [B] are subsets of equivalence classes of concrete contexts modulo $\approx_{\mathcal{F}(A)}$. More precisely, if C is a concrete context such that $C\{A\} \approx B$, then arrow $[C]_{[A]}: [A] \rightarrow [B]$ is the equivalence class $[C]_{[A]} \triangleq \{C' \mid C'\{A\} \approx B \text{ and } C' \approx_{\mathcal{F}(A)} C\}$. Notice that in general $C \approx_{\mathcal{F}(A)} C'$ does not imply $C\{A\} \approx C'\{A\}$. The definition is well given because if $A \approx A'$, then $C\{A\} \approx C\{A'\}$. An arrow $[C]_{[A]}$ will be denoted also by $[C]_A$.

It is easy to check that **Term** is a category, by defining arrow composition as context composition: $[C']_{C\{A\}} \circ [C]_A = [C' \circ C]_A$. The correctness of this definition follows from the above Fact. The associativity of arrow composition comes from the properties of substitutions. Finally, the identity arrow of an object A is the equivalence class $[\bullet]_A$: the above Fact states that identities behave correctly.

Category **CTerm** is the full subcategory of **Term** having elements of $\mathcal{CA}$ as objects. Similarly, **WfTerm** is the full subcategory of **Term** having elements of $\mathcal{WfA}$ as objects, i.e., all well-formed terms. ♦

We are now ready to extend functions $\mathcal{Gr}$ and $\mathcal{CT}$ to arrows.

3.17 Definition *(functors $\mathcal{Gr}$: Term → FGraph and $\mathcal{CT}$: FGraph → CTerm)*

1) We extend function $\mathcal{Gr}$ to arrows, yielding a functor $\mathcal{Gr}$: **Term** → **FGraph**. Let $[C]_A: [A] \rightarrow [C\{A\}]$ be an arrow of **CTerm**. Then $\mathcal{Gr}([C]_A): \mathcal{Gr}(A) \rightarrow \mathcal{Gr}(C\{A\})$ is defined inductively on the structure of C. Intuitively, $\mathcal{Gr}$ maps $[\bullet]_A$ to the identity morphism of $\mathcal{Gr}(A)$; if C applied to A produces $A \mid A'$, then $\mathcal{Gr}([C]_A)$ is the inclusion morphism from $\mathcal{Gr}(A)$ to $\mathcal{Gr}(A \mid A')$; and if C applied to A yields $A[\Phi]$, then $\mathcal{Gr}([C]_A)$ is a graph morphism that applies the substitution Φ to all the variables of $\mathcal{Gr}(A)$. More formally, $\mathcal{Gr}([C]_A): \mathcal{Gr}(A) \rightarrow \mathcal{Gr}(C\{A\})$ is defined inductively as follows:

- $\mathcal{Gr}([\bullet]_A) = id_A: \mathcal{Gr}(A) \rightarrow \mathcal{Gr}(A)$
- $\mathcal{Gr}([C \mid A']_A) = in_{C\{A\}|A'} \circ \mathcal{Gr}([C]_A): \mathcal{Gr}(A) \rightarrow \mathcal{Gr}(C\{A\} \mid A')$
  where $in_{C\{A\}|A'}: \mathcal{Gr}(C\{A\}) \rightarrow \mathcal{Gr}(C\{A\} \mid A')$ is the obvious graph inclusion.
- $\mathcal{Gr}([C[\Phi]]_A) = k_{C\{A\},\Phi} \circ \mathcal{Gr}([C]_A): \mathcal{Gr}(A) \rightarrow \mathcal{Gr}(C\{A\}[\Phi])$
  where $k_{B,\Phi}: \mathcal{Gr}(B) \rightarrow \mathcal{Gr}(B[\Phi])$ is defined as $(k_{B,\Phi})_N(a) = \Phi(a)$ for each $a \in N_B$, and $(k_{B,\Phi})_E(\underline{p(a_1, \ldots, a_n)}) = \underline{q(b_1, \ldots, b_n)}$ for all $\underline{p(a_1, \ldots, a_n)} \in E_A$, if $\Phi(p) = q$, and $\Phi(a_i) = b_i$ (it is easy to check that $k_{B,\Phi}$ is indeed a graph morphism).

It is routine to check that $\mathcal{Gr}$ is indeed a functor. Functor $\mathcal{CGr}$ is defined as the restriction of $\mathcal{Gr}$ to the full subcategory **CTerm** of **Term**, thus $\mathcal{CGr}$: **CTerm** → **FGraph**.

2) Similarly, we extend function $\mathcal{CT}$ to arrows, yielding a functor $\mathcal{CT}$: **FGraph** → **CTerm**. If $h = \langle h_N, h_E \rangle: G \rightarrow H$, then $\mathcal{CT}(h): \mathcal{CT}(G) \rightarrow \mathcal{CT}(H)$ is defined as

$$\mathcal{CT}(h) \triangleq [\ \bullet[h_N \cup h_E] \mid \mathcal{CT}(H)]_{\mathcal{T}(G)},$$

The definition is well given because $\mathcal{CT}(h)\{\mathcal{CT}(G)\} \approx \mathcal{CT}(H)$: indeed, if A is an atom of $\mathcal{CT}(G)$, then $A[h_N \cup h_E]$ is an atom of $\mathcal{CT}(H)$, and it is absorbed by the idempotence of the | operator. Again, the fact that $\mathcal{CT}$ is a functor is routine check. ♦

It is worth explaining here the restriction to *concrete* contexts in the definition of category **Term**. This choice is motivated by the fact that, although contexts including restriction have a precise interpretation in the category of graphs, the two functors just defined could not be extended consistently to arbitrary contexts. If fact, while each concrete context C is associated with an arrow from $\mathcal{Gr}(A)$ to $\mathcal{Gr}(C\{A\})$, a context like $\bullet\backslash x$ would induce naturally an arrow $\mathcal{Gr}([\bullet\backslash x]): \mathcal{Gr}(A\backslash x) \rightarrow \mathcal{Gr}(A)$ in a contravariant way, which is the obvious inclusion. Although they are kept outside the categorical structure, non-concrete contexts will be needed for the results presented later in Section 4.

3.18 Theorem *($\mathcal{CT}$ is a full and faithful embedding with right adjoint $\mathcal{CGr}$)*

Functor $\mathcal{CT}$: **FGraph** → **CTerm** is full and faithful, and is the left adjoint to $\mathcal{CGr}$: **CTerm** → **FGraph**.

Proof. $\mathcal{CT}$ is full. Let $[C]_{\mathcal{CT}(G)}$: $\mathcal{CT}(G) \to \mathcal{CT}(H)$ be an arrow of **CTerm** and let Φ_C be the substitution associated with C, as in Definition 3.14. Define h: $G \to H$ as $h = \langle h_N, h_E \rangle$, where h_N is the restriction of Φ_C to the node variables appearing free in $\mathcal{CT}(G)$, and similarly $h_E \triangleq \Phi_C \mid_{\mathcal{E} \cap \mathcal{F}(\mathcal{CT}(G))}$. Then $\mathcal{CT}(h) = [\,\bullet[h_N \cup h_E] \mid \mathcal{CT}(H)]_{\mathcal{T}(G)} = [C]_{\mathcal{CT}(G)}$ because $C \approx_{\mathcal{F}(\mathcal{CT}(G))} \bullet[h_N \cup h_E] \mid \mathcal{CT}(H)$.

$\mathcal{CT}$ is faithful. Let h, h': $G \to H$ be such that $\mathcal{CT}(h) = \mathcal{CT}(h')$. This implies that $\bullet[h_N \cup h_E] \mid \mathcal{CT}(H) \approx_{\mathcal{CT}(G)} \bullet[h'_N \cup h'_E] \mid \mathcal{CT}(H)$, and thus h = h'.

Finally, it is routine to check that $\mathcal{CT}$ is the left adjoint to $\mathcal{CGr}$. The unit component η_G: $G \to \mathcal{CGr}(\mathcal{CT}(G))$ is the isomorphism mentioned in Fact 3.13. The counit component ε_A: $\mathcal{CT}(\mathcal{CGr}(A)) \to A$ is the context $\bullet[\Phi]$, where Φ is the identity on $\mathcal{N}$, and renames every edge of the form $\underline{p(a_1,\ldots,a_n)}$ to p. ♦

3.3 Well-formed terms as partially abstract graphs

In this section we extend the correspondence among open terms and concrete graphs to a similar correspondence among well-formed terms and partially abstract graphs. We first show how to associate a pa-graph $\mathcal{PaG}(A)$ with each term A of the algebra $\mathcal{A}$. Then in Theorem 3.21 we prove that the definition is correct for well-formed terms, in the sense that equivalent terms are associated with the same pa-graph. This does not hold in general for non-well-formed terms. Next we show how to extract from every pa-graph X a well-formed term $\mathcal{WfT}(X)$. Finally $\mathcal{PaG}$ and $\mathcal{WfT}$ are extended to arrows, and an adjunction similar to the one presented above is proved to hold. A relevant fact, that will be exploited in Section 4, is correspondence between the application of a concrete context to a well-formed term and a morphism of pa-graphs, i.e., a pushout in category **FGraph**.

The pa-graph associated with a term A, $\mathcal{PaG}(A)$, will be determined by indicating a representative of the equivalence class (see Definition 2.9), i.e., a specific monomorphism denoted by $m(A)$: $\mathcal{Gr}(A) \rightarrowtail \mathcal{Abs}(A)$. $\mathcal{Gr}(A)$, the concrete part of $\mathcal{PaG}(A)$, is defined as in Definition 3.10. On the other hand, $\mathcal{Abs}(A)$, the abstract part of $\mathcal{PaG}(A)$, will be defined in most cases as a colimit object, thus up to isomorphism. We remind that if h is an injective morphism, then [h] denotes the unique pa-graph containing h.

3.19 Definition *(from terms to pa-graphs: function $\mathcal{PaG}$: $\mathcal{A} \to$ |PA-FGraph|)*

The partially abstract graph $\mathcal{PaG}(A) \triangleq [m(A): \mathcal{Gr}(A) \rightarrowtail \mathcal{Abs}(A)]$ associated with a term A is defined by induction as follows. The definition exploits the functor $\mathcal{Gr}$: **Term** $\to$ **FGraph** of Definition 3.17.

- If A is an atom, i.e., A = 0, A = a, or A = $p(a_1, \ldots, a_n)$, then $\mathcal{PaG}(A) = [id_{\mathcal{Gr}(A)}]$.
- If A = B | C, $\mathcal{PaG}(B) = [m(B): \mathcal{Gr}(B) \rightarrowtail \mathcal{Abs}(B)]$ and $\mathcal{PaG}(C) = [m(C): \mathcal{Gr}(C) \rightarrowtail \mathcal{Abs}(C)]$, then $m(A)$ is obtained as in the next diagram. Graph $\mathcal{Gr}(A)$ (= $\mathcal{Gr}$(B | C)) is the union of graphs $\mathcal{Gr}(B)$ and $\mathcal{Gr}(C)$ (see Definition 3.10), with the obvious injection morphisms in_1 and in_2. $\mathcal{Abs}(A)$ is defined as the colimit object of the resulting diagram in **Graph**; and $m(A)$: $\mathcal{Gr}(A) \rightarrowtail \mathcal{Abs}(A)$ is the corresponding morphism (which is injective by Proposition 2.8).

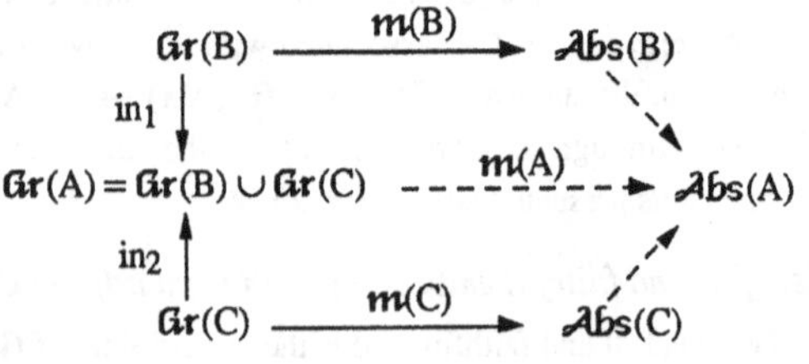

It is worth stressing that $in_1 = Gr([\cdot \mid C]_B)$ and $in_2 = Gr([\cdot \mid B]_C)$, respectively.

- If $A = B[\Phi]$ and $PaG(B) = [m(B): Gr(B) \rightarrowtail Abs(B)]$, then $m(A): Gr(A) \rightarrowtail Abs(A)$ is defined as the bottom arrow in the following pushout diagram in category **Graph**. Morphism $m(A)$ is injective by Proposition 2.8.

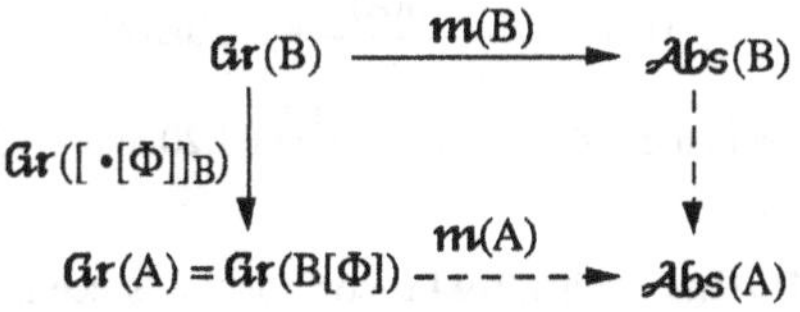

$$
\begin{array}{ccc}
Gr(B) & \xrightarrow{\ m(B)\ } & Abs(B) \\[2pt]
\Big\downarrow{\scriptstyle Gr([\ \cdot[\Phi]]_B)} & & \Big\downarrow \\[6pt]
Gr(A) = Gr(B[\Phi]) & \dashrightarrow{\ m(A)\ } & Abs(A)
\end{array}
$$

- Finally, if $A = B\backslash x$, and $PaG(B) = [m(B): Gr(B) \rightarrowtail Abs(B)]$, then $Abs(A) \triangleq Abs(B)$, and $Gr(A) = Gr(B\backslash x)$. The arrow $m(A): Gr(A) \rightarrowtail Abs(A)$ is obtained as the composition of $m(B)$ with the obvious injection in: $Gr(B\backslash x) \rightarrowtail Gr(B)$.

$$
Gr(B\backslash x) \xrightarrow{\ in\ } Gr(B) \xrightarrow{\ m(B)\ } Abs(B\backslash x) = Abs(B)
$$
$$
\underbrace{\hspace{7cm}}_{m(B\backslash x)}
$$
$\blacklozenge$

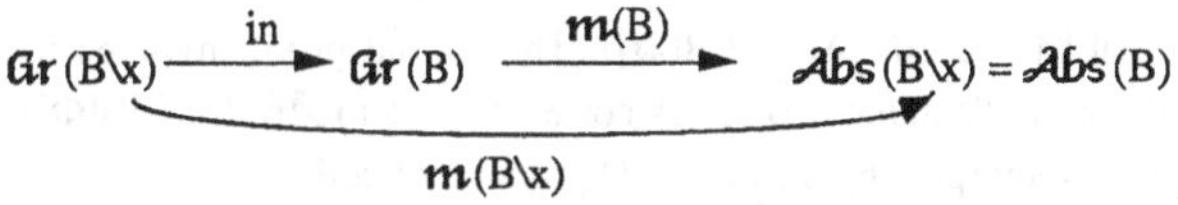

3.20 Lemma *(open terms correspond to classes of isomorphisms)*

1) If A is open, then $PaG(A)$ is the class of isomorphisms $[id_{Gr(A)}]$.
2) If A and B are open, $PaG(A) = PaG(B)$ iff $Gr(A) = Gr(B)$.

Proof. 1) If A is an atom, the statement holds by definition. If $A = B \mid C$ (resp. $A = D[\Phi]$) with B, C (resp. D) open, then by induction hypothesis and by Proposition 2.8 it follows that $PaG(A)$ is a class of isomorphisms. Then the statement follows by observing that if $X = [h: G \rightarrowtail H]$ is a pa-graph, and h is an isomorphism, then $X = [id_G]$; in fact, by definition of pa-graph, $id_G = h^{-1} \circ h \in X$. Finally, if $A = B\backslash x$ with B open and $x \notin F(B)$, the statement follows by induction hypothesis since $Gr(A) = Gr(B)$.

2) The *only if part* is obvious. For the *if part*, by point 1) we have $PaG(A) = [id_{Gr(A)}] = [id_{Gr(B)}] = PaG(B)$. $\blacklozenge$

3.21 Theorem *(PaG is well-defined of WfA)*

If $A, B \in WfA$ and $A \approx B$, then $PaG(A) = PaG(B)$.

Proof. By Proposition 3.11, equivalent terms are associated with the same concrete graph. Thus it is sufficient to verify that the abstract part of the pa-graphs associated with the lhs and the rhs of each axiom are isomorphic. We check this for all the axioms in turn.

C-l: $PaG(A \mid B) = PaG(B \mid A)$. The colimit construction defining object $Abs(A \mid B)$ is symmetric in the two arguments.

A-l: $PaG((A \mid B) \mid C) = PaG(A \mid (B \mid C))$. It is easy to check that both $Abs((A \mid B) \mid C)$ and $Abs(A \mid (B \mid C))$ are colimit objects of the diagram including $m(A)$, $m(B)$, $m(C)$, and the obvious inclusions of $Gr(A)$, resp. $Gr(B)$, resp. $Gr(C)$ into $Gr(A) \cup Gr(B) \cup Gr(C)$. Thus they are isomorphic by the universal property of colimits.

I-l: $PaG(A \mid 0) = PaG(A)$. Since $Gr(0) = \emptyset$ is initial in **Graph** and $PaG(0) = [id_\emptyset]$, then the following is a colimit diagram. Thus $Abs(A) \cong Abs(A \mid \emptyset)$.

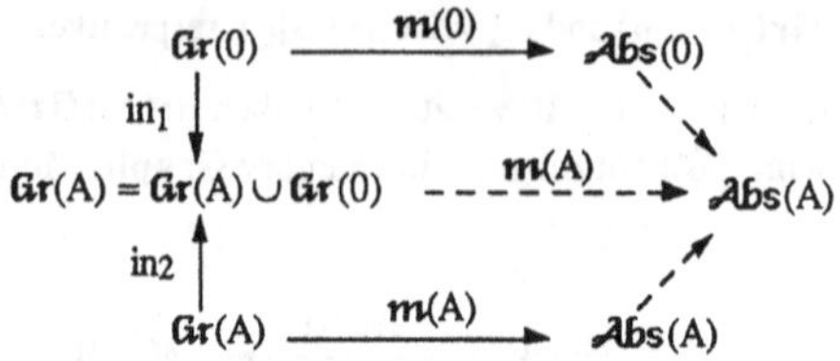

D-|, MAP: For each axiom the statement follows by Lemma 3.20 and Proposition 3.11, since both sides are open.

COMP-[]: PaG(A[Φ][Ψ]) = PaG(A[Ψ∘Φ]). By Proposition 2.8, since the composition of two pushout squares is again a pushout square.

C-\, EL-\: PaG(A\x\y) = PaG(A\y\x); PaG(A\x) = PaG(A), if x is not free in A. Indeed, the abstract graphs associated with the lhs and rhs of the axioms are identical (restriction does not change them by Definition 3.19).

DIS-[]: PaG((A | B)[Φ]) = PaG(A[Φ] | B[Φ]). The formal proof involves the construction of a polyhedron as depicted below. Then the two vertex corresponding to Abs((A | B)[Φ]) and to Abs(A[Φ] | B[Φ]) can be shown to be isomorphic by the universal property of colimits.

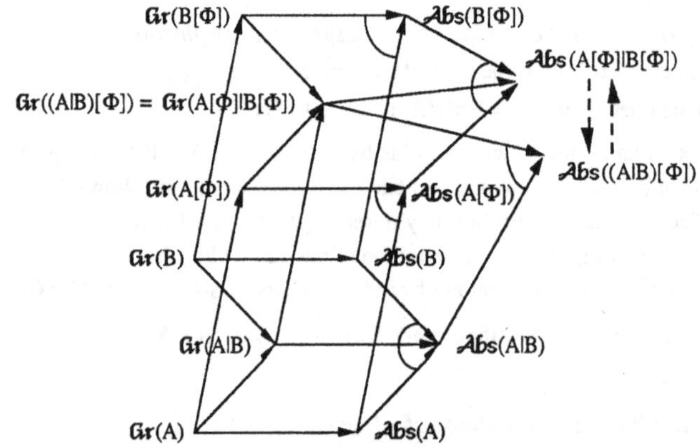

FAC-\: PaG(A\x | B) = PaG((A | B)\x), if x is not free in B. It is easy to show that square (1) in the following diagram is a pushout, assuming that x is not in Gr(B), and that both A and A\x are well-formed. Then the universal property of colimits can be used to prove the existence and uniqueness of the dashed arrows, and thus that Abs(A\x | B) ≅ Abs((A | B)\x) under the given condition.

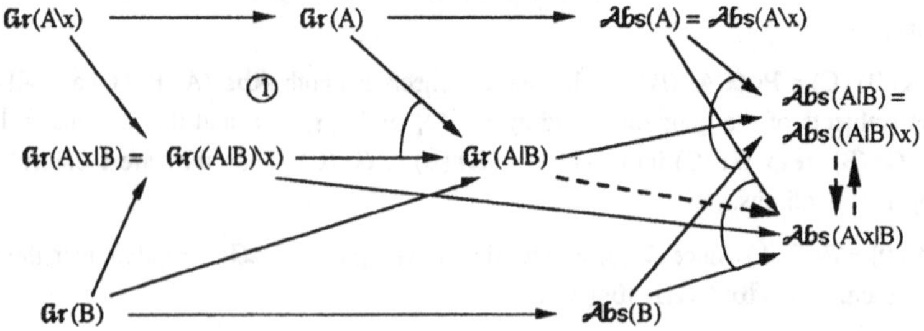

SWAP: $\mathcal{PaG}(A\backslash x[\Phi]) = \mathcal{PaG}(A[\Phi]\backslash\Phi(x))$, if $\forall y \in \mathcal{F}(A\backslash x)$, $\Phi(y) \neq \Phi(x)$. Using the side condition and the well-formedness of both A and A\x, it could be shown that square (1) below is a pushout. Then as usual the two colimit objects are isomorphic by the universal property.

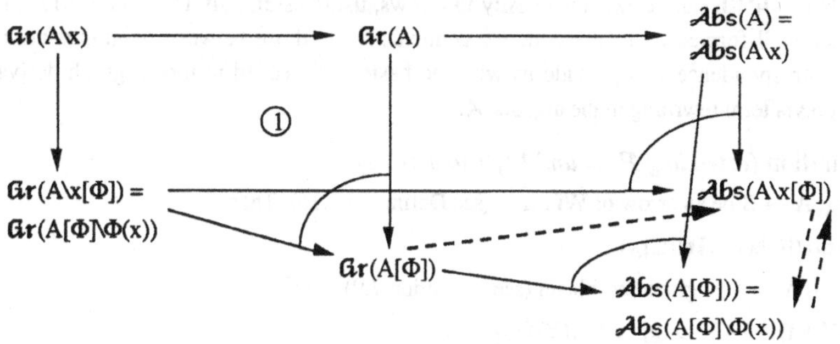

♦

The next definition shows how to associate with each pa-graph X a well-formed term $\mathcal{WfT}(X)$, such that $\mathcal{PaG}(\mathcal{WfT}(X)) \equiv X$.

3.22 Definition *(form pa-graphs to well-formed terms. $\mathcal{WfT}$: |PA-FGraph| $\to \mathcal{WfA}$)*
Let $X = [h: G \rightarrowtail H]$ be a pa-graph. Then $\mathcal{WfT}(X)$, the *well-formed term representing* X is defined as

$$\mathcal{WfT}(X) \triangleq (\mathcal{CT}(H)\backslash x_1\backslash \ldots \backslash x_n)[h^{-1}]$$

where $\{x_1, \ldots, x_n\}$ is the set of items of H which are not in the image of G through h (i.e., $\{x_1, \ldots, x_n\} = \{x \in H \mid \neg\exists\, y \in G.\ x = h(y)\}$, and h^{-1} improperly denotes the substitution such that $h^{-1}(y) = x$ if $h(x) = y$, and $h^{-1}(y) = y$ otherwise (it is well defined because h is injective). $\mathcal{WfT}(X)$ is well-formed, because if a node is restricted then it is not in the image of h; thus every edge *e* connected to that node cannot be in the image of h, and therefore *e* is restricted as well. ♦

3.23 Fact *(properties of $\mathcal{WfT}$)*

1) $\mathcal{WfT}$ is well defined, i.e., if $h: G \rightarrowtail H$ and $k: G \rightarrowtail K$ are in a pa-graph X, then

$$(\mathcal{CT}(H)\backslash x_1\backslash \ldots \backslash x_n)[h^{-1}] \approx (\mathcal{CT}(K)\backslash y_1\backslash \ldots \backslash y_m)[k^{-1}],$$

where $\{x_1, \ldots, x_n\}$ (resp. $\{y_1, \ldots, y_m\}$) is the set of items of H (resp. of K) which are not in the image of G through h (resp. through k).

2) $\mathcal{PaG}(\mathcal{WfT}(X)) \equiv X$.

Proof outline. 1) By Definition 3.1 there exists an isomorphism $f: K \to H$ such that $f \circ k = h$. Thus

$(\mathcal{CT}(H)\backslash x_1\backslash \ldots \backslash x_n)[h^{-1}] = (\mathcal{CT}(H)\backslash x_1\backslash \ldots \backslash x_n)[k^{-1} \circ f^{-1}] = (\mathcal{CT}(H)\backslash x_1\backslash \ldots \backslash x_n)[f^{-1}][k^{-1}] =$
$= (\mathcal{CT}(H)[f^{-1}]\backslash f^{-1}(x_1)\backslash \ldots \backslash f^{-1}(x_n))[h^{-1}]$, because f is an isomorphism,
$= (\mathcal{CT}(K)\backslash f^{-1}(x_1)\backslash \ldots \backslash f^{-1}(x_n))[h^{-1}]$, and $\{f^{-1}(x_1), \ldots, f^{-1}(x_n)\}$ are exactly the items of K which are not in the image k.

2) Let $X = [h: G \rightarrowtail H]$ be a pa-graph and $\mathcal{WfT}(X) = (\mathcal{CT}(H)\backslash x_1\backslash \ldots \backslash x_n)[h^{-1}]$ be as in the previous definition. Then since $\mathcal{Gr}(\mathcal{CT}(H)) \cong H$ (by Fact 3.13), we have that $\mathcal{Gr}(\mathcal{CT}(H)\backslash x_1\backslash \ldots\backslash x_n) \cong G$, because $\{x_1, \ldots, x_n\}$ are all the items of H which are not in the image of G through the monomorphism h. On the other hand, $\mathcal{Abs}(\mathcal{CT}(H)\backslash x_1\backslash \ldots\backslash x_n) = \mathcal{Abs}(\mathcal{CT}(H)) \cong H$, because by Definition 3.19 the restriction does not change the abstract part. Thus we have $\mathcal{PaG}(\mathcal{CT}(H)\backslash x_1\backslash \ldots\backslash x_n) \equiv X$. Finally, the statement follows because $\mathcal{Gr}(\bullet[h^{-1}])$ is an isomorphism, and thus by Proposition 2.8 and Definition 3.19 $X \equiv \mathcal{PaG}(\mathcal{CT}(H)\backslash x_1\backslash \ldots\backslash x_n)$

$\cong \mathcal{PaG}((\mathcal{CT}(H)\backslash x_1\backslash\ldots\backslash x_n)[h^{-1}]) = \mathcal{PaG}(\mathcal{WfT}(X))$. It is not difficult to check that, under the given hypotheses, the concrete part of $\mathcal{PaG}(\mathcal{WfT}(X))$ is exactly $\mathcal{Gr}(\mathcal{CT}(G))$, and that the mentioned isomorphism is based of the isomorphism between G and $\mathcal{Gr}(\mathcal{CT}(G))$, mentioned in Fact 3.13. ♦

Function $\mathcal{PaG}$ and $\mathcal{WfT}$ can be extended easily to arrows, using essentially Definition 3.17. Since the arrows between well-formed terms are concrete contexts, while those between pa-graphs are pushout squares, this correspondence will provide us with the basic tool needed to model graph derivation in category **Graph** via term rewriting in the algebra $\mathcal{A}$.

3.24 Definition *(extending $\mathcal{PaG}$ and $\mathcal{WfT}$ to arrows)*

1) Let $[C]_A: A \to B$ be an arrow of **WfTerm** (see Definition 3.16). Then

$$\mathcal{PaG}([C]_A) \triangleq \mathcal{Gr}([C]_A).$$

2) Let $f: X \to Y$ be a pa-graph morphism (see Definition 2.9).Then

$$\mathcal{WfT}(f) \triangleq [\ \bullet[f_N \cup f_E] \mid \mathcal{CT}(\mathcal{G}(Y))]_{\mathcal{WfT}(X)},$$

where $\mathcal{G}(Y)$ denotes the concrete part of Y. It must be stressed that if $X = [h: G \to H]$ and $Y = [k: L \to K]$, then f is an arrow from G to L by definition, and $\mathcal{WfT}(f) = \bullet[f_N \cup f_E] \mid \mathcal{CT}(\mathcal{G}(Y))]_{\mathcal{WfT}(X)} = \bullet[f_N \cup f_E] \mid \mathcal{CT}(L)]_{\mathcal{CT}(G)} = \mathcal{CT}(f)$ (see Definition 3.17).♦

3.25 Theorem *($\mathcal{PaG}$ is a functor: concrete contexts correspond to pushouts)*

Let A and B be well-formed terms and C be a concrete context such that $C\{A\} \approx B$. Then $\mathcal{PaG}([C]_A)$: $\mathcal{PaG}(A) \to \mathcal{PaG}(B)$ is an arrow of **PA-FGraph.** Moreover, $\mathcal{PaG}$ preserves identities and arrow composition. Thus, it is a functor $\mathcal{PaG}$: **WfTerm** $\to$ **PA-FGraph.**

Proof. Let $\mathcal{PaG}(A) = [\mathfrak{m}(A): \mathcal{Gr}(A) \rightarrowtail \mathcal{Abs}(A)]$ and $\mathcal{PaG}(B) = [\mathfrak{m}(B): \mathcal{Gr}(B) \rightarrowtail \mathcal{Abs}(B)]$. By the definition of pa-graph morphisms, we have to show that $\mathcal{PaG}([C]_A)$ is an arrow from $\mathcal{Gr}(A)$ to $\mathcal{Gr}(B)$ (and this is obvious because $\mathcal{PaG}([C]_A) = \mathcal{Gr}([C]_A)$ by definition, and $\mathcal{Gr}$ is a functor), and that there exists and arrow $g: \mathcal{Abs}(A) \to \mathcal{Abs}(B)$ making the following square a pushout in **FGraph**.

$$\begin{array}{ccc}
\mathcal{Gr}(A) & \xrightarrow{\mathfrak{m}(A)} & \mathcal{Abs}(A) \\
{\scriptstyle \mathcal{Gr}([C]_A)}\downarrow & & \downarrow{\scriptstyle g} \\
\mathcal{Gr}(B) & \xrightarrow{\mathfrak{m}(B)} & \mathcal{Abs}(B)
\end{array}$$

We prove the existence of such an arrow g by induction on the structure of C.

- If $C = \bullet$, we have $B = C\{A\} = A$, thus $\mathcal{Gr}(A) = \mathcal{Gr}(B)$ and $\mathcal{PaG}(A) = \mathcal{PaG}(B)$. Therefore by Definition 2.9 there exists an isomorphism, say g, such that $g \circ \mathfrak{m}(A) = \mathfrak{m}(B)$. The resulting square is a pushout, like every square having two parallel isomorphisms.

- Suppose now that $C = C' \mid B$. Consider the next colimit diagram, which defines $\mathcal{PaG}(C'\{A\} \mid B)$ by Definition 3.19.

First, note that $\mathfrak{m}(B)$ is an isomorphism by Lemma 3.20 because context $C'|B$ is concrete by hypothesis. Then it is easy to show that square (1) is a pushout. In fact, for any graph G and arrows f, k such that $k \circ \mathcal{Gr}([\bullet \mid B]) = f \circ \mathfrak{m}(C'\{A\})$, there is a unique arrow from $\mathcal{Abs}(C'\{A\} \mid B)$ to G. This is due to the universal property of $\mathcal{Abs}(C'\{A\} \mid B)$, since there is an obvious arrow from $\mathcal{Abs}(B)$ to G making everything commutative, namely $k \circ \mathcal{Gr}([C'\{A\} \mid \bullet]) \circ \mathfrak{m}(B)^{-1}$.

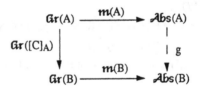

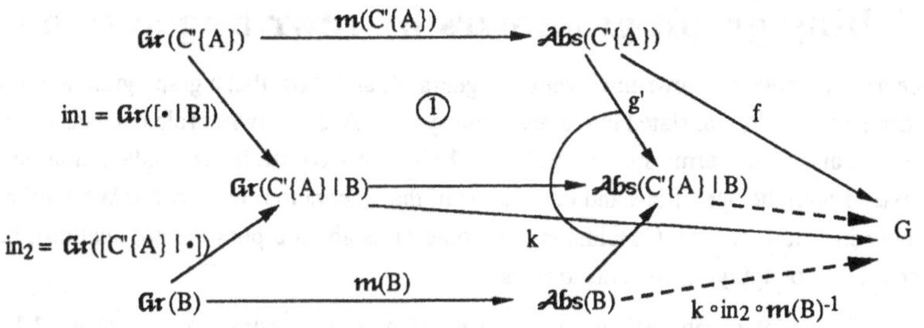

Let g' be the arrow making square (1) a pushout. Then the statement follows by induction hypothesis, by exploiting the fact that the composition of two pushouts is a pushout (Proposition 2.8), and the fact that (• | B) ∘ C' = C' | B.

• Finally, if C = C'[Φ], the statement follows directly by the definition of $\mathcal{PaG}$(A[Φ]) (see Definition 3.19), and by an inductive argument like the one above.

The fact that $\mathcal{PaG}$ preserves identities and arrow composition follows by the corresponding properties of functor $\mathcal{Gr}$. ♦

3.26 Proposition *(functor $\mathcal{WfT}$: PA-FGraph → WfTerm is well defined)*

If X and Y are pa-graphs and f: X → Y is a pa-graph morphism, then $\mathcal{WfT}$(f): $\mathcal{WfT}$(X) → $\mathcal{WfT}$(Y) is an arrow of **WfTerm**. Moreover, $\mathcal{WfT}$ preserves identities and arrow composition. Thus, it is a functor $\mathcal{WfT}$: **PA-FGraph** → **WfTerm**.

Proof outline. By Definition 3.16, one has to show that $\mathcal{WfT}$(f){$\mathcal{WfT}$(X)} ≈ $\mathcal{WfT}$(Y). Let X = [h: G ↣ H] and Y = [k: L ↣ K]. Then by definition f: G → L, and there exists an arrow g: H → K such that the square ‹h, g, f, k› is a pushout. By exploiting the existence of g and the equation $f \circ h^{-1} = k^{-1} \circ g$, it is possible to prove that $\mathcal{WfT}$(f){$\mathcal{WfT}$(X)} ≙ (•[$f_N \cup f_E$] | $\mathcal{CT}$(L)){($\mathcal{CT}$(H)\x$_1$\...\x$_n$)[h^{-1}]} ≈ $\mathcal{WfT}$(Y) = ($\mathcal{CT}$(K)\g(x$_1$)\...\g(x$_n$))[k^{-1}]. The rest of the proof is routine. ♦

3.27 Theorem *($\mathcal{WfT}$ is a full and faithful embedding with right adjoint $\mathcal{PaG}$)*

Functor $\mathcal{WfT}$: **PA-FGraph** → **WfTerm** is full and faithful. Moreover, $\mathcal{PaG}$: **WfTerm** → **PA-FGraph** is its right adjoint.

Proof. $\mathcal{WfT}$ is full. Let $[C]_{\mathcal{WfT}(X)}$: $\mathcal{WfT}$(X) → $\mathcal{WfT}$(Y) be an arrow of **WfTerm**, with X = [h: G ↣ H] and Y = [k: L ↣ K], and let Φ_C be the substitution associated with C, as in Definition 3.14. Define f: G → L as f = ‹f_N, f_E›, with $f_N = \Phi_C |_{N_G}$, and $f_E = \Phi_C |_{E_G}$. Then f: X → Y is a pa-graph morphism, and $\mathcal{WfT}$(f) = $[C]_{\mathcal{WfT}(X)}$. We first have to how that there is an arrow g: H → K such that ‹h, g, f, k› is a pushout square. This follows by the fact that $\mathcal{PaG}$ is a functor, and that $\mathcal{PaG}$($\mathcal{WfT}$(X)) ≡ X. Next we have $\mathcal{WfT}$(f) ≙ [•[$f_N \cup f_E$] | $\mathcal{CT}$(L)]$_{\mathcal{WfT}(X)}$ =$[C]_{\mathcal{WfT}(X)}$, because C $\approx_{\mathcal{WfT}(X}$ •[$f_N \cup f_E$] | $\mathcal{CT}$(L), by the definition of f.

$\mathcal{WfT}$ is faithful. Let f, f': X → Y be two pa-graph morphisms such that $\mathcal{WfT}$(f) = $\mathcal{WfT}$(f'). By definition, this implies that •[$f_N \cup f_E$] | $\mathcal{CT}$($\mathcal{G}$(Y)) $\approx_{\mathcal{WfT}(X)}$ •[$f'_N \cup f'_E$] | $\mathcal{CT}$($\mathcal{G}$(Y)), and thus f = f'.

Finally, as for the case of concrete graphs, it is routine to check that $\mathcal{WfT}$ is the left adjoint to $\mathcal{PaG}$. The unit component η_X: X → $\mathcal{PaG}$($\mathcal{WfT}$(X)) is the isomorphism mentioned in Fact 3.23. The counit component ε_A: $\mathcal{WfT}$($\mathcal{PaG}$(A)) → A is the context •[Φ], where Φ is the identity on the free node variables, and renames every free edge variable of the form $\underline{p(a_1,...,a_n)}$ to p. ♦

4 Modelling graph grammars by rewriting systems

We introduce here the notion of rewriting over the algebra $\mathcal{A}$, and show that a graph grammar as defined in Section 2.1 can be faithfully translated into a rewriting system. A term rewrite rule over $\mathcal{A}$ has the form R: $A \rightarrow B$, where A and B are terms such that $\mathcal{F}(A) = \mathcal{F}(B)$. This constraint is fundamental, since it will allow us to extend both the operations and the axioms of the algebra to rewrite rules. We shall not explore this opportunity in detail (see the Conclusions for some hints about a possible application), but we will exploit the possibility of applying contexts to rules.

Using the formulation of graph derivations in terms of partially abstract graphs (Fact 2.12) and the correspondence between terms and pa-graphs explored in the last section, we propose a translation of graph productions into (well-formed) term rewrite rules over $\mathcal{A}$. Since concrete contexts applied to terms represent pushouts, their application to rewrite rules models faithfully the double pushout construction which defines a direct derivation. This allows us to prove a result of soundness and completeness of the term representation of a graph production. The sequential composition of rewriting steps gives rise to computations, which, as expected, faithfully model graph derivations.

4.1 Definition *(term rewrite rules, term rewriting systems, rewriting, computations)*

A *(term) rewrite rule R* (over $\mathcal{A}$) is a pair of terms $R = \langle A, B \rangle$ (written R: $A \rightarrow B$) such that $\mathcal{F}(A) = \mathcal{F}(B)$. The *set of free variables of R*, $\mathcal{F}(R)$, is defined as $\mathcal{F}(R) \triangleq \mathcal{F}(A)$. A rewrite rule R: $A \rightarrow B$ is *well-formed* if A and B are well-formed, and $\mathcal{G}r(A) = \mathcal{G}r(B)$ (which implies $\mathcal{F}(A) = \mathcal{F}(B)$). A *(term) rewriting system* $\mathcal{RS}$ is a finite set $\mathcal{RS} = \{R_i: A_i \rightarrow B_i\}_{i \leq n}$ of rewrite rules. $\mathcal{RS}$ is *well-formed* if all R_i are well-formed.

Given a term A' and a rule R: $A \rightarrow B$, *A' rewrites to B'*, written $A' \rightarrow_R B'$, if there exists a context (possibly not concrete) C such that $C\{A\} \approx A'$ and $C\{B\} \approx B'$. Given a rewriting system $\mathcal{RS} = \{R_i\}$, a *computation from A to B over* $\mathcal{RS}$ is a sequence of rewritings $A = A_0 \rightarrow_{R_0} A_1 \rightarrow_{R_1} \dots \rightarrow_{R_{n-1}} A_n = B$ such that $R_i \in \mathcal{RS}$. ♦

4.2 Definition *(from graph grammars to well-formed term rewriting systems)*

Let $p = (L \overset{l}{\leftarrow} K \overset{r}{\rightarrow} R)$ be a graph production, as defined in Definition 2.2. Then the *well-formed rewrite rule representing p*, $\mathcal{R}(p)$, is defined as $\mathcal{R}(p)$: $\mathcal{W}f\mathcal{T}([l]) \rightarrow \mathcal{W}f\mathcal{T}([r])$ (see Definition 3.24). Clearly, $\mathcal{R}(p)$ is well-formed. If $P = \{p_i\}_{i \leq n}$ is a set of graph productions, the *well-formed term rewriting system representing P* is $\mathcal{R}(P) \triangleq \{\mathcal{R}(p_i) \mid p_i \in P\}$. ♦

4.3 Theorem *(soundness and completeness of $\mathcal{R}(P)$)*

Let $p = (L \overset{l}{\leftarrow} K \overset{r}{\rightarrow} R)$ be a graph production and $\mathcal{R}(p)$: $\mathcal{W}f\mathcal{T}([l]) \rightarrow \mathcal{W}f\mathcal{T}([r])$ be its associated rule. Then:

Soundness. If $A \rightarrow_{\mathcal{R}(p)} B$, then $\mathcal{A}bs(A) \Rightarrow_p \mathcal{A}bs(B)$.

Completeness. If $G \Rightarrow_p H$ then $\mathcal{W}f\mathcal{T}([d]) \rightarrow_{\mathcal{R}(p)} \mathcal{W}f\mathcal{T}([b])$, where d and b are as in

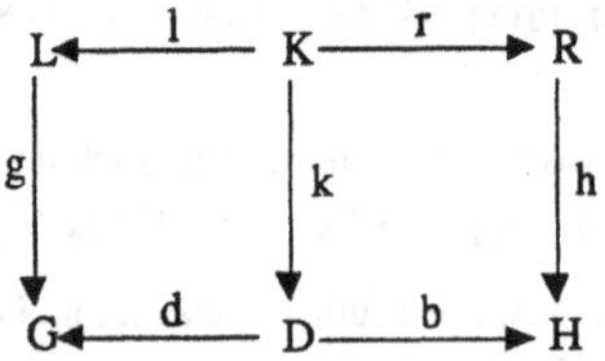

Weak completeness. If $G \Rightarrow_p H$ then $\mathcal{W}f\mathcal{T}([G]) \rightarrow_{\mathcal{R}(p)} \mathcal{W}f\mathcal{T}([H])$.

Proof. *Soundness.* If A $\rightarrow_{\mathcal{R}(p)}$ B then there exists a context C such that C{$\mathcal{W}f\mathcal{T}$([l])} ≈ A and C{$\mathcal{W}f\mathcal{T}$([r])} ≈ B. Let K ≜ $\mathcal{F}$($\mathcal{W}f\mathcal{T}$([l])) (= $\mathcal{F}$($\mathcal{W}f\mathcal{T}$([r])), by Definition 4.2). By Fact 2.12, in order to prove that $\mathcal{A}bs$(A) $\Rightarrow_p$ $\mathcal{A}bs$(B) it is sufficient to find a graph morphism k such that k: [l] → $\mathcal{P}a\mathcal{G}$(A) and k: [r] → $\mathcal{P}a\mathcal{G}$(B) are pa-morphisms.

If C is concrete, let k' ≜ $\mathcal{P}a\mathcal{G}$($[C]_K$). Then since $[C]_K$: $\mathcal{W}f\mathcal{T}$([l]) → A and $[C]_K$: $\mathcal{W}f\mathcal{T}$([r]) → B are arrows of **WfTerm** and $\mathcal{P}a\mathcal{G}$ is a functor, we have that k' = $\mathcal{P}a\mathcal{G}$($[C]_K$): $\mathcal{P}a\mathcal{G}$($\mathcal{W}f\mathcal{T}$([l])) → $\mathcal{P}a\mathcal{G}$(A) and k' = $\mathcal{P}a\mathcal{G}$($[C]_K$): $\mathcal{P}a\mathcal{G}$($\mathcal{W}f\mathcal{T}$([r])) → $\mathcal{P}a\mathcal{G}$(B) are arrows in **PA-FGraph**. Then the statement follows by composing k' with the unit of the adjunction $\eta_{[l]}$: [l] ≡ $\mathcal{P}a\mathcal{G}$($\mathcal{W}f\mathcal{T}$([l])), i.e., k ≜ k' ∘ $\eta_{[l]}$.

If C is not concrete, then by using the same techniques as in Proposition 3.6 it can be shown that there is a K-equivalent context C" = C'$\backslash x_1 \ldots \backslash x_n$, with C' concrete, such that C"{$\mathcal{W}f\mathcal{T}$([l])} ≈ A and C"{$\mathcal{W}f\mathcal{T}$([r])} ≈ B. Therefore we have that A$\backslash x_1 \ldots \backslash x_n$ $\rightarrow_{\mathcal{R}(p)}$ B$\backslash x_1 \ldots \backslash x_n$ using the concrete context C'. Since C' is concrete one has $\mathcal{A}bs$(A$\backslash x_1 \ldots \backslash x_n$) $\Rightarrow_p$ $\mathcal{A}bs$(B$\backslash x_1 \ldots \backslash x_n$), and thus $\mathcal{A}bs$(A) $\Rightarrow_p$ $\mathcal{A}bs$(B) follows because by Definition 3.19 $\mathcal{A}bs$(A) = $\mathcal{A}bs$(A$\backslash x_1 \ldots \backslash x_n$) and $\mathcal{A}bs$(B) = $\mathcal{A}bs$(B$\backslash x_1 \ldots \backslash x_n$).

Completeness. In the given hypotheses, let C ≜ $\mathcal{CT}$(k) (as in Definition 3.17). Then we have that C{$\mathcal{W}f\mathcal{T}$([l])} ≈ $\mathcal{W}f\mathcal{T}$([d]) because C = $\mathcal{CT}$(k) = $\mathcal{W}f\mathcal{T}$(k) by Definition 3.24, and because $\mathcal{W}ft$ is a functor, and similarly C{$\mathcal{W}f\mathcal{T}$([r])} ≈ $\mathcal{W}f\mathcal{T}$([b]). Thus $\mathcal{W}f\mathcal{T}$([d]) $\rightarrow_{\mathcal{R}(p)}$ $\mathcal{W}f\mathcal{T}$([b]).

Weak completeness. Let C ≜ $\mathcal{CT}$(k) as for completeness, and let C' ≜ •$\backslash x_1 \backslash \ldots \backslash x_n$, where $\{x_1, \ldots, x_n\}$ = $\mathcal{F}$($\mathcal{CT}$(D)). Then C'∘C{$\mathcal{W}f\mathcal{T}$([l])} ≈ C'{$\mathcal{W}f\mathcal{T}$([d])} ≈ $\mathcal{W}f\mathcal{T}$([∅ ↣ G]) = $\mathcal{W}ft$([G]) (using Notation 2.10), because C' restricts all the free items of D; similarly, C'∘C{$\mathcal{W}f\mathcal{T}$([r])} = $\mathcal{W}f\mathcal{T}$([H]). Thus $\mathcal{W}ft$([G]) $\rightarrow_{\mathcal{R}(p)}$ $\mathcal{W}ft$([H]). ♦

For what concerns *concrete* contexts, the last result is clearly a straightforward consequence of the properties of the embedding of pa-graphs into well-formed terms presented in Section 3.3. For contexts which are not concrete, their application does not correspond to double pushouts. Nevertheless, the soundness result states that also using non-concrete contexts, every abstract graph which can be derived through the rule representing a graph production is also derivable by a direct application of the production itself. This is due to the fact that a restriction may delete an item from the context graph D (and possibly from the gluing graph K as well) but does not affect the starting and the ending graph of the direct derivation.

The completeness result stresses the fact that modeling a direct derivation G $\Rightarrow_p$ H through the application of the rewrite rule $\mathcal{R}$(p) via context C, all the items which are not affected by the direct derivation (i.e., the items of the context graph D) are preserved as free variables of the rule C{$\mathcal{R}$(p)}. This property does not hold for the representation of graph derivations via term rewriting proposed in [BC87], and could be used to rephrase in our algebraic framework the Church-Rosser properties of graph derivations, as hinted in the Conclusions.

We conclude this section with a result similar to the one presented in [BC87], showing that the representation of a graph grammar as a term rewriting system over $\mathcal{A}$ has the same generative power. To this aim, we need to use the weaker form of completeness introduced above, showing by the way that non-concrete contexts are necessary in our framework. In fact, in order to extend the completeness result to computations, we should show that whenever two direct graph derivations can be composed into a derivation sequence, then the corresponding term rewritings can be sequentially composed as well. But two (well-formed) rewritings A $\rightarrow_R$ B and B' $\rightarrow_{R'}$ D can be sequentially composed if B ≈ B', i.e., if $\mathcal{P}a\mathcal{G}$(B) = $\mathcal{P}a\mathcal{G}$(B'): this means that also the concrete parts of the two rewritings (corresponding to the context graphs) must be identical. This requirement is too strong when modeling graph derivations, for which it would be sufficient that $\mathcal{A}bs(B) = \mathcal{A}bs(B')$, ignoring the concrete part. Thus, in order to have a faithful representation of graph derivations via term computations, the concrete part of the rules should be forgotten through the application of restriction operators, as done in Theorem 4.3 for weak completeness.

4.4 Proposition *(derivations as computations: soundness and completeness)*

Let $P = \{p_i\}$ be a set of graph productions, and $\mathcal{R}(P) = \{\mathcal{R}(p_i) \mid p_i \in P\}$ be its associated well-formed rewriting system. Then:

1) If $G_0 \Rightarrow_{p_1} G_1 \Rightarrow_{p_2} \dots \Rightarrow_{p_n} G_n$ is a graph derivation, then there exists a computation $A_0 \rightarrow_{\mathcal{R}(p_1)} A_1 \rightarrow \dots \rightarrow_{\mathcal{R}(p_n)} A_n$ such that every A_i is closed, and $\mathbf{PaG}(A_i) \equiv [G_i]$ for all $0 \leq i \leq n$.

2) If $B_0 \rightarrow_{\mathcal{R}(p_1)} B_1 \rightarrow \dots \rightarrow_{\mathcal{R}(p_n)} B_n$ is a computation, then $\mathcal{A}bs(B_0) \Rightarrow_{p_1} \mathcal{A}bs(B_1) \Rightarrow_{p_2} \dots \Rightarrow_{p_n} \mathcal{A}bs(G_n)$ is a derivation.

Proof. 1) By weak completeness (Theorem 4.3), since $G_{i-1} \Rightarrow_{p_i} G_i$ implies that $\mathcal{W}f\mathcal{T}([G_{i-1}]) \rightarrow_{\mathcal{R}(p_i)} \mathcal{W}f\mathcal{T}([G_i])$, and recalling that $\mathcal{W}f\mathcal{T}([G]) \equiv [G]$ by Fact 3.23.

2) Immediate, by soundness. ♦

5 Conclusions and future work

In this paper we presented an axiomatization of partially abstract graphs, that is, of suitable classes of monomorphisms which may be interpreted as graphs having both a concrete part and an abstract part (i.e., defined up to isomorphism). Morphisms between pa-graphs are pushout squares. We showed that the basic notions of the algebraic theory of graph grammars (instantiated to a suitable category of graphs) can be rephrased in a natural way using partially abstract graphs.

The terms of the algebra we have proposed are built over a small set of operators, including parallel composition, substitution application, and restriction. The free variables of a term represent the concrete part of the associated pa-graph: thus *concrete* terms (i.e., terms without restrictions) axiomatize concrete graphs. By equipping the algebra of terms with a categorical structure (where the arrows are equivalence classes of monadic contexts), we showed that there is a full and faithful embedding (with a right adjoint) of the category of pa-graphs (resp. graphs) into the category of well-formed terms (resp. concrete terms).

By exploiting this embedding we were able to show that rewriting (in the sense of term rewriting systems) in the algebra of pa-graphs models in a sound and complete way the double pushout construction, which is the keystone of the algebraic theory of graph grammars. In particular we showed that also graph productions having non-discrete gluing graphs can be represented as term rewrite rules without loss of information, unlike a similar approach proposed in [BC87].

This possibility could be exploited in order to capture in our framework the Church-Rosser properties of graph derivations, which exploits the possibility of having non-discrete gluing graphs. In short, if two direct derivation (p, g): $G \Rightarrow H$ and (p', g'): $G \Rightarrow H'$ are 'parallel independent' (see Definition 2.7), the Church-Rosser property [Eh79] states that they can be applied in either order to G producing isomorphic graphs X and X', i.e., there are derivation sequences $G \Rightarrow_p H \Rightarrow_{p'} X$ and $G \Rightarrow_{p'} H' \Rightarrow_p X'$. Moreover, the so-called 'Parallelism Theorem' states that in this case a 'parallel' production p+p' may be applied to G as well, producing a graph X" isomorphic to X.

These results can be formulated in our algebraic framework by applying to the term rewriting systems over the algebra of pa-graphs some standard techniques, which have been firstly developed for an algebraic presentation of Petri Nets in [MM88], and then have been generalized to arbitrary term rewriting systems [Me90] and to structured transition systems [Co90]. Shortly, given an equational signature $\langle\Sigma, E\rangle$ and a term rewriting system over a ‹Σ, E›-algebra, one has to extend the operations and the axioms of the algebra to the rewrite rules and to the computations as well, adding some 'functoriality' axioms which state that all the operations of the algebra distribute with respect to the sequential composition. The result of these definitions is that computations are equipped with a rich, modular proof system, and often the resulting equivalence classes of computations can be shown to be a truly concurrent semantics of the term rewriting system one started with.

The case of Petri Nets can be considered as a paradigmatical example. As defined in [MM88], Petri nets are naturally equipped with a monoidal operation representing parallel composition. When extended to computations, the monoidal operation means that two computations can 'run in parallel'. The fundamental axiom imposed by the extension of the operation to computations is the functoriality of the monoidal operation (here denoted by '+'), i.e.,

$$(t \,;\, t') + (s \,;\, s') \;=\; (t + s) \,;\, (t' + s')$$

As pointed out in [MM88], this single axiom captures a basic fact about concurrency, i.e., that the parallel composition of two independent computations can be broken into a sequence of parallel compositions of their elementary steps. Thanks to this axiom, it has been shown that the equivalence classes of computations of a net are a kind of 'non-sequential processes', which are indeed the most widely accepted truly concurrent semantics of nets.

In order to apply these general techniques to our rewriting systems, first one has to extend all the operations and the axioms of the algebra of pa-graphs to rewrite rules. As hinted at the beginning of Section 4, this is possible because rewrite rules, like terms, are equipped with a set of free variables. This step provides a rich calculus of rewrite rules; in particular, 'parallel' rules of the form R | R' are generated as well, and they can be used to model faithfully the parallel productions of a graph grammar.

Next, the operations should be extended to computations, by adding one functoriality axiom for each operation of the algebra. This is possible, indeed, because it can be observed that in a computation $\rho = (A_0 \rightarrow_{R_0} A_1 \rightarrow_{R_1} \dots \rightarrow_{R_{n-1}} A_n)$, the set of free variables is identical for each term A_i, and thus we can define set of free variables of ρ as $\mathcal{F}(\rho) = \mathcal{F}(a_i)$ for any $0 \le i \le n$. The functoriality axioms include that for parallel composition (similar to the one above for Petri nets), and those for substitution application and restrictions, of the form $(\rho \,;\, \sigma)[\Phi] = \rho[\Phi] \,;\, \sigma[\Phi]$ and $(\rho \,;\, \sigma)\backslash x = \rho\backslash x \,;\, \sigma\backslash x$, respectively, for all computations ρ, σ, all substitutions Φ, and all variables x.

The overall result of these definitions is a rich, modular proof system which can be used to prove that the term rewriting computations representing the derivation sequences $G \Rightarrow_p H \Rightarrow_{p'} X$, $G \Rightarrow_{p'} H' \Rightarrow_p X'$, and $G \Rightarrow_{p+p'} X''$, respectively, are all equivalent under the hypothesis of parallel independence. Moreover, as for Petri nets, the equivalence classes of computations induced by the axioms of the algebra of pa-graphs and by the axioms of functoriality should provide a truly concurrent semantics for graph grammars.

6 References

[BC87] M. Bauderon, B. Courcelle, *Graph Expressions and Graph Rewritings*, Mathematical System Theory **20**, 1987, pp. 83-127.

[CER79] V. Claus, H. Ehrig, G. Rozenberg, (Eds.) *Proceedings of the 1st International Workshop on* Graph-Grammars and Their Application to Computer Science and Biology, LNCS 73, 1979.

[CMREL91] A. Corradini, U. Montanari, F. Rossi, H. Ehrig, M. Löwe, *Logic Programming and Graph Grammars*, in [EKR91].

[Co90] A. Corradini, *An Algebraic Semantics for Transition Systems and Logic Programming*, Ph.D. Thesis TD-8/90, Dipartimento di Informatica, Università di Pisa, March '90.

[Eh79] H. Ehrig, *Introduction to the Algebraic Theory of Graph-Grammars*, in [CER79], pp. 1-69.

[EHKP91a] H. Ehrig, A. Habel, H.-J. Kreowski, F. Parisi-Presicce, *From Graph Grammars to High-Level Replacement Systems*, in [EKR91].

[EHKP91b] H. Ehrig, A. Habel, H.-J. Kreowski, F. Parisi-Presicce, *Parallelism and Concurrency in High-Level Replacement Systems*, Technical Report n• 90/35, Technische Universität Berlin. To appear in Mathematical Structures in Computer Science, 1991.

[EKMRW81] H. Ehrig, H.-J. Kreowski, A. Maggiolo-Schettini, B. Rosen, J. Winkowski, *Transformation of Structures: an Algebraic Approach*, Mathematical System Theory **14**, 1981, pp. 305-334.

[EKR91] H. Ehrig, H.-J. Kreowski, G. Rozenberg, (Eds.) *Proceedings of the 4th International Workshop on Graph-Grammars and Their Application to Computer Science*, LNCS, 1991.

[ENR83] H. Ehrig, M. Nagl, G. Rozenberg, (Eds.) *Proceedings of the 2nd International Workshop on Graph-Grammars and Their Application to Computer Science*, LNCS 153, 1983.

[ENRR87] H. Ehrig, M. Nagl, G. Rozenberg, A. Rosenfeld, (Eds.) *Proceedings of the 3rd International Workshop on Graph-Grammars and Their Application to Computer Science*, LNCS 291, 1987.

[EPS73] H. Ehrig, M. Pfender, H.J. Schneider, *Graph-grammars: an algebraic approach*, Proc, IEEE Conf. on Automata and Switching Theory, 1973, pp. 167-180.

[Ha91] A. Habel, *Collage Grammars*, in [EKR91].

[HK87] A. Habel, H.-J. Kreowski, *May we introduce to you: hyperedge replacement*, in [ENRR87], pp. 15-26.

[HKP88] A. Habel, H-J. Kreowski, D. Plump, *Jungle evaluation*, in Proc. Fifth Workshop on Specification of Abstract Data Types, LNCS 332, 1988, pp. 92-112.

[Ke91] J.R. Kennaway, *Graph rewriting in some categories of partial morphisms*, in [EKR91].

[Me90] J. Meseguer, *Rewriting as a Unified Model of Concurrency*, Proc. CONCUR '90, LNCS 458, 1990, pp. 384-400. Full version as Tech. Rep. SRI-CSL-90-02, SRI International, February '90.

[MM88] J. Meseguer, U. Montanari, *Petri Nets are Monoids: A New Algebraic Foundation for Net Theory*, Proc. Logics In Computer Science, Edinburgh, 1988, pp. 155-164. Also in Information and Computation, **88** (2), 1990, pp. 105-155.

[MR91] U. Montanari, F. Rossi, *True Concurrency in Concurrent Constraint Programming*, Proc. International Logic Programming Symposium, MIT Press, 1991.

[PEM87] F. Parisi-Presicce, H. Ehrig and U. Montanari, *Graph Rewriting with Unification and Composition*, in [ENRR87], pp. 496-514.

[PP90] F. Parisi-Presicce, *A Rule Based Approach to Modular System Design*, Proc. 12th Int. Conf. Software Engineering, March 1990.

Dataflow Networks are Fibrations

Eugene W. Stark*
Department of Computer Science
State University of New York at Stony Brook
Stony Brook, NY 11794 USA
(stark@cs.sunysb.edu)

Abstract

Dataflow networks are a paradigm for concurrent computation in which a collection of concurrently and asynchronously executing processes communicate by sending messages over FIFO message channels. In a previous paper, we showed that dataflow networks could be represented as certain spans in a category of automata, or more abstractly, in a category of domains, and we identified some universal properties of various operations for building networks from components. Not all spans corresponded to dataflow processes, and we raised the question of what might be an appropriate categorical characterization of those spans that are "dataflow-like." In this paper, we answer this question by obtaining a characterization of the dataflow-like spans as *split right fibrations*, either in a 2-category of automata or a 2-category of domains. This characterization makes use of the theory of fibrations in a 2-category developed by Street. In that theory, the split right fibrations are the algebras of a certain doctrine (or 2-monad) R on a category of spans. For the 2-categories we consider, R has a simple interpretation as an "input buffering" construction.

1 Introduction

Dataflow networks [4, 5] are a paradigm for concurrent computation in which a collection of concurrently and asynchronously executing processes communicate by sending messages over FIFO message channels. *Determinate* dataflow networks compute continuous functions from input message histories to output message histories, and have a well-understood theory. Less developed is the theory of *indeterminate* or non-functional networks. These more general networks are especially interesting because they exhibit both concurrency and indeterminacy, and insight gained from their study will likely contribute to a better overall understanding of these two concepts.

This paper is part of a research program aimed at finding the correct algebraic setting for the study of indeterminate dataflow networks. We wish to view dataflow networks as the elements of an algebra whose operations represent ways to build networks from

*Research supported in part by NSF Grant CCR-8902215.

components, and we would like to understand fully the notions of behavioral equivalence that are appropriate in this context. For some time, we have been studying a particular automata-theoretic model for dataflow networks, in an attempt to identify whatever useful algebraic structure might be present. Based on the progress we have made so far [8, 9, 10, 13], a general structure appears to be emerging. However, all is not yet completely clear, and it continues to be difficult to identify and separate the important structure from the incidental artifacts of the model.

In a previous paper [9] we showed that a dataflow network with input "ports" X and output ports Y could be represented as a span from FX to FY (*i.e.* a diagram $FY \xleftarrow{g} A \xrightarrow{f} FX$) in a finitely complete category **Auto** of concurrent automata. Here F is a suitable functor that associates "objects of inputs" FX and FY in **Auto** with finite sets of ports X and Y. We showed that various constructions, corresponding intuitively to ways of composing smaller networks into larger ones, could be defined in terms of limits in **Auto**. In particular, the operation of "feeding back" outputs to inputs was defined in terms of equalizers. We also showed that dataflow networks could be modeled more abstractly as spans in a category of **EvDom** of "conflict event domains," and that this model is related to the more concrete automaton model by a coreflection. Consequently, operations defined in terms of limits are preserved in the passage from the more concrete model to the more abstract version.

At the end of the previous paper, we noted several interesting properties, valid in the domain-theoretic model, of spans corresponding to dataflow processes, and we raised the question of what might be the correct categorical characterization of the "dataflow-like" spans. A proper answer to this question would be prerequisite to the construction of a fully categorical theory of dataflow networks. In the present paper, we obtain a characterization of dataflow-like spans as *split right fibrations*, either in a 2-category of automata, or in a 2-category of domains. The fact that essentially the same characterization holds in both cases lends credence to the idea that it is in fact the correct categorical notion. Further support comes from an intuitive interpretation of the definition of fibration. Fibrations in a 2-category are defined to be the algebras of a certain "doctrine," or 2-monad. In the present situation, this doctrine corresponds to the construction "compose with an input buffer." Thus, the dataflow-like spans are those spans that are algebras of the input buffering doctrine.

The theory of fibrations was first developed in terms of concrete constructions on categories [3]. Then, Street [14, 15], building on work of Gray [2], showed that this theory has a bicategorical formulation, which can be applied not only to the 2-category **Cat**, but to any bicategory with sufficient completeness properties. Here, we examine how the theory applies to the category **Auto** of automata and the category **EvOrd** of "conflict event orderings," which is equivalent to the category **EvDom** of our previous paper. These categories have 2-categorical structure we have not exploited until now. As a technical matter, the 2-categories **Auto** and **EvOrd** do not quite have the necessary completeness properties (existence of comma objects and certain 2-pullbacks), so our results are complicated somewhat by the necessity of enlarging them to 2-categories **AutoWk**, of "automata and weak morphisms," and **EvOrdWk**, of "conflict event orderings and sup-preserving maps." We determine the structure of the split right fibrations in each of the

2-categories **AutoWk** and **EvOrdWk**. Our main results are: (1) the dataflow-like spans in **Auto** are exactly those that are split right fibrations in **AutoWk** with an **Auto**-morphism as cleavage, and (2) the dataflow-like spans in **EvOrd** are exactly those that are split right fibrations in **EvOrdWk** with an **EvDom**-morphism as cleavage.

Some further comments are in order concerning the general directions envisioned for this research. We would like very much to be able to define a notion of "dataflow model," which would permit both the study of the algebra of network-forming operations, and the comparison of different dataflow models by means of homomorphisms. In the author's opinion, the evidence at present suggests quite strongly that a dataflow model ought to be a certain kind of bicategory, whose objects are "types," whose arrows are "processes," and whose 2-cells are morphisms of processes. Composition of arrows would correspond to "sequential composition" of processes, in which the output of one process becomes the input of another. The bicategory would also be equipped with a functorial tensor product $\otimes$, which would provide a way to form the "parallel composition" $p \otimes p' : X \otimes X' \to Y \otimes Y'$ of processes $p : X \to Y$ and $p' : X' \to Y'$. It ought to be possible to define other operations on processes using bicategorical constructions, although in some cases (in particular the feedback operation) this remains problematic. At present, we have several examples of this type of bicategorical model for dataflow networks, but the correct general formulation seems elusive. The "bicategories of spans" of Carboni and Walters [1] represent an example of the structure we might hope to find.

We hope that the organization chosen for the rest of this paper will make the motivations from the intended application area clear both to category theorists and to computer scientists, and will hold the attention, at least for a little while, of computer scientists unfamiliar with 2-categories. To this end, we have kept Section 2 of the paper almost completely free of 2-categories. Instead, we define the ordinary categories **Auto** of automata and **AutoWk** of automata and weak morphisms, ignoring their 2-categorical structure, and we show how certain spans in these categories model processes that consume inputs and produce outputs. We show that these "monotone automata" are in fact the algebras of an "input buffering monad" on a category of spans in **AutoWk**. This result prepares the connection, made in Section 3, with fibrations in **AutoWk**. For this, the use of 2-categories is necessary, and the reader is referred to [6, 14, 15] for the basic terminology and notations. In Section 4, we apply the theory to the 2-category **EvOrdWk** of domains and obtain a similar characterization of the dataflow-like spans in **EvOrdWk**.

Finally, a comment on notation. In this paper, fx or $f(x)$ denotes the application of a mapping f to its argument x. Compositions are written in reverse diagrammatic order, so that gf denotes f followed by g. We extend this convention to all types of composition. For example, if $e_1, e_2, \ldots, e_n$ are elements of a set E, then the string "e_1 followed by e_2 followed by ... followed by e_n" will be denoted $e_n \ldots e_2 e_1$.

2 Trace Automata

In this section, we review the automata-theoretic model for dataflow networks presented in [9], along with its associated intuition. The objects of this model are certain spans in a category **Auto** of "trace automata." We develop various properties of these spans, to motivate the idea that they ought to be examples of fibrations.

2.1 Concurrent Alphabets and Traces

A *concurrent alphabet* is a pair $(E, \|)$, where E is a set, and $\|$ is a symmetric, irreflexive relation on E, called the *concurrency relation*. If $e \| e'$, then we say that e and e' are *concurrent* or that they *commute*. A set $U \subseteq E$ is called *commuting* if $e \| e'$ for all $e, e' \in U$ with $e \neq e'$, and we use $\text{Comm}(E)$ to denote the set of all *finite* commuting subsets of E. Intuitively, if E is the set of basic observable actions of interest for some system, then $\text{Comm}(E)$ is the set of all possible instantaneous occurrences that might be observed during an execution of that system. If $U, V \in \text{Comm}(E)$, then U and V are called *orthogonal*, and we write $U \perp V$, if $U \cup V \in \text{Comm}(E)$ and $U \cap V = \emptyset$.

A *morphism* from a concurrent alphabet E to a concurrent alphabet E' is a function $h : \text{Comm}(E) \to \text{Comm}(E')$, such that

1. $h(\emptyset) = \emptyset$.
2. If $U \cup V \in \text{Comm}(E)$, then $h(U) \cup h(V) \in \text{Comm}(E')$, and $h(U \setminus V) = h(U) \setminus h(V)$.

The symbol $\setminus$ denotes set difference. It can be shown [9] that if $U \cup V \in \text{Comm}(E)$, then $h(U \cup V) = h(U) \cup h(V)$, so that a morphism of concurrent alphabets is uniquely determined by its action on singleton sets. We often use this fact in defining particular morphisms. It can also be shown $U \perp V$ implies $f(U) \perp f(V)$ — as a special case we see that $e \| e'$ implies $f(\{e\}) \cap f(\{e'\}) = \emptyset$. Let **Alph** denote the category of concurrent alphabets and their morphisms. The identity morphism on E is just the identity function from $\text{Comm}(E)$ to $\text{Comm}(E)$. Composition of morphisms is function composition.

The *product* $E \otimes E'$ of concurrent alphabets E and E' is the concurrent alphabet whose set of elements is the disjoint union of the sets of elements of E and E', and whose concurrency relation extends those of E and E' by making each element of E concurrent with each element of E'. Equipping $E \otimes E'$ with the restriction maps

$$\begin{aligned} - \cap E : \quad & E \otimes E' \to E \\ - \cap E' : \quad & E \otimes E' \to E' \end{aligned}$$

makes $E \otimes E'$ into a categorical product.

Suppose E is a concurrent alphabet. The *free partially commutative monoid* generated by E is obtained by factoring the monoid of finite sequences of elements of E by the least congruence that relates ee' and $e'e$ whenever $e \| e'$. We use E^* to denote this monoid, whose elements are called *traces* [7]. We shall find it convenient in the sequel to identify a set $U = \{e_1, \ldots, e_n\} \in \text{Comm}(E)$ with the corresponding trace $e_n \ldots e_1 \in E^*$. In this way we give meaning to expressions such as $U_m U_{m-1} \ldots U_1$, where $U_i \in \text{Comm}(E)$ for $1 \leq i \leq m$.

2.2 Trace Automata

"Trace automata" are transition systems whose transition labels are drawn from a concurrent alphabet. To capture the idea that "concurrent transitions commute," the definition of trace automata includes the requirement that if two transitions with concurrent labels are enabled in the same state, then they can be executed in either order with the same effect.

Formally a *trace automaton* (or more simply, an *automaton*) is a four-tuple

$$A = (E, Q, T, q_I),$$

where

- E is a concurrent alphabet of *actions*.
- Q is a set of *states*, with $q_I \in Q$ a distinguished *initial state*.
- $T : Q \times E \to Q$ is a partial function, called the *transition map*, such that whenever $e \| e'$, $T(q,e) = r$, and $T(q,e') = r'$, then there exists $s \in Q$ with $T(r,e') = s = T(r',e)$.

We often write $e : q \to r$ or $q \xrightarrow{e} r$ to assert that $T(q,e) = r$, and call the triples (q,e,r) such that $q \xrightarrow{e} r$ the *transitions* of A. Condition (3) in the above definition embodies the intuitive idea that the order of occurrence of concurrent actions is immaterial.

A *computation sequence* for a trace automaton A is a finite sequence of transitions of the form:

$$q_0 \xrightarrow{e_1} q_1 \xrightarrow{e_2} \dots \xrightarrow{e_n} q_n.$$

Each computation sequence γ determines a corresponding trace $\mathrm{tr}(\gamma) = e_n e_{n-1} \dots e_1 \in E^*$. Two computation sequences are *equivalent* if they have the same trace, and the equivalence classes of computation sequences are called the *computations* of A.

A *morphism* from a trace automaton A to a trace automaton A' is a pair of maps $h = (h_a, h_s)$, where $h_a : E \to E'$ is a morphism of concurrent alphabets, and $h_s : Q \to Q'$ is a function, such that:

- $h_s(q_I) = q'_I$.
- For all $q,\ r \in Q$ and $e \in E$, if $e : q \to r$ in A and $h_a(e) = \{e_1, e_2, \dots, e_n\}$, then $T'(h_s(q), e_i)$ is defined for all i with $1 \le i \le n$, and A' has a computation sequence

  $$q_0 \xrightarrow{e_1} q_1 \xrightarrow{e_2} \dots \xrightarrow{e_n} q_n.$$

  with $q_0 = h_s(q)$ and $q_n = h_s(r)$.

The above notion of morphism was introduced in [9], where it was also shown that the resulting category **Auto** has reasonable properties. (In particular, **Auto** has finite limits, which are created by the forgetful functor from **Auto** to the product of **Alph** and the category **Set**$_*$ of pointed sets.) The intuition behind the definition is that finite commuting sets of transitions are what a trace automaton can do in a single instantaneous step, and these are the things that ought to be preserved by their morphisms. The reason why finite commuting sets of transitions, rather than single transitions, are what correspond to instantaneous steps, can be seen by considering a finite product $A = A_1 \times \dots \times A_n$ of automata. We wish to think of A as representing a system $\{A_1, \dots, A_n\}$, executing concurrently and independently. In such a system, there is the possibility of the simultaneous occurrence of a transition from each of the A_i. Formally, this is reflected in the requirement that a collection of morphisms $h_i : B \to A_i$ $(1 \leq i \leq n)$ induce a unique morphism $h : B \to A$. In order for h to be a morphism, we must allow morphisms to map single transitions of B to finite commuting sets of transitions of A.

The category **Alph** is isomorphic to a full reflective subcategory of **Auto** via the functor that takes each concurrent alphabet E to the one-state automaton with alphabet E, having a transition for each of its actions. Since the embedding of **Alph** in **Auto** is right-adjoint to the forgetful functor from **Auto** to **Alph**, it preserves limits. In the sequel, it will be convenient for us to identify a concurrent alphabet E with the corresponding one-state automaton.

2.3 Monotone Automata

We wish to discuss automata that consume inputs and produce outputs. Therefore, if X and Y are concurrent alphabets, we define an *automaton from X to Y* to be a span $Y \xleftarrow{g} A \xrightarrow{f} X$ from X to Y in **Auto**. Intuitively, if $Y \xleftarrow{g} A \xrightarrow{f} X$ is an automaton from X to Y, then we think of X as an "object of inputs," whose transitions represent the possible inputs that A might consume. Similarly, we regard Y as an "object of outputs," whose transitions represent the possible outputs that A might produce. If $Y \xleftarrow{g} A \xrightarrow{f} X$ and $Y \xleftarrow{g'} A' \xrightarrow{f'} X$ are automata from X to Y, then an *arrow of spans* from A to A' is a morphism $h : A \to A'$ such that $g'h = g$ and $f'h = f$. Let **Auto**(X,Y) denote the category of automata from X to Y, with arrows of spans as morphisms.

We wish to single out automata with the following properties:

1. They are always prepared to consume arbitrary input.

2. The act of consuming input can never cause enabled output transitions to become disabled, although it may cause more output transitions to become enabled.

In previous papers, we have formalized such automata using variants of the following definition: an automaton A from X to Y is *monotone* iff the alphabet E of A is (up to isomorphism) of the form $Z \otimes X$, with $f_a : E \to X$ the restriction to X, with $g_a e = \emptyset$ for all $e \in X$, and in addition with the following condition satisfied:

(Receptivity) For all states q of A, and all actions $e \in X$, there exists a transition $q \xrightarrow{e} r$ of A.

The "port automata" defined in [9], are special cases of monotone automata, in which X and Y are concurrent alphabets whose elements represent the transmission of data values over finite sets of "ports," and E is required to be of the form $Y \otimes Z \otimes X$, with g the restriction to Y. We use the more general monotone automata here because they have a clean characterization in terms of fibrations.

The next result, easily proved from the definitions, gives the essential properties of monotone automata. Statement (1) is a generalization of receptivity to arbitrary traces, rather than single actions. Statement (2) states that under certain conditions, a transition of A in state q can be "pushed out" along an input computation sequence, to yield a transition of A from the final state of that computation sequence.

Proposition 2.1 *Suppose A is a monotone automaton from X to Y. Then*

1. *For all states q of A and all traces x in X^*, there exists a computation sequence of A, starting from state q and having trace x. Moreover, any two such computation sequences arrive in the same final state, which we denote by $q \cdot x$.*

2. *Suppose $q \xrightarrow{e} r$ is a transition of A, and x and x' are traces in X^*, such that $x'e = ex$. Then A also has a transition $q \cdot x \xrightarrow{e} q \cdot x'$.*

2.4 The Input Buffering Monad

Suppose $Y \xleftarrow{g} A \xrightarrow{f} X$ is an automaton from X to Y. The *input buffering construction* is a way to obtain a monotone automaton $Y \xleftarrow{g'} BA \xrightarrow{f'} X$ by "composing A with an input buffer." Formally, if $A = (E, Q, T, q_{\mathrm{I}})$, then $BA = (E', Q', T', q'_{\mathrm{I}})$, where:

- $E' = E \otimes X$.
- $Q' = Q \times X^*$. That is, the states of BA are pairs (q, x), where q is a state of A, and x is a trace in X^*. The initial state q'_{I} of BA is the pair $(q_{\mathrm{I}}, \epsilon)$, where ϵ is the empty trace.
- The transition map $T' : Q' \times E' \to Q'$ of BA is defined as follows:

  1. If $e \in X$, then $T'((q, x), e) = (q, ex)$.
  2. If $e \in E$, then $T'((q, x), e)$ is defined iff A has a transition $q \xrightarrow{e} r$ and there exists x' with $x = x'(fe)$, in which case $T'((q, x), e) = (r, x')$.

Intuitively, a state of BA is a pair (q, x) consisting of a state q of A and an "input buffer" x. There are two types of transitions of BA: *input* transitions (case (1) above), in which input $e \in X$ arrives and is appended to the tail of the input buffer, and *noninput* transitions (case (2) above), in which A performs a transition $q \xrightarrow{e} r$, absorbing any necessary input fe from the head of the input buffer.

Define $f' : BA \to X$ to take (q, x) to the unique state of X, and to take $e \in X$ to $\{e\}$ and $e \in E$ to $\emptyset$. Define $g' : BA \to Y$ to take (q, x) to the unique state of Y, to take $e \in X$ to $\emptyset$, and $e \in E$ to ge. It is straightforward to verify that f' and g' are morphisms of automata. We have the following result:

Proposition 2.2 *For any automaton A from X to Y, the automaton BA from X to Y is a monotone automaton. Moreover, the map taking A to BA extends in an obvious way to a functor $B : \mathbf{Auto}(X, Y) \to \mathbf{Auto}(X, Y)$.*

In fact, more can be shown. For each automaton A from X to Y, there is a morphism $\mu_A : BBA \to BA$. Intuitively, μ_A is the morphism that collapses two tandem input buffers into one, hiding actions corresponding to the transfer of input between the two buffers. Formally, the map μ_A takes a state $((q, x), x')$ of BBA to the state $(q, x'x)$ of BA. To define the behavior of μ_A on actions, observe that the alphabet of BBA is of the form $(E \otimes X) \otimes X$, and the alphabet of BA is of the form $E \otimes X$. The map μ_A takes $e \in E$ to $\{e\}$, it takes e in the "outer copy" of X to $\{e\}$, and it takes e in the "inner copy" of X to $\emptyset$.

Proposition 2.3 *The maps $\mu_A : BBA \to BA$ are morphisms of automata, which are the components of a natural transformation $\mu : BB \to B$ that satisfies the associative law $\mu \circ (B\mu) = \mu \circ (\mu B)$.*

If A is a monotone automaton from X to Y, then we may also obtain a morphism $h_A : BA \to A$. To see this, note that the alphabet E of A has the form $Z \otimes X$, and the alphabet of BA has the form $(Z \otimes X) \otimes X$. Let $h_A : BA \to A$ take each state (q, x) of BA to the state $q \cdot x$ of A (see Proposition 2.1), each action e in Z to $\{e\}$, each action e in the "outer copy" of X to $\{e\}$, and each action e in the "inner copy" of X to $\emptyset$.

Proposition 2.4 *If A is a monotone automaton from X to Y, then the map $h_A : BA \to A$ is an arrow of spans. Moreover, $h_A \circ \mu_A = h_A \circ Bh_A$.*

Proof – If h_A is a morphism of automata, then the fact that it is an arrow of spans is immediate from the definitions. To see that h_A is a morphism of automata, suppose $(q, x) \xrightarrow{e} (q', x')$ is a transition of BA. If e is an input action (*i.e.* it is in the outer copy of X), then $x' = ex$ and $q' = q$. In this case, A has a transition $q \cdot x \xrightarrow{e} q \cdot ex$, which is then the image of $(q, x) \xrightarrow{e} (q', x')$ under h_A. If e is a noninput action in the inner copy of X, then $x'e = x$ and A has a transition $q \xrightarrow{e} q'$. By Proposition 2.1, A also has a transition $q \cdot x'e \xrightarrow{e} q' \cdot x'$ which is then the image of $(q, x) \xrightarrow{e} (q', x')$ under h_A. Finally, if $e \in Z$, then $x' = x$ and A has a transition $q \xrightarrow{e} q'$. By Proposition 2.1, A also has a transition $q \cdot x \xrightarrow{e} q' \cdot x'$, which is the image of $(q, x) \xrightarrow{e} (q', x')$ under h_A. ∎

Proposition 2.3 states that B is almost the underlying functor of a monad, with the natural transformation μ as multiplication. Proposition 2.4 states that every monotone automaton from X to Y almost carries a structure of B-algebra. What is missing to make B an actual monad, and the monotone automata its actual algebras, is the monad unit, which would be a natural transformation $\eta : 1 \to B$. The components $\eta_A : A \to BA$ of such a natural transformation almost exist. More precisely, the map η_A would have to take a state q of A to the state (q, ϵ) of BA, and an action $e \in E$ to the disjoint sum $\{e\} + fe \in \mathrm{Comm}(E \otimes X)$. The only problem is that such a map η_A need not be a morphism.

One can see what goes wrong by considering the case $A = Z \otimes X$, with $f : A \to X$ the restriction to X and $g : A \to Y$ the zero map. In this case, A has just one state $*$, so up to isomorphism BA has as its states the traces in X^*, and the alphabet of BA is $(Z \otimes X) \otimes X$. The map η_A would have to take the unique state $*$ of A to the empty trace ϵ, and each action $a \in X$ to the set $\{a, a'\} \in \mathrm{Comm}((Z \otimes X) \otimes X)$, where a' is in the "inner copy" of X and a is in the "outer copy." Since A has a transition $a : * \to *$ whenever $a \in X$, for η_A to be a morphism, BA would have to have transitions from state ϵ for both actions a and a'. In fact, BA has transitions $\epsilon \xrightarrow{a} a \xrightarrow{a'} \epsilon$, but no transition for action a' in state ϵ. Intuitively, BA has the capability of accepting input a in one step and then processing it on the next step, but not of doing both in a single step. There are two ways around this problem: we can enlarge the class of objects of **Auto** to include automata capable of the behavior required of BA, or we can weaken the properties required of morphisms **Auto**, so as to include all the maps η_A. Ultimately, the first approach is probably the correct way to proceed, but the definition of a suitably general class of automata (such as the "concurrent transition systems" of [11, 12]) introduces an additional layer of abstraction that would tend to obscure the ideas we wish to convey here. We therefore follow the second approach in this paper.

Thus, a *weak morphism* from a trace automaton A to a trace automaton A' is a pair of maps $h = (h_a, h_s)$, where $h_a : E \to E'$ is a morphism of concurrent alphabets, and $h_s : Q \to Q'$ is a function, such that:

- $h_s(q_I) = q'_I$.
- For all q, $r \in Q$ and $e \in E$, if $e : q \to r$ in A, then for some enumeration $\{e_1, e_2, \ldots, e_n\}$ of $h_a(e)$, the automaton A' has a computation sequence

$$q_0 \xrightarrow{e_1} q_1 \xrightarrow{e_2} \ldots \xrightarrow{e_n} q_n.$$

  with $q_0 = h_s(q)$ and $q_n = h_s(r)$.

The change is that we have deleted the requirement that $T'(h_s(q), e_i)$ be defined for all i with $1 \leq i \leq n$.

Let **AutoWk** denote the category of trace automata and weak morphisms. This category is not as nice as the category **Auto**: although **AutoWk** does have finite products, which are constructed the same way as in **Auto**, it fails to have all equalizers. In addition, there is no way to extend the coreflection between **Auto** and **EvDom**, exhibited in [9], to a coreflection between **AutoWk** and some expansion of **EvDom**. For these reasons, we shall have to work with both categories **Auto** and **AutoWk**.

Proposition 2.5 *The functor B extends to an endofunctor of* **AutoWk**(X,Y). *The maps $\eta_A : A \to BA$ are weak morphisms of trace automata, which are the components of a natural transformation $\eta : 1 \to B$ that satisfies the unit laws $\mu \circ B\eta = 1 = \mu \circ \eta B$. Thus, the triple (B, η, μ) is a monad in* **AutoWk**(X,Y).

We now arrive at the main result of this section.

Theorem 1 *Suppose A is an automaton from X to Y. Then A is a monotone automaton iff there is a* **Auto**-*morphism $h_A : BA \to A$ enriching A with a structure of B-algebra.*

Proof – Suppose $Y \xleftarrow{g} A \xrightarrow{f} X$ is a monotone automaton from X to Y. Let the map $h_A : BA \to A$ be as in Proposition 2.4. It is a straightforward use of the definitions to check that $h_A \circ \eta_A = 1$ and $h_A \circ \mu_A = h_A \circ (Bh_A)$, so that h_A is a B-algebra structure on A.

Conversely, suppose $h_A : BA \to A$ is an **Auto**-morphism enriching the span $Y \xleftarrow{g} A \xrightarrow{f} X$ with a structure of B-algebra. Now, the alphabet of BA is of the form $E \otimes X$, with f the restriction to X. Suppose $e \in X$. Because $fh_A = f' : BA \to A$, we must have $fh_A e = f'e = \{e\}$. There must therefore exist some particular $e' \in h_A e$ with $fe' = \{e\}$. So, for each $e \in X$, there exists $e' \in E$ with $fe' = \{e\}$. Moreover, if $e_1 \| e_2$, then $h_A e_1 \perp h_A e_2$, so we must have $e'_1 \| e'_2$. Conversely, if $e'_1 \| e'_2$, then $\{e_1\} = fe'_1 \perp fe'_2 = \{e_2\}$, so $e_1 \| e_2$. Thus, $E \simeq (E \backslash \{e' : e \in X\}) \otimes \{e' : e \in X\}$, with f the restriction on the second factor. Since the second factor is isomorphic to X, we have shown that the alphabet E of

A has the form required of a monotone automaton. To prove that A has the receptivity property, observe that given e' corresponding to $e \in X$, applying h_A to the transition $(q, \epsilon) \xrightarrow{e} (q, e)$ of BA gives a computation $h_A e : q \to r$ of A. Using the fact that $e' \in h_A e$ and the assumption that h_A is a morphism, not just a weak morphism, we see that e' is enabled for A in state q, yielding the required transition $q \xrightarrow{e'} q \cdot e$ of A. ∎

3 Fibrations between Automata

The definition of monotone automaton given in Section 2, although successful at capturing our intuition, and having a number of interesting consequences as well, is not categorical, and is therefore unsuitable as a basis for a category-theoretic study of the properties of dataflow-like networks, viewed as spans in a category of automata. The purpose of this section is to show that monotone automata can be characterized categorically as certain *split right fibrations* in a suitable 2-category of automata.

In the theory of fibrations in a 2-category developed by Street [14, 15], the (two-sided) split fibrations from X to Y in a 2-category $\mathbf{K}$ are defined to be the algebras of a certain "doctrine" (or "2-monad") M on the 2-category of spans from X to Y in $\mathbf{K}$. The structure map for an M-algebra is called a "cleavage" for the underlying span. The one-sided "left" and "right" fibrations (essentially corresponding to what were earlier called split fibrations and split op-fibrations) are also algebras of doctrines L and R, which compose according to certain distributive laws to form M.

The connection between the abstract theory of fibrations and the concrete definitions we have given so far is made when one realizes that in the automata-theoretic case, the functor R is essentially the input buffering construction B, defined concretely for trace automata in the previous section, which takes an automaton A from X to Y and composes it with an input buffer to yield a "free X-input-buffered automaton" BA from X to Y. It then follows from Theorem 1 that the monotone automata from X to Y are exactly those automata from X to Y that are split right fibrations in **AutoWk** having an **Auto**-morphism as cleavage. Dually, it is possible to identify the doctrine L as an "output buffering construction," however we do not develop that idea further in this paper.

3.1 AutoWk as a 2-Category

Suppose A is an automaton. Let A^* be the category whose objects are the states of A, and whose arrows are its computations (computation sequences modulo trace equivalence). Empty computations serve as identities and computations are composed by concatenation. We call A^* the *computation category* of A. It can be shown [12] that: (1) A^* has no nontrivial isomorphisms, (2) every arrow of A^* is both epi and mono, (3) every span $s \xleftarrow{g} q \xrightarrow{f} r$ that can be completed to a commuting square has a pushout. Moreover, if $f : A \to B$ is a morphism of **AutoWk**, then f determines a pushout-preserving functor $f^* : A^* \to B^*$, in such a way that the map taking A to A^* becomes a functor $(\text{-})^* : \mathbf{AutoWk} \to \mathbf{Cat}$. Specifically, f^* takes each state q of A^* (which is nothing more than a state q of A) to the state fq of B^*. The action of f^* on arrows of A^* (that is, on computations of A) is determined by the fact that it is to be a functor from A^* to B^*

(hence it must preserve empty computations and concatenation of computations), and that it takes each single-transition computation $\gamma : q \xrightarrow{e} r$ of A to its image $f\gamma$ in B.

The category **AutoWk** can be made into a 2-category in such a way that $(\ \text{-}\ )^* :$ **AutoWk** $\to$ **Cat** becomes a 2-functor. Specifically, if $f, g : A \to B$ in **AutoWk**, then a 2-cell from f to g is a natural transformation $\tau : f^* \Rightarrow g^*$. Identity 2-cells, along with vertical and horizontal composition, are inherited from **Cat**, and the interchange law holds because it does in **Cat**. The subcategory **Auto** becomes a 2-category in the same way.

It should be emphasized here that although **Auto** is finitely complete as a 1-category, it is not the case that all limits are 2-limits. In particular, it is not the case that every pullback is a 2-pullback. Similarly, even for the cases in which ordinary limits exist in **AutoWk**, these need not be 2-limits. We shall see, though, that enough 2-pullbacks do exist in **AutoWk** to satisfy our needs.

3.2 Comma Objects

A *comma object* [14] for an opspan $Y \xrightarrow{g} A \xleftarrow{f} X$ from X to Y in a 2-category **K** is a span $Y \xleftarrow{d_0} g/f \xrightarrow{d_1} X$ from X to Y, together with a 2-cell $\lambda : gd_0 \Rightarrow fd_1$, such that composition with λ yields a 2-natural isomorphism

$$\mathbf{K}(\ \text{-}\ , g/f) \simeq \mathbf{K}(\ \text{-}\ , g)/\mathbf{K}(\ \text{-}\ , f).$$

Here for each object X the expression $\mathbf{K}(X, g)/\mathbf{K}(X, f)$ denotes the usual comma category of the functors $\mathbf{K}(X, g)$, $\mathbf{K}(X, f)$. An equivalent elementary description of this situtation is that, for each span $Y \xleftarrow{u_0} B \xrightarrow{u_1} X$ from X to Y in **K**, we have the following properties:

1. For every 2-cell $\sigma : gu_0 \Rightarrow fu_1$, there exists a unique $h : B \to g/f$, such that $\sigma = \lambda h$.

2. Given 2-cells $\xi : d_0 h \Rightarrow d_0 h'$ and $\eta : d_1 h \Rightarrow d_1 h'$ such that $\lambda h' \cdot g\xi = f\eta \cdot \lambda h$, then there exists a unique 2-cell $\phi : h \Rightarrow h'$ such that $\xi = d_0\phi$, $\eta = d_1\phi$.

Lemma 3.1 *The 2-category* **AutoWk** *has a comma object for every opspan.*

Proof – Given an opspan $Y \xrightarrow{g} A \xleftarrow{f} X$ from X to Y in **AutoWk**, define an automaton g/f as follows:

- The alphabet of actions of g/f is the product $Y \otimes X$.

- The states of g/f are arrows $gr \xrightarrow{\gamma} fq$ of A^*, or more precisely, triples (r, γ, q) with r a state of Y, q a state of X, and $\gamma : gr \to fq$ an arrow of A^*. The initial state of g/f is the identity computation on the initial state q_I of A.

- Suppose $gr \xrightarrow{\gamma} fq$ and $gr' \xrightarrow{\gamma'} fq'$ are states of g/f. There are two cases in which there are transitions of g/f from γ to γ':

    1. In case $q' = q$, then the transitions from γ to γ' are the transitions $r \xrightarrow{e} r'$ of Y such that $\gamma'(ge) = \gamma$ in A^*:

$$\begin{array}{ccc} gr & \xrightarrow{\gamma} & fq \\ {\scriptstyle ge}\downarrow & & \downarrow{\scriptstyle 1} \\ gr' & \xrightarrow[\gamma']{} & fq' \end{array}$$

2. In case $r' = r$, then the transitions from γ to γ' are the transitions $q \xrightarrow{e} q'$ of X such that $\gamma' = (fe)\gamma$:

$$\begin{array}{ccc} gr & \xrightarrow{\gamma} & fq \\ {\scriptstyle 1'}\downarrow & & \downarrow{\scriptstyle fe} \\ gr' & \xrightarrow[\gamma']{} & fq' \end{array}$$

These are the only types of transitions that g/f has. One may verify that g/f satisfies the commutativity condition required of an automaton.

Define maps $d_0 : g/f \to Y$ and $d_1 : g/f \to X$ as follows:

- d_0 takes a state $gr \xrightarrow{\gamma} fq$ of g/f to the state r of Y, and restricts the alphabet of g/f to the Y component.
- d_1 takes a state $gr \xrightarrow{\gamma} fq$ of g/f to the state q of X, and restricts the alphabet of g/f to the X component.

It is easily seen that d_0 and d_1 are morphisms of trace automata. The automaton g/f is also equipped with a 2-cell $\lambda : gd_0 \Rightarrow fd_1$, which associates with each state $gr \xrightarrow{\gamma} fq$ of g/f the computation γ of A.

We claim that g/f, equipped with the maps d_0 and d_1 and the 2-cell λ, is a comma object in **AutoWk** for the opspan $Y \xrightarrow{g} A \xleftarrow{f} X$. To see this, suppose $Y \xleftarrow{u_0} B \xrightarrow{u_1} X$ is a span in **AutoWk**.

1. Suppose we are given a 2-cell $\sigma : gu_0 \Rightarrow fu_1$. With each state q of D, this 2-cell (which is actually a natural transformation from $(gu_0)^*$ to $(fu_1)^*$) associates a computation $\gamma_q : gu_0q \to fu_1q$ of A, in other words with each state q of D is associated a state of g/f.

   Define the map $h : B \to g/f$ to take each state q of B to the corresponding state γ_q of g/f, and each action e of B to the disjoint sum $ge + fe$, which is a commuting set of actions of g/f. Clearly, if h is a weak morphism of automata, then it is the unique such morphism such that $\sigma = \lambda h$.

   To see that h is a weak morphism of automata, suppose we are given a transition $q \xrightarrow{e} r$ of B. This transition determines the following commuting diagram in A^*:

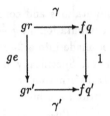

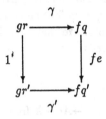

$$
\begin{array}{ccc}
fu_1q & \xrightarrow{fu_1e} & fu_1r \\
\gamma_q \Big\uparrow & & \Big\uparrow \gamma_r \\
gu_0q & \xrightarrow[gu_0e]{} & gu_0r
\end{array}
$$

which factors as follows:

$$
\begin{array}{ccccc}
fu_1q & \xrightarrow{fu_1e} & fu_1r & \xrightarrow{1} & fu_1r \\
\gamma_q \Big\uparrow & & \Big\uparrow & & \Big\uparrow \gamma_r \\
gu_0q & \xrightarrow[1]{} & gu_0q & \xrightarrow[gu_0e]{} & gu_0r
\end{array}
$$

Since the left and right squares clearly determine computations of g/f, their composite does too. This composite computation is the image of the transition $q \xrightarrow{e} r$ under h. (A map h behaving in this way is in general not a morphism of automata, but rather only a weak morphism.)

2. Suppose we are given 2-cells $\xi : d_0h \Rightarrow d_0h'$ and $\eta : d_1h \Rightarrow d_1h'$, such that $\lambda h' \cdot g\xi = f\eta \cdot \lambda h$. For each state q of B, let ϕ_q denote the computation of g/f corresponding to the following commuting square in A^*:

$$
\begin{array}{ccc}
d_1hq & \xrightarrow{f\eta q} & d_1h'q \\
\lambda hq \Big\uparrow & & \Big\uparrow \lambda h'q \\
d_0hq & \xrightarrow[g\xi q]{} & d_0h'q
\end{array}
$$

To see that this square does in fact determine a computation of g/f, use the same factoring trick as in (1) above. The map ϕ that assigns to each state q of B the corresponding computation ϕ_q of g/f, is now easily seen to be the unique 2-cell $\phi : h \Rightarrow h'$ such that $\xi = d_0\phi$, $\eta = d_1\phi$. ∎

The alphabet of actions of a comma object g/f of an opspan from X to Y is the product $Y \otimes X$. Let us call the actions in the Y component *output* actions, and those in the X component *input* actions. This terminology suggests an intuitive interpretation of g/f as kind of "nondeterministic X to Y transducer," which accepts as input a sequence of actions of X, records some information about these actions internally in the form of a computation of automaton A, and then outputs this information, perhaps after some delay, in the form of a sequence of actions of Y. Of special interest is the comma object ΦA for the opspan $A \xrightarrow{1} A \xleftarrow{1} A$ from A to A. We interpret ΦA as an "A-buffer."

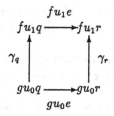

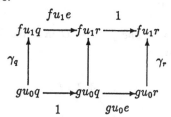

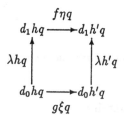

The factoring trick used in the proof above represents an important property of comma objects in **AutoWk**, which can be formalized as follows:

Lemma 3.2 *Every 2-cell γ between morphisms from B to g/f has a unique* input/output factorization; *that is, γ factors uniquely as $\xi\eta$ with $d_0\eta$ and $d_1\xi$ both identity 2-cells.*

As noted above, **AutoWk** does not even have all ordinary pullbacks, let alone all 2-pullbacks. Fortunately, though, the 2-pullbacks that we need for the notion of fibration to make sense actually do exist:

Lemma 3.3 *Suppose $f : A \to C$, and $g : B \to C$ in* **AutoWk**. *Then all three indicated 2-pullbacks exist in* **AutoWk**.

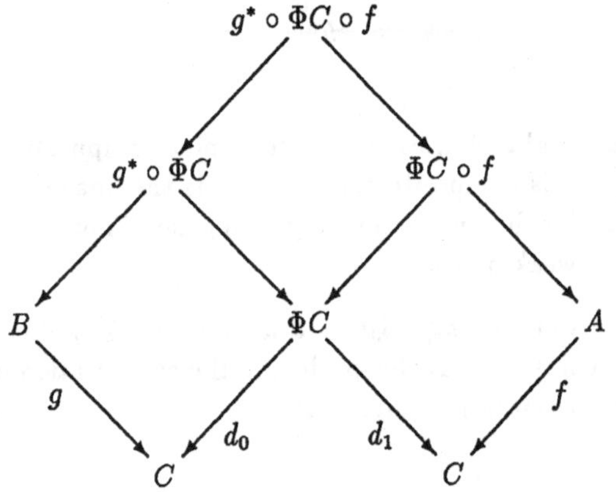

The proof of this result relies heavily on the input/output factorization property of computations of ΦC.

3.3 The Input Buffering Doctrine

We are now in a position to apply the theory of fibrations to **AutoWk**. Given concurrent alphabets X and Y, let **Span**(X,Y) denote the 2-category of spans in **AutoWk** from X to Y. Suppose $Z \xleftarrow{g} A \xrightarrow{f} X$ is a span from X to Z and $Y \xleftarrow{k} B \xrightarrow{h} Z$ is a span from Z to Y. If the following diagram is a 2-pullback,

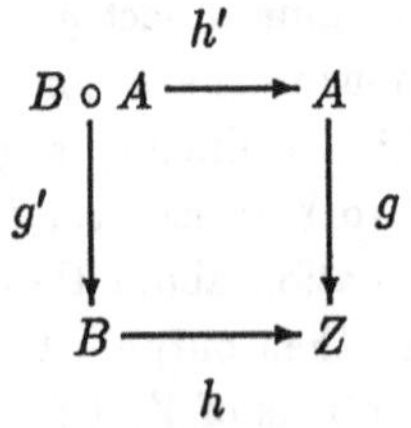

then the span $Y \xleftarrow{kg'} B \circ A \xrightarrow{fh'} X$ from X to Y is called the *composite* of B and A. Composition on the right with the comma object ΦX, viewed as a span from X to X, yields an endo-2-functor

$$R : \mathbf{Span}(X, Y) \to \mathbf{Span}(X, Y).$$

Explicitly, if A is a span from X to Y, then RA is the span $A \circ \Phi X$ from X to Y. In view of our intuitive interpretation of ΦX as an "X-buffer," we may interpret R as an "input buffering construction," which takes a span from X to Y and places it in tandem with an input buffer, producing an "input-buffered" span from X to Y.

The 2-functor R comes equipped with 2-natural transformations $\eta : 1 \to R$ and $\mu : RR \to R$ making (R, η, μ) a monad in **2Cat** (also called a *2-monad* or a *doctrine*). The component ηA of η at a span $Y \xleftarrow{g} A \xrightarrow{f} X$ from X to Y is the unique arrow of spans $A \to RA$ whose composition with the projection $RA \to A$ is 1_A and whose composition with the projection $RA \to \Phi X$ is the map $i_f : A \to \Phi X$ corresponding to the identity 2-cell on $f : A \to X$. The component μA of μ at the span A is the unique arrow of spans $1_A \circ c : RRA \to RA$ induced by the map $1_A : A \to A$ and the map $c : \Phi X \circ \Phi X \to \Phi X$ corresponding to the composite 2-cell:

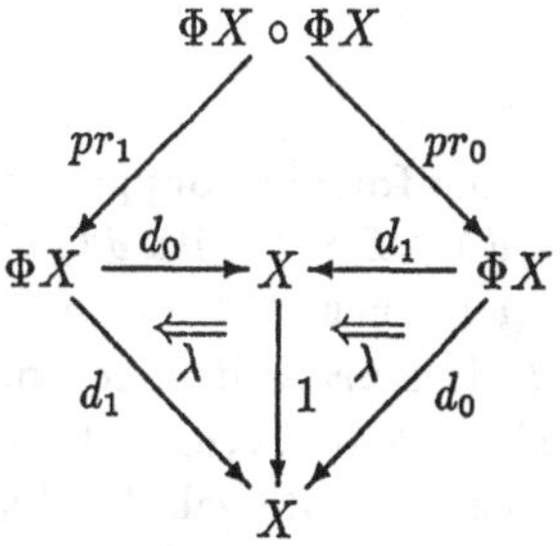

The existence of R, η, and μ, and the fact that they form a doctrine, follows automatically from general properties of comma objects and 2-pullbacks (see [14]). The concrete form taken by these data in the 2-category **AutoWk** is given by the following result, which is easily proved by working through the definitions.

Lemma 3.4 *Suppose X and Y are one-state trace automata. Then the doctrine (R, η, μ), regarded as an ordinary monad on the category* $\mathbf{Span}(X, Y)$*, is nothing but the input buffering monad (B, η, μ) defined in Section 2.*

A span A from X to Y is called a *split right fibration* if it admits a structure of algebra for the doctrine R. This means that there exists an arrow of spans $h_A : RA \to A$ such that $h_A \circ (\eta A) = 1$ and $h_A \circ (Rh_A) = h_A \circ (\mu A)$. The structure map h_A is called a *right cleavage* for A. It is a consequence of the general theory that if a span A is a split right fibration, then the cleavage h_A is left-adjoint to the map $\eta A : A \to RA$, hence is determined uniquely up to an invertible 2-cell. In the case of **AutoWk**, there are no invertible 2-cells other than identities, so h_A is unique. Street also defines a more general notion of *right fibration*, which is a span A that bears a structure of *pseudo-R* algebra. For a pseudo R-algebra, the above diagrams are required to commute only up to invertible 2-cell, rather than exactly. Again, since **AutoWk** has no nontrivial invertible 2-cells, there is no difference between a right fibration and a split right fibration.

The following is the main result of this section. It is a direct consequence of Theorem 1 and Lemma 3.4.

Theorem 2 *Suppose X and Y are one-state automata. Then an automaton from X to Y is monotone iff it is a split right fibration in* **AutoWk** *having an* **Auto***-morphism as cleavage.*

4 Fibrations between Domains

Although the fact that the monotone automata coincide exactly with the split right fibrations in **AutoWk** provides some evidence that the latter is the correct categorical notion, stronger evidence comes from the fact that essentially the same coincidence occurs for domains of computations of monotone automata and split right fibrations in a suitably defined 2-category of domains. The purpose of this section is to develop these results.

4.1 Conflict Event Orderings

We begin by recalling some definitions from [9]. Suppose $D = (D, \sqsubseteq)$ is a partially ordered set. An *interval* of D is a pair $(d, d') \in D \times D$ with $d \sqsubseteq d'$. A *prime* (or *covering*) interval is an interval (d, d') with $d \sqsubset d'$ and such that for no $d'' \in D$ do we have $d \sqsubset d'' \sqsubset d'$. We say that an interval $I = (d, d')$ is $\sqcup$-*prime* if there exists a finite set of prime intervals $\{(d, d_1), \ldots, (d, d_n)\}$ such that $d' = \bigsqcup\{d_1, \ldots, d_n\}$. Call two intervals $I = (d_0, d_1)$ and $J = (d_0', d_1')$ *coinitial* if $d_0 = d_0'$. Coinitial intervals $I = (d_0, d_1)$ and $J = (d_0, d_1')$ are called *consistent* if the set $\{d_1, d_1'\}$ has an upper bound. Call coinitial $\sqcup$-prime intervals I and J *orthogonal* if they are consistent, and there is no prime interval (d_0, d_1'') with $d_1'' \sqsubseteq d_1$ and $d_1'' \sqsubseteq d_1'$. We say that D is *finitely consistently complete* if every finite subset U of D having an upper bound, has a supremum $\bigsqcup U$. If D is finitely consistently complete, and if intervals $I = (d_0, d_1)$ and $J = (d_0, d_1')$ are consistent, then let $I \setminus J$ denote the interval $(d_1', d_1 \sqcup d_1')$.

Suppose D is a nonempty, finitely consistently complete poset with the following additional property:

1. $I \setminus J$ is a prime interval whenever I and J are distinct, consistent prime intervals.

We may then define $\equiv$ to be the least equivalence relation on prime intervals of D, such that $I \equiv I \setminus J$ whenever I and J are distinct and consistent.

A *conflict event ordering* is a nonempty, finitely consistently complete poset D having property (1) above and in addition having the properties:

2. $I \equiv J$ implies $I = J$, whenever I and J are coinitial prime intervals.

3. If I, I', J, J' are prime intervals such that $I \equiv I'$, $J \equiv J'$, I and J are coinitial, and I' and J' are coinitial, then I and I' are consistent iff J and J' are consistent.

A *weak morphism* from a conflict event ordering D to a conflict event ordering D' is a function $f : D \to D'$ that preserves all finite suprema existing in D. A *morphism* from D to D' is a weak morphism $f : D \to D'$ with the additional properties:

1. If I is a $\sqcup$-prime interval of D, then $f(I)$ is a $\sqcup$-prime interval of D'.
2. If $\sqcup$-prime intervals I, J are orthogonal in D, then $f(I)$ and $f(J)$ are orthogonal in D'.

Let **EvOrd** denote the category of conflict event orderings and morphisms, and let **EvOrdWk** denote the category of conflict event orderings and weak morphisms. Note that the map taking a poset to its ideal completion determines an equivalence of categories between **EvOrd** and the category **EvDom** of conflict event domains defined in [9]. The adjoint maps a conflict event domain to its finite basis. Since we shall have no need for morphisms that map finite elements to infinite elements, we prefer here to dispense with infinite elements entirely, and work with the category **EvOrd** instead of **EvDom**.

We showed in [9] that the category **EvOrd** is finitely complete. We also showed the following:

Proposition 4.1 *If A is an automaton, then the set HA of (finite) computations of A from its initial state, partially ordered by prefix, is a conflict event ordering. Moreover, the map taking A to HA extends to a functor H :* **Auto** $\to$ **EvOrd**, *which is right-adjoint to a full and faithful embedding, with unit an isomorphism.*

We add that although the functor H extends to a functor from **AutoWk** to **EvOrdWk**, the adjunction does not.

4.2 EvOrdWk as a 2-Category

The categories **EvOrd** and **EvOrdWk** have partially ordered homs, with strict, monotone composition. Hence they are actually 2-categories. Although **EvOrd** is finitely complete as a 1-category, it is not finitely 2-categorically complete, for essentially the same reasons as for **Auto**.

Proposition 4.2 *The 2-category* **EvOrdWk** *has a comma object for every opspan.*

Proof – A comma object g/f for an opspan $Y \xrightarrow{g} D \xleftarrow{f} X$ in **EvOrdWk** is $g/f = \{(b,a) : gb \sqsubseteq fa\}$, ordered componentwise, and equipped with the evident projections and 2-cell. It is necessary to verify that g/f is a conflict event ordering—this can be done by a direct check of the axioms, using the fact that f and g preserve finite suprema. ∎

The same result concerning 2-pullbacks holds for **EvOrdWk** as for **AutoWk**:

Proposition 4.3 *Lemma 3.3 holds for* **EvOrdWk**.

Next, we describe the doctrine R on spans in **EvOrdWk** and obtain a characterization of its algebras, the split right fibrations in **EvOrdWk**. The object map of R takes a span $Y \xleftarrow{g} D \xrightarrow{f} X$ to the span $Y \xleftarrow{g'} RD \xrightarrow{f'} X$, where

$$RD \simeq \{(d, x) \in D \times X : fd \sqsubseteq x\},$$

the map f' takes (d, x) to $x \in X$, and g' takes (d, x) to $gd \in Y$. The unit $\eta : 1 \to R$ has components $\eta_D : D \to RD$ that take d to (d, fd). The multiplication $\mu : RR \to R$ has components $\mu_D : RRD \to RD$ that take $((d, x), x')$ to (d, x').

Theorem 3 *A span* $Y \xleftarrow{g} D \xrightarrow{f} X$ *in* **EvOrdWk** *is a split right fibration iff the following condition holds:*

- *For all* $d \in D$ *and all* $x \geq fd$, *there exists an element* $d \sqcup x$ *of* D, *which is the least* $d' \geq d$ *with* $fd' \geq x$. *Moreover,* $f(d \sqcup x) = x$ *and* $g(d \sqcup x) = gd$.

Proof – Suppose the condition. Let $h : RD \to D$ be the map taking $(d, x) \in RD$ to $d \sqcup x \in D$. It is easy to check that h preserves finite suprema, hence is a weak morphism. Since $f(d \sqcup x) = x$ and $g(d \sqcup x) = gd$, it follows that h is an arrow of spans. Also, if $fd \sqsubseteq x \sqsubseteq x'$ then $d \sqcup x' = (d \sqcup x) \sqcup x'$, so $h \circ \mu D = h \circ Rh$. Finally, $h(\eta_D(d)) = h(d, fd) = d$, so h is a right cleavage for D.

Conversely, suppose $h : RD \to D$ is a right cleavage for D. Then h is left-adjoint to $\eta D : D \to RD$, with counit the identity. It follows by properties of adjunctions that for all $(d, x) \in RD$, the element $h(d, x)$ of D is the least $d' \geq d$ with $fd' \geq x$. Since h is an arrow of spans, we have also $f(h(d, x)) = x$ and $g(h(d, x)) = gd$. ∎

4.3 Connection with Automata

In this section, we show that the "unwinding functor" H, which takes each automaton A to the poset HA of its computations from the initial state, preserves and reflects split right fibrations. This gives a connection, as in [9], between an "operational" semantics of dataflow networks, defined in terms of automata, and a more "denotational," order-theoretic semantics. The proofs of the results in this section involve a detailed examination of the structure of the posets of computations of monotone automata. It ought to be possible to prove at least some of these results categorically, though at present it is not clear to the author what satisfactory categorical versions of the proofs would look like.

For the techniques to prove the following result, the reader is referred to [12]:

Lemma 4.4 *Suppose* $Y \xleftarrow{g} A \xrightarrow{f} X$ *is a monotone automaton. For each computation* $\gamma : q_I \to r$ *of* A *and trace* $x \in X^*$ *with* $Hf\gamma \sqsubseteq x$, *there exists a least computation* $\gamma \sqcup x$ *such that* $\gamma \sqsubseteq \gamma \sqcup x$ *and* $x \sqsubseteq Hf(\gamma \sqcup x)$. *Moreover,* $Hf(\gamma \sqcup x) = x$, $Hg(\gamma \sqcup x) = Hg\gamma$, *and the map taking* (γ, x) *to* $\gamma \sqcup x$ *is an* **EvOrd**-*morphism from* $R(HA)$ *to* HA.

The following result makes use of technical properties of conflict event orderings. A complete proof would be rather lengthy, so we just sketch the main ideas.

Lemma 4.5 *Let X and Y be concurrent alphabets. Suppose $HY \xleftarrow{g} D \xrightarrow{f} HX$ is a span in* **EvOrd**, *and suppose $h_D : RD \to D$ is an* **EvOrd**-*morphism that is also an R-algebra structure on D. Then there exists a monotone automaton $Y \xleftarrow{g'} A \xrightarrow{f'} X$, and an order-isomorphism $\phi : HA \to D$, such that $g\phi = Hg'$ and $f\phi = Hf'$.*

Proof – (Sketch) We first observe the following facts about the poset D:

1. Every prime interval (d, d') in D satisfies exactly one of the following two conditions:

   (a) $d' = h_D(d, fd')$, with (fd, fd') a prime interval of HX.

   (b) $fd = fd'$.

   Call intervals of type (a) *input* intervals, and those of type (b) *noninput* intervals.

2. If I is an input interval $(d, h_D(d, x))$ and J is a noninput interval (d, d'), then $I \perp J$. Moreover, $I \setminus J = (d', h_D(d', x))$ and $J \setminus I = (h_D(d, x), h_D(d', x))$.

3. If $I = (d, d')$ is a noninput prime interval, and $fd' \sqsubseteq x$, then $(h_D(d, x), h_D(d', x))$ is also a noninput prime interval.

   The automaton A is then constructed as follows:

- The alphabet of A is $E \otimes X$, where E is the set of all $\equiv$-equivalence classes of noninput intervals of D and $[I] \| [J]$ in E iff there exist $I' = (d, d') \in [I]$ and $J' = (d, d'') \in [J]$ such that $I' \perp J'$.

- The states of A are the elements of RD, with $(\perp, \epsilon)$ as the initial state.

- The transitions of A are of two types:

  1. If $e \in X$, then for all states (d, x) of A there is a transition $(d, x) \xrightarrow{e} (h_D(d, ex), ex)$ of A.

  2. If $[I] \in E$, then A has a transition $(d, x) \xrightarrow{[I]} (d', x)$ whenever $(d, d') \in [I]$.

Verification that A satisfies the commutativity condition, hence is an automaton, requires a case analysis on the various ways in which actions of A can be concurrent. These arguments make use of the properties of D stated above, plus the hypothesis that h_D is an R-algebra structure on D.

Let the map $f' : A \to X$ take each state of A to the unique state of X, and on actions, let f' be the restriction to X. Let the map $g' : A \to Y$ take each state of A to the unique state of Y. On actions, let g' be the map that takes $e \in X$ to $\emptyset$ and takes each $[I] \in E$, where $I = (d, d')$, to the trace $gd' \setminus gd \in Y^*$, which must be in Comm(Y) because the **EvOrd**-morphism g preserves $\sqcup$-prime intervals. It then follows from the definitions that A has the receptivity property, hence the span $Y \xleftarrow{g'} A \xrightarrow{f'} X$ is a monotone automaton.

To complete the proof, one may check that the map $\phi : HA \to D$ that takes each computation $\gamma : (\perp, \epsilon) \to (d, x)$ of A to $d \in D$, is an isomorphism of spans in **EvOrdWk**, from the span HA to the span D. The verification of this fact involves the observation that prime intervals (d, d') in D correspond exactly to transitions of A from (d, fd) to (d', fd'), and thus the computation sequences of A from state $(\bot, \epsilon)$ correspond to covering chains from $\bot$ in D. Moreover, prime intervals (d, d') and (d, d'') in D are orthogonal iff the corresponding transitions of A are for commuting actions. These facts allow us to prove that the map ϕ is in fact an arrow of spans in **EvOrdWk**, with an inverse that is also an arrow of spans in **EvOrdWk**. ▮

We can now answer the question raised at the end of our previous paper [9], concerning a characterization of the dataflow-like spans in **EvOrd**.

Theorem 4 *Suppose X and Y are concurrent alphabets. A span D from HX to HY in* **EvOrd** *is HA for some monotone automaton A iff D is a split right fibration in* **EvOrdWk**, *having an* **EvOrd**-*morphism as cleavage.*

5 Conclusion

We have shown that spans arising as behaviors of dataflow networks can be characterized in terms of split right fibrations, both in a 2-category of automata and a 2-category of domains. This characterization should make it possible to give categorical proofs that this class of spans is closed under network-forming operations, such as parallel and sequential composition, and feedback. We hope also that it will facilitate the continuity proofs required in the development of a semantics for recursively defined networks. There remains, however, the problem of understanding the correct way to formulate the universal properties satisfied by the feedback operation.

It is a bit annoying that our characterizations had to be stated in terms of the 2-categories **AutoWk** and **EvOrdWk** and their sub-2-categories **Auto** and **EvOrd**. For intuitive reasons, though, it seems necessary that the results be stated in this way. In this paper, we tried to make the simplest extensions to the 2-categories **Auto** and **EvOrd** that would show the connection with fibrations. Perhaps a cleaner (though less concrete) formulation of the results might be achieved by making **AutoWk** and **EvOrdWk** much larger, and then giving categorical characterizations of **Auto** and **EvOrd** as sub-2-categories. For example, we expect that the 2-category **Cts** of "concurrent transition systems" [11, 12] would be a suitable replacement for **AutoWk**.

References

[1] A. Carboni and R. F. C. Walters. Cartesian bicategories I. *Journal of Pure and Applied Algebra*, 49:11–32, 1987.

[2] J. W. Gray. Fibred and cofibred categories. In *Proc. Conference on Categorical Algebra at La Jolla*, pages 21–83, Springer-Verlag, 1966.

[3] A. Grothendieck. Catégories fibrées et descente. In *Séminaire de Géométrie Algébrique de l'Institute des Hautes Études Scientifiques, Paris 1960/61 (SGA 1)*, pages 145–194, Springer-Verlag, 1971.

[4] G. Kahn. The semantics of a simple language for parallel programming. In J. L. Rosenfeld, editor, *Information Processing 74*, pages 471–475, North-Holland, 1974.

[5] G. Kahn and D. B. MacQueen. Coroutines and networks of parallel processes. In B. Gilchrist, editor, *Information Processing 77*, pages 993–998, North-Holland, 1977.

[6] G. M. Kelly and R. H. Street. Review of the elements of 2-categories. In *Lecture Notes in Mathematics 420*, pages 75–103, Springer-Verlag, 1974.

[7] A. Mazurkiewicz. Trace theory. In *Advanced Course on Petri Nets*, GMD, Bad Honnef, September 1986.

[8] P. Panangaden and E. W. Stark. Computations, residuals, and the power of indeterminacy. In T. Lepisto and A. Salomaa, editors, *Automata, Languages, and Programming*, pages 439–454, Springer-Verlag. Volume 317 of *Lecture Notes in Computer Science*, 1988.

[9] E. W. Stark. Compositional relational semantics for indeterminate dataflow networks. In *Category Theory and Computer Science*, pages 52–74, Springer-Verlag. Volume 389 of *Lecture Notes in Computer Science*, Manchester, U. K., 1989.

[10] E. W. Stark. Concurrent transition system semantics of process networks. In *Fourteenth ACM Symposium on Principles of Programming Languages*, pages 199–210, January 1987.

[11] E. W. Stark. Concurrent transition systems. *Theoretical Computer Science*, 64:221–269, 1989.

[12] E. W. Stark. Connections between a concrete and abstract model of concurrent systems. In *Fifth Conference on the Mathematical Foundations of Programming Semantics*, Springer-Verlag. *Lecture Notes in Computer Science*, New Orleans, LA, 1990. (to appear).

[13] E. W. Stark. A simple generalization of Kahn's principle to indeterminate dataflow networks. In M. Z. Kwiatkowska, M. W. Shields, and R. M. Thomas, editors, *Semantics for Concurrency, Leicester 1990*, pages 157–176, Springer-Verlag, 1990.

[14] R. H. Street. Fibrations and Yoneda's lemma in a 2-category. In *Lecture Notes in Mathematics 420*, pages 104–133, Springer-Verlag, 1974.

[15] R. H. Street. Fibrations in bicategories. *Cahier de Topologie et Géometrie Différentielle*, XXI-2:111–159, 1980.

Applications of the Calculus of Trees to Process Description Languages

P. Degano* S. Kasangian† S. Vigna†

"Other maps are such shapes, with their islands and capes!
But we've got our brave captain to thank"
(So the crew would protest) "that he's bought *us* the best—
A perfect and absolute blank!"

(*Lewis Carroll*
The Hunting of the Snark
Fit the Second, 4)

Abstract

Bénabou's notion of *motor* is extended to cover labelled finite trees. Operations on them are defined that permit to easily define the semantics for a finite concurrent calculus. Then, suitable motors that constructively define canonical representatives for strong and observational congruence based on the notion of bisimulation are introduced in a clean and straightforward way. This enables us to provide the calculus with a fully abstract semantics up to the above congruences.

1 Introduction

In lectures held at the Mathematics Department and at the Computer Science Department of the University of Milano [Bén90], Jean Bénabou introduced the concept of *motor*, a simple algebraic structure which reflects very well the recursive nature of finite forests and provides a uniform, clean and elegant treatement of them. The novelty of Bénabou's approach is that his new algebraic structure has induction as a "built-in" feature, which allows to make extensive use of the *initial* motor (in a suitable category) with its powerful universal properties. That is, using very simple categorical tools Bénabou is able to characterize many attributes of forests (for instance, the number of nodes, of leaves etc.).

In a previous paper [KV91], two of the authors developed Bénabou's calculus for *labelled* trees, instead of unlabelled forests. Basically, a labelled tree is built with a set of endofunctions, one for each label, and with the monoid operator $\oplus$, interpreted

*Dipartimento di Matematica, Università di Parma, Via M. d'Azeglio 85, I-43100 Parma PR, Italy, Fax: +39-521-205350, and Dipartimento di Informatica, Università di Pisa, Corso Italia 40, I-56125 Pisa PI, Italy, Fax: +39-50-510226; email: uucp: degano@di.unipi.it.

†Dipartimento di Matematica, Università di Milano, Via Saldini 50, I-20133 Milano MI, Italy, Fax: +39-2-230346; email: uucp: kasan@imiucca.unimi.it.

as branching operator (such a structure is called a Σ-*motor*)[1]. The authors introduced and studied several examples of counting functions, both *qualitative* and *quantitative*. They showed that maps which should have been defined through recursive definitions involving several steps could be described in just one gulp, simply assigning a *correct* motor structure to the codomain, dispensing with clumsy manipulations of strings, terms, graphs or whatever else. Moreover, the powerful inductive structure allows for easy and, above all, *natural* proofs which can be carried on using only two very simple concepts, i.e., equivariance and preservation of the product.

The present paper aims at systematically applying Σ-motors to Process Description Languages. Some of the concepts we formalize here are by no means new, in particular those of Sections 2-4; indeed definitions for them can be found in [Mil89, Win84, BK85]. Our first contribution is that all definitions are *invariant*, in the sense that they are derived from the universal property of the initial Σ-motor, only. Nonetheless, the calculus for trees that we use has the same shape of the algebras of trees used so far, and thus it is as natural as the latter. Moreover, it is much more powerful and general. Indeed, the existence of the initial Σ-motor in any distributive category with countable sums [KV] guarantees a wide range of applicability of our techniques (for the relevance of these categories to Computer Science, see, e.g., [Wal89]). The link between the invariant aspects and the presentation of trees is given by a unique decomposition theorem (see Section 2.2).

The operations we present in Sections 2 and 3 are just those we need for defining through the initial Σ-motor a denotational semantics for a finite CCS-like concurrent calculus (Section 4) with prefixing, restriction, relabelling, nondeterministic choice and parallel composition.

The more specific result of the paper concerns the constructive definition of canonical representatives for trees up to strong and observational congruence (and for weak equivalence, too) [Mil89]. It suffices to define a suitable Σ-motor $\mathbb{M}^c_\Sigma$, the operations of which take care in quite a natural way of the well-known reduction rules, e.g., those named by Bergstra and Klop *arc reduction* and *removal of a non-initial deterministic τ-step* [BK85]. The elements of the resulting Σ-motor are trees, and *not* equivalence classes; furthermore, all the Σ-motor operations directly act on and always produce such canonical representatives. Therefore, it becomes straightforward to define a semantics for the finite concurrent calculus considered above that associates to all observationally congruent processes a unique tree. An immediate consequence is that this semantics is fully abstract *in se* [Mil75]. As a matter of fact, the definition of this more abstract semantics is got from the more concrete one of Section 4, simply by substituting the motor $\mathbb{M}^c_\Sigma$ and its operations for the initial Σ-motor. This shows the friendliness of Σ-motors, and the ease in accomplishing complex tasks with them. Indeed, only very recently the problem of constructing canonical representatives for observational equivalences has been solved independently by Rutten [Rut90], who defines an *ad hoc* technique for calculating them. Until our proposals, only the existence of a canonical representative for each observational equivalence class was known [BK85, Cas87, MS89]. Also, a purely syntactic approach, based on rewriting systems, is proposed in [IN90]. The latter reduces a finite CCS process term to its normal form, through a cute rewriting strategy.

---

[1]The authors thank André Arnold for having pointed out that the notion of motor and Σ-motor have already appeared under the name of *binoids* in [PQ68], where the authors used them to define a generalization of regular languages.

A further development of the results presented here will cover a truly concurrent approach to the semantics of Process Description Languages (see, e.g., [DD89, Win87, Pet80, Pra86, DDM90]). We conjecture that the above sketched results can be easily transferred to this more descriptive domains, thus vindicating the generality of Σ-motors.

On the other side, we have no satisfactory results when trees are infinite, although many definitions can be lifted to this case.

2 Fundamentals

Let us recall some of the basic definitions that we gave in our previous paper [KV91], to which we refer for a full and detailed account.

Definition 2.1 *Given an alphabet* Σ, *a* Σ*-motor is a tuple* $\langle M, \oplus, 0, f\rangle$, *where* $\langle M, \oplus, 0\rangle$ *is a monoid, and*

$$f : \Sigma \times M \longrightarrow M$$

is a map. $\mathbf{M}_\Sigma$ *will denote the initial object of* Σ-**Mot** *(the category of* Σ*-motors).*

The definition above generalizes Bénabou's notion of *motor*, that we recover when $\Sigma = \{*\}$.

If $\alpha, \beta, \gamma, \ldots$ are elements of Σ, and x is an element of M, we write $\alpha(x)$ (or even αx) for $f(\alpha, x)$, according to the usage in R-module theory and Ω-group theory. Indeed, f should be viewed as an *action* of Σ on M, like the external product in an R-module or in an Ω-group. The axioms for a Σ-motor as given in Definition 2.1 are in fact a weakening of the axioms for a left R-module or a left Ω-group. For the same reason we write α for $\alpha(0)$ (one-dimensional vectors can be identified with the elements of the ground ring). If needed, we will feel free to use

$$\sum_k \sigma_k x_k$$

to denote indexed sums, for we will use almost exclusively abelian structures. Furthermore, while dealing with actions f which actually depend only on M, we will write only the map $M \longrightarrow M$, understanding its composition with the canonical projection $\pi_M : \Sigma \times M \longrightarrow M$.

By cartesian closedness, an alternative definition could be given using an Σ-indexed family of endomaps of M: The resulting structure, which is basically a one-sorted algebra given by a monoid and by $|\Sigma|$ unary operators, is a slight weakening of Milner's RST's [Mil80]. The interplay between these two points of view turned out to be extremely productive. Introducing a new Σ-motor, we will often prefer to state the definition in terms of endofunctions, because the R-module-like notation permits a more readable description of f. The definition of abelian Σ-motor is obvious.

Let us notice also that since the usual interpretation of the sum of processes is commutative, we will always use an initial abelian Σ-motor $\mathbf{M}_\Sigma$ (for sake of generality, however, in this section we will use the same symbol for the non-abelian initial Σ-motor).

2.1 Morphisms to N and Σ^*

In [KV91] we made some heuristic considerations that enabled us to exhibit $\mathbf{M}_\Sigma$ as a "true" object of labelled finite trees. We recall also some useful definitions of maps in N and Σ^* that will give the flavour of our constructions.

Example 2.1 *Let $T \subseteq \Sigma$. Call ν_T the unique morphism from $\mathbf{M}_\Sigma$ to $\langle \mathrm{N}, +, 0, f' \rangle$, where*

$$\sigma n = \begin{cases} n+1 & \text{if } \sigma \in T \\ n & \text{otherwise} \end{cases}$$

Intuitively, ν_T counts the number of arcs with labels in T. Indeed, we will see that every tree t can be written uniquely as $\alpha_1 x_1 \oplus \alpha_2 x_2 \oplus \cdots \oplus \alpha_k x_k$. Since ν_T is a Σ-motor map, we have

$$\nu_T(t) = \nu_T(\alpha_1 x_1 \oplus \alpha_2 x_2 \oplus \cdots \oplus \alpha_k x_k) = j + \nu_T(x_1) + \nu_T(x_2) + \cdots + \nu_T(x_k),$$

where j is the number of occurrences of elements of T in $\{\alpha_1, \alpha_2, \ldots, \alpha_k\}$, and the meaning of this expression is the usual recursive definition of the number of arcs. Interesting special cases are of course $T = \Sigma$ (all arcs) and $T = \{\alpha\}$ (occurrences of a chosen label).

Example 2.2 *Let $T \subseteq \Sigma$. Call ∂_0^T the unique morphism from $\mathbf{M}_\Sigma$ to $\langle \mathrm{N}, +, 0, f' \rangle$, where*

$$\sigma n = \begin{cases} 1 \vee n & \text{if } \sigma \in T \\ 0 \vee n = n & \text{otherwise} \end{cases}$$

($\vee$ denotes the supremum of two natural numbers).

Again, ∂_0^T counts the number of arcs with labels in T which are not followed (at any depth) by other arcs with labels in T. Indeed, when two trees are joined, these numbers get added. When a new arc prefixes a tree t with no arcs labelled in T, you count just one if the label of the arc was in T. ∂_0^Σ clearly counts all the terminal arcs (hence the leaves). In other words, this is the frontier of T.

Example 2.3 *Let $T \subseteq \Sigma$. Call ∂_1^T the unique morphism from $\mathbf{M}_\Sigma$ to $\langle \mathrm{N}, +, 0, f' \rangle$, where*

$$\sigma n = \begin{cases} 1 & \text{if } \sigma \in T \\ 0 & \text{otherwise} \end{cases}$$

It is obvious that ∂_1^T counts the number of arcs immediately under the root (i.e., the branching degree) which have labels in T. Just like above, ∂_1^Σ counts them all.

Example 2.4 *Let $T \subseteq \Sigma$. Call ρ_T the unique morphism from $\mathbf{M}_\Sigma$ to $\langle \mathrm{N}, +, 0, f' \rangle$, where*

$$\sigma n = \begin{cases} n+1 & \text{if } \sigma \in T \\ 0 & \text{otherwise} \end{cases}$$

This morphism counts the number of arcs occurring in paths labelled only in T. In other words, the number of places reachable moving only through T-actions.

If now we change also the monoid operation in N, we can obtain some other interesting maps. For instance,

Example 2.5 *Let $T \subseteq \Sigma$. Call h_T the unique morphism from $\mathbf{M}_\Sigma$ to $\langle \mathbf{N}, \vee, 0, f' \rangle$, where*

$$\sigma n = \begin{cases} n+1 & \textit{if } \sigma \in T \\ 0 & \textit{otherwise} \end{cases}$$

This map is very important, and counts the *maximum length* of a path starting from the root with labels only in T. In fact, if we join two trees t and t', we must take as h_T of the sum the longest path, i.e., $h_T(t) \vee h_T(t')$. If $T = \Sigma$ we get the usual definition of tree *height.*

Now, we give a flash on how one of the previous examples can be *refined.* We defined some interesting maps landing in $\mathbf{N}$, which provide a *quantitative* evaluation of some properties of trees in $\mathbf{M}_\Sigma$. It would be interesting to have also some kind of *qualitative* evaluations. We use (with a slight abuse of notation) the same letters as before, and we write $\cdot$ for the concatenation in Σ^*, and ε for the empty word.

Example 2.6 *Let $T \subseteq \Sigma$. Call ν_T the unique morphism from $\mathbf{M}_\Sigma$ to $\langle \Sigma^*, \cdot, \varepsilon, f' \rangle$, where*

$$\sigma x = \begin{cases} \sigma \cdot x & \textit{if } \sigma \in T \\ x & \textit{otherwise} \end{cases}$$

ν_T produces the ordered list of the arcs we would meet if we scanned completely a tree with the *preorder traversal* (using the usual recursive algorithm) [Knu73]. However, only labels in T would appear in the list. Notice that simply by inverting the concatenation (i.e., $\sigma x = x \cdot \sigma \; \forall \sigma \in T$) we could have got the *postorder traversal* [Knu73].

2.2 Some algebraic properties

We list here without proof some of the algebraic properties that we investigated in [KV91], and which will be useful in the present paper.

2.2.1 Unique factorization

Definition 2.2 *Let $(\Sigma \times -) : \mathbf{Sets} \longrightarrow \mathbf{Sets}$ be the "product by Σ" functor (with the trivial action on morphisms). We define*

$$\Lambda_\Sigma = (-)^* \circ (\Sigma \times -) \circ U = (\Sigma \times U(-))^* : \mathbf{Mon} \longrightarrow \mathbf{Mon}$$

We give a short description of how Λ_Σ works. It maps a monoid M into the free monoid generated by the elements of the set $\Sigma \times U(M)^2$, i.e., to the set of strings of the form

$$\langle \sigma_1, x_1 \rangle \langle \sigma_2, x_2 \rangle \cdots \langle \sigma_n, x_n \rangle$$

where $\sigma_i \in \Sigma$ and $x_i \in M$.

The intuitive meaning of this operation can be better understood if, in particular, we apply it to $\mathbf{M}_\Sigma$: in this case, we should think of the x_i's as trees which are freely copied in $|\Sigma|$ different ways and then freely concatenated. Our interest will be in showing that this construction *leads again to* $\mathbf{M}_\Sigma$.

Now, we want to extend Λ_Σ to a functor Σ-$\mathbf{Mot}$$\longrightarrow$$\Sigma$-$\mathbf{Mot}$, still keeping the same name (with a slight abuse of notation).

[2] Recall that $U(M)$ is the underlying set of M.

Proposition 2.1 *Λ_Σ can be made into a functor from* Σ-**Mot** *to* Σ-**Mot** *by composition with the forgetful functor* Σ-**Mot** $\longrightarrow$ **Mon**, *and by endowing the monoid $\Lambda_\Sigma(M)$ with the map*

$$\eta \circ (1_\Sigma \times U(\phi(f))),$$

where ϕ is the natural bijection of the adjunction $(-)^* \dashv U$, η is the insertion of generators

$$\eta : \Sigma \times U(M) \longrightarrow U((\Sigma \times U(M))^*)$$

and f is the action of M.

We can now state the following

Theorem 2.1 $\mathbf{M}_\Sigma \cong \Lambda_\Sigma(\mathbf{M}_\Sigma)$. *Therefore, every tree $t \in \mathbf{M}_\Sigma$ can be uniquely written as a sum $\sigma_1 x_1 \oplus \sigma_2 x_2 \oplus \cdots \oplus \sigma_n x_n$.*

This theorem shows clearly the recursive nature of $\mathbf{M}_\Sigma$. The isomorphism, actually, will map a tree t, uniquely written as $\sigma_1 x_1 \oplus \sigma_2 x_2 \oplus \cdots \oplus \sigma_n x_n$, in the concatenation $\langle \sigma_1, x_1\rangle\langle \sigma_2, x_2\rangle \cdots \langle \sigma_n, x_n\rangle$. In other words, the label of the copy "remembers" the fact that the arc was labelled[3].

Let us notice that, in the abelian case, the order is not determined, so we can only write t as $\sigma_1 x_1 \oplus \sigma_2 x_2 \oplus \cdots \oplus \sigma_n x_n$ up to the order of the addenda. However, we can also write a formal sum $\sum_j k_j \sigma_j x_j$, where $k_j \in \mathbb{N}$, $\sigma_j \in \Sigma$, $x_j \in \mathbf{M}_\Sigma$, and $i = j \Longrightarrow \sigma_i x_i = \sigma_j x_j$.

2.3 Some operations

We conclude recalling some unary and binary operations which will be also useful to define the semantics of concurrent languages (like TCSP, CCS, ACP, MEIJE,...).

2.3.1 Pruning

The pruning of a tree is the abstract counterpart of the corresponding botanic operation: we want to cut out every branch *starting* with a label in a given subset T. This operation will be used to give meaning to the usual *restriction* operator.

Definition 2.3 *Let $T \subseteq \Sigma$. We call p_T the unique morphism in* Σ-**Mot** *from* $\mathbf{M}_\Sigma$ *to* $\mathbf{M}' = \langle \mathbf{M}_{\Sigma\backslash T}, \oplus, 0, f'\rangle$, *where $\mathbf{M}_{\Sigma\backslash T}$ is the initial $\Sigma\backslash T$-motor, $\oplus$ its monoid operation, and*

$$f'(\sigma, t) = \begin{cases} 0 & \textit{if } \sigma \in T \\ \sigma t & \textit{otherwise} \end{cases}$$

We must also notice that $\mathbf{M}_{\Sigma\backslash T}$ is endowed with a unique "injection morphism" i_T in $\mathbf{M}_\Sigma$ (you need only to restrict suitably the action). And, what really matters, for the universality of $\mathbf{M}_{\Sigma\backslash T}$ in $\Sigma\backslash T$-**Mot**, we have $p_T \circ i_T = \mathbf{1}_{\mathbf{M}_{\Sigma\backslash T}}$, i.e., the following diagram commutes:

[3] An immediate consequence is that we can prove a property of t assuming that the claim is true for the x_i's.

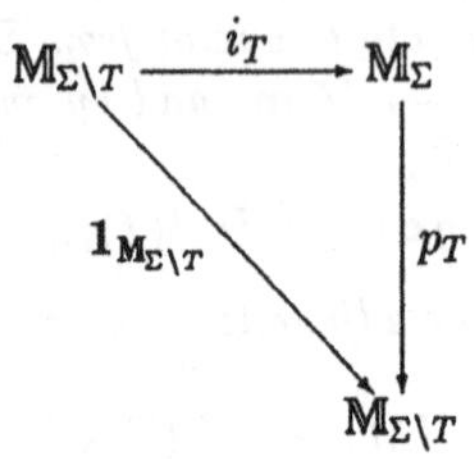

We will freely write i_ω and p_ω for $i_{\{\omega\}}$ and $p_{\{\omega\}}$.

2.3.2 Tree leaf product

We introduce now a binary operation on trees which can be useful to represent the *sequential composition* of processes (see [BK85]). A copy of a tree is attached to each *leaf* of another one.

Definition 2.4 *Let $t, t' \in \mathbb{M}_\Sigma$, and $t = \sum_k \sigma_k x_k$. The leaf product $t \mathbin{\textcircled{;}} t'$ is inductively defined by*

$$\begin{aligned} 0 \mathbin{\textcircled{;}} t &= t \\ t \mathbin{\textcircled{;}} t' &= \sum_k \sigma_k x_k \mathbin{\textcircled{;}} t' = \sum_j \sigma_j (x_j \mathbin{\textcircled{;}} t') \end{aligned}$$

It is easy to see the product $\textcircled{;}$ is associative, but not distributive; thus, $\langle \mathbb{M}_\Sigma, \textcircled{;}, 0, f\rangle$ is a Σ-motor and the map $\partial_0' = \partial_0 \vee 1$ in $\langle \mathbb{N}, *, 1, 1_{\mathbb{N}}\rangle$ (i.e., $t \overset{\partial_0'}{\mapsto} \partial_0(t) \vee 1$) is a morphism in Σ-**Mot**.

2.3.3 Relabelling

Finally, we show how to manage the *relabelling*. Given two alphabets Σ and Σ', we have the following

Definition 2.5 *Given a relabelling map $\xi : \Sigma \longrightarrow \Sigma'$, the relabelling arrow R_ξ is the unique morphism in* Σ-**Mot** *from $\mathbb{M}_\Sigma$ to $\mathbb{M}' = \langle \mathbb{M}_{\Sigma'}, \oplus, 0, f'(\xi(-), -)\rangle$.*

3 Synchronization of trees

We define now a quite general *synchronization* operation on two trees, whose rules are defined through a synchronization algebra S, taken from [Win84]. Through this operation, it is immediate to define the semantics of the usual synchronization operators of Process Description Languages such as CCS, TCSP, MEIJE, SCCS, etc. The next section will describe the first one.

Definition 3.1 *A synchronization algebra is an algebra $S = \langle L, \bullet, *, \omega\rangle$ where L is a set of labels such that $L \backslash \{*, \omega\} \neq \emptyset$ and $\bullet$ is a binary commutative associative operation on S which satisfies*

1. $\forall \sigma \in L \quad \sigma \bullet \omega = \omega$

2. $* \bullet * = *$

3. $\forall \sigma, \sigma' \in L \quad \sigma \bullet \sigma' = * \Rightarrow \sigma = *$

Basically, a product is inductively defined that keeps track even of failed synchronizations, and then the resulting tree is pruned at each failure ω. It is clear that $L = \Sigma \cup \{*, \omega\}$ is always used as underlying set of our synchronization algebra (we suppose of course $\Sigma \cap \{*, \omega\} = \emptyset$)

Definition 3.2 *Let $S = \langle L, \bullet, *, \omega\rangle$ be a synchronization algebra, $t, t' \in \mathbf{M}_L$, $t = \sum_j \alpha_j x_j$ and $t' = \sum_k \beta_k y_k$. The product $t \mid t'$ is defined by induction on $h(t) + h(t')$ as*

$$\begin{aligned} 0 \mid 0 &= 0 \\ t \mid t' &= \sum_j (\alpha_j \bullet *)(x_j \mid t') \oplus \sum_{j,k} (\alpha_j \bullet \beta_k)(x_j \mid y_k) \oplus \sum_k (* \bullet \beta_k)(t \mid y_k) \end{aligned}$$

Proposition 3.1 *The product $\mid$ is associative and commutative.*

Notice that in the proof of the associativity $* \bullet * = *$ and the commutativity of $\oplus$ are crucial.

Proposition 3.2 *The product $\mid$ is left and right distributive on $\oplus$.*

Now we define the product we are really interested in by pruning every subtree starting with the label ω. Since no action can be labelled by $*$ or ω, we can think of embedding the trees of $\mathbf{M}_\Sigma$ in $\mathbf{M}_{\Sigma \cup \{\omega\}}$, performing the operation $\mid$ and then pruning on ω[4].

Definition 3.3 *Let $t, t' \in \mathbf{M}_\Sigma$, and let $S = \langle L, \bullet, *, \omega\rangle$ be a synchronization algebra. The product $t \parallel t'$ is defined by*

$$t \parallel t' = p_\omega(i_\omega(t) \mid i_\omega(t'))$$

In a sense, the pruning does not "interfere" with the synchronization, since

Proposition 3.3 *The product $\parallel$ is associative and commutative.*

4 A finite calculus and its semantics

We will now provide a finite concurrent calculus with the "obvious" denotational semantics expressed through $\mathbf{M}_\Sigma$. We take a CCS-like language [Mil89] without recursion, which would produce infinite trees. The syntax of our processes is

$$P \longrightarrow \mathsf{NIL} \mid \alpha.P \mid P + P \mid P\backslash T \mid P[\phi] \mid P|P,$$

where Σ has an involution $(\bar{\ })$ with unique fixed point τ, $\alpha \in \Sigma$, $T \subseteq \Sigma$ and $\phi : \Sigma \longrightarrow \Sigma$. As usual, α is interpreted as the prefixing operator, $+$ denotes nondeterministic choice,

[4]The last sychronization algebra axiom guarantees that the only divisor of $*$ is $*$ itself, so that no $*$ can arise by a $|$-sychronization

$\backslash T$ is restriction, $[\phi]$ is relabelling and $|$ is the operator for synchrony/asynchrony. The definition of its semantics is

$$\begin{aligned}
[\![\mathsf{NIL}]\!] &= 0 \\
[\![\alpha.P]\!] &= \alpha([\![P]\!]) \\
[\![P+P']\!] &= [\![P]\!] \oplus [\![P']\!] \\
[\![P\backslash T]\!] &= p_{T\cup\overline{T}}([\![P]\!]) \\
[\![P[\phi]]\!] &= R_\phi([\![P]\!]) \\
[\![P|P']\!] &= [\![P]\!] \parallel [\![P']\!]
\end{aligned}$$

The synchronization algebra of $\|$ is defined to be $S = \langle \Sigma \cup \{*, \omega\}, \bullet, *, \omega \rangle$, where $\alpha \bullet * = \alpha$, $\alpha \bullet \beta = \omega$ if $\beta \neq \overline{\alpha}$, and $\alpha \bullet \overline{\alpha} = \tau$ (recall that $\overline{\overline{\alpha}} = \alpha$). It is straightforward verifying that the above semantics agrees with the classical one [Mil89]. The reader can easily figure out how similar semantics can be defined for other calculus for concurrency, e.g., for ACP, where the operation ③ should be used to interpret the sequential composition ;.

5 Strong bisimulation and strong equivalence

We are now going to discuss in our framework the well known concept of *bisimulation*, introduced by Park [Par81] and Milner [Mil80]. We will refer for the definitions to the more recent [Mil89].

Historically, the basic ingredient for the description of the bisimulation is the concept of *derivation*. Given two trees t and t', $t \xrightarrow{\alpha} t'$ (t' is the *α-derivative* of t) if in the unique decomposition of t there is a term $\alpha t'$ (not necessarily unique).

Definition 5.1 *A binary relation $R \subseteq U(\mathbf{M}_\Sigma) \times U(\mathbf{M}_\Sigma)$*[5] *is a* strong bisimulation *if $(t, u) \in R$ implies, for all $\sigma \in \Sigma$,*

- *Whenever $t \xrightarrow{\sigma} t'$ then, for some u', $u \xrightarrow{\sigma} u'$ and $(t', u') \in R$*
- *Whenever $u \xrightarrow{\sigma} u'$ then, for some t', $t \xrightarrow{\sigma} t'$ and $(t', u') \in R$*

The greatest such relation is an equivalence relation and is called strong equivalence *($\sim$).*

It is trivial verifying that $\sim$ is a congruence of Σ-motors.

However, an algebraically minded approach to bisimulation has been given by Milner in [Mil84], consisting in a set of equations whose induced congruence classes coincide with those of R. In particular, strong equivalence (in the finite case) is induced by the idempotence equational law $t \oplus t = t$.

5.1 Canonical representatives

Since it is more comfortable to deal with canonical representatives rather than with equivalence classes, we now apply the "initial object" approach to finding them for the strong

[5] Recall that $U(\mathbf{M}_\Sigma)$ is the underlying set of $\mathbf{M}_\Sigma$.

equivalence. We will use in the first place the equational characterization of strong bisimulation, since its algebraic flavour fits naturally with our approach. Then, the fact that the formulation of Definition 5.1 admits the same canonical representatives will exhibit idempotence as a sound and complete axiomatization of $\sim$. Let us define a suitable motor:

Definition 5.2 *Let* $\mathbf{M}_\Sigma^s = \langle U(\mathbf{M}_\Sigma), 0, \oplus^s, f^s\rangle$, *where* $f^s(\sigma, t) = \sigma t$, *and if for* $t, t' \neq 0$ *we have* $t \oplus t' = \sum_j k_j \sigma_j x_j$, *then* $t \oplus^s t$ *is defined to be* $\sum_j \sigma_j x_j$, *while* $t \oplus^s 0 = t$.

In other words, if $t = \alpha_1 x_1 \oplus \alpha_2 x_2 \oplus \cdots \oplus \alpha_j x_j$ and $t' = \beta_1 y_1 \oplus \beta_2 y_2 \oplus \cdots \oplus \beta_k y_k$, $t \oplus^s t'$ can be defined by the following process: consider the sum $t \oplus t' = \alpha_1 x_1 \oplus \alpha_2 x_2 \oplus \cdots \oplus \alpha_j x_j \oplus \beta_1 y_1 \oplus \beta_2 y_2 \oplus \cdots \oplus \beta_k y_k$, and for each pair of equal terms, delete one of them until no more duplicates are present. Notice that the order in which the terms are picked up for deletion is irrelevant, because $\mathbf{M}_\Sigma$ is abelian. For the same reason, $\oplus^s$ is associative. The following theorem describes in just one gulp the entire constructive process of finding the canonical representatives. Note that we are *not* taking any kind of quotient.

Theorem 5.1 *The unique map*

$$\mathbf{M}_\Sigma \xrightarrow{!_s} \mathbf{M}_\Sigma^s$$

yields canonical representatives for the congruence induced by $t \oplus t = t$.

Proof. Consider the quotient of $\mathbf{M}_\Sigma$ with respect to the congruence induced by the idempotence equational law $t \oplus t = t$, that we will call $\mathbf{M}_\Sigma/\mathcal{I}$: we have a unique map in **Σ-Mot**

$$p : \mathbf{M}_\Sigma \longrightarrow \mathbf{M}_\Sigma/\mathcal{I},$$

namely the projection on the quotient. Now, if we consider the epi-mono factorization of the map $\mathbf{M}_\Sigma \xrightarrow{!_s} \mathbf{M}_\Sigma^s$ as

$$\mathbf{M}_\Sigma \xrightarrow{e} !_s(\mathbf{M}_\Sigma) \xrightarrow{m} \mathbf{M}_\Sigma^s,$$

$!_s(\mathbf{M}_\Sigma)$ can be easily seen to satisfy the idempotence law, since an element $t \in !_s(\mathbf{M}_\Sigma)$ does not contain duplicates by construction, and computing $t \oplus^s t$ gives trivially t as result. For the universal property of the quotient, the map $\mathbf{M}_\Sigma \xrightarrow{e} !_s(\mathbf{M}_\Sigma)$ factors through $\mathbf{M}_\Sigma/\mathcal{I}$, i.e., there exists a unique e' such that the following diagram commutes:

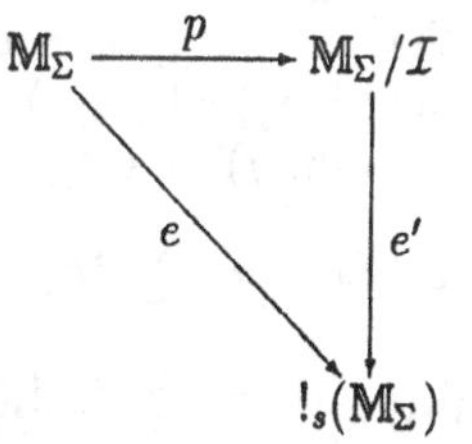

and consequently, by composition with m, this one commutes too:

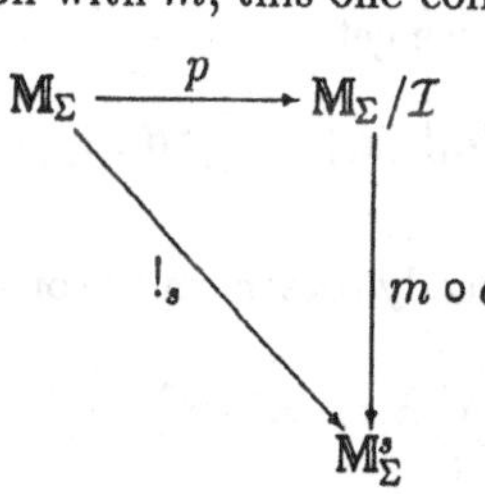

so that if two trees are strongly bisimilar, they have the same image in $\mathbf{M}^s_\Sigma$, i.e., $p(t) = p(t') \Longrightarrow !_s(t) = !_s(t')$.

Consider now the map p as a (set) map $\mathbf{M}^s_\Sigma \longrightarrow \mathbf{M}_\Sigma/\mathcal{I}$. Trivially, $p(0) = 0$ and $p(\sigma^s t) = p(\sigma t) = \sigma p(t)$ (because $f = f^s$). Moreover, $p(t \oplus^s t') = p(t \oplus t')$, because deleting duplicated terms does not change the equivalence class of $t \oplus t'$. Additionally, $p(t \oplus t') = p(t) \oplus p(t')$, because p is a Σ-motor morphism from $\mathbf{M}_\Sigma \longrightarrow \mathbf{M}_\Sigma/\mathcal{I}$. But then p is a Σ-motor map from $\mathbf{M}^s_\Sigma$ to $\mathbf{M}_\Sigma/\mathcal{I}$, so that by initiality the following diagram commutes:

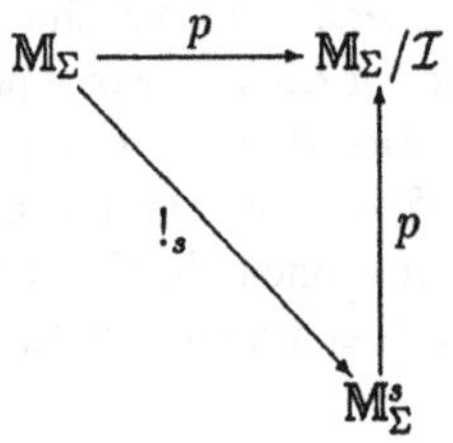

Thus, $!_s(t) = !_s(t')$ implies $p(!_s(t)) = p(!_s(t'))$ and $p(t) = p(t')$. □

Let us now prove the following

Theorem 5.2 *The unique map*

$$\mathbf{M}_\Sigma \xrightarrow{!_s} \mathbf{M}^s_\Sigma$$

yields canonical representatives for the relation $\sim$.

Proof. Since $\sim$ is a congruence, $\mathbf{M}_\Sigma/\sim$ is endowed with naturally induced Σ-motor operations. Thus, we have a unique map in Σ-**Mot**

$$p : \mathbf{M}_\Sigma \longrightarrow \mathbf{M}_\Sigma/\sim .$$

With the same notations of the previous theorem, in order to prove that $!_s$ factors through p we have to prove that $p(t) = p(u) \Longrightarrow !_s(t) = !_s(u)$. But $p(t) = p(u)$ means that (writing t as $\sum_i \alpha_i x_i$ and u as $\sum_j \beta_j y_j$ with $1 \leq i \leq m$ and $1 \leq j \leq n$) there are two maps $a : \underline{m} \longrightarrow \underline{n}$ and $b : \underline{n} \longrightarrow \underline{m}$ such that $\alpha_i = \beta_{a(i)}$, $x_i \sim y_{a(i)}$, $\beta_j = \alpha_{b(j)}$ and $y_j \sim x_{b(j)}$. By induction, we have $!_s(x_i) = !_s(y_{a(i)})$ and $!_s(y_j) = !_s(x_{b(j)})$. Multiplying both members by α_i (resp. β_j) and summing in $\mathbf{M}^s_\Sigma$ on i (resp. j) we get

$$\sum_i{}^s \alpha_i !_s(x_i) = \sum_i{}^s \beta_{a(i)} !_s(y_{a(i)})$$
$$\sum_j{}^s \alpha_{b(j)} !_s(x_{b(j)}) = \sum_j{}^s \beta_j !_s(y_j)$$

If we now add these two equations, we get

$$\sum_i{}^s \alpha_i !_s(x_i) = \sum_j{}^s \beta_j !_s(y_j)$$

because any re-indexed term was already present in the original sum. Since $!_s$ is a Σ-motor morphism, we get

$$!_s(\sum_i \alpha_i x_i) = !_s(\sum_j \beta_j y_j)$$

Consider now the map p as a (set) map $\mathbf{M}^s_\Sigma \longrightarrow \mathbf{M}_\Sigma/\sim$. As in the proof of Theorem 5.1, p is a Σ-motor map from $\mathbf{M}^s_\Sigma$ to $\mathbf{M}_\Sigma/\sim$, so that by initiality the following diagram commutes:

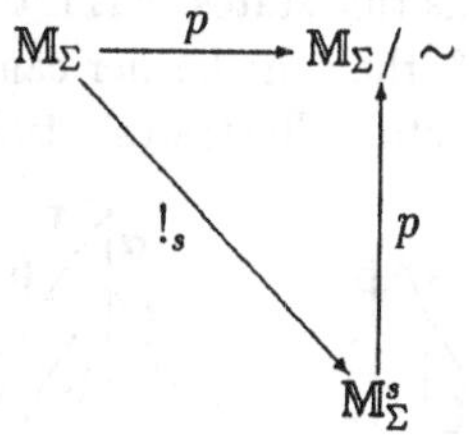

Thus, $!_s(t) = !_s(t')$ implies $p(!_s(t)) = p(!_s(t'))$ and $p(t) = p(t')$. □

Corollary 5.1 *The idempotence equational law is a sound and complete axiomatization for $\sim$.*

6 Observational congruence

We now discuss the most interesting concept, namely the *observational congruence*, which takes account for the silent moves τ. As a matter of fact, we proceed analogously to the previous section; besides, the structure of the proofs is essentially the same. The equational characterization of observational congruence [HM85] follows:

Definition 6.1 *The* observational congruence *($=_\mathcal{A}$) is defined as the congruence generated by the following set $\mathcal{A}$ of equations:*

1. $\alpha\tau t = \alpha t$
2. $t \oplus \tau t = \tau t$
3. $\alpha(t \oplus \tau t') \oplus \alpha t' = \alpha(t \oplus \tau t')$
4. $t \oplus t = t$

The starting point of our treatment is [BK85], where the authors define five confluent rules to reduce a labelled transition system to a canonical representative of its observational congruence class.

The first obstruction is of course that in [BK85] the rules are applied (in any order) for an arbitrarily large number of times, and the process stops only when no rule can be applied again. Instead, the recursive nature of $\mathbf{M}_\Sigma$ induces a constructive and finite process of reduction. The order in which the rules must be applied is completely determined by the definition of a suitable Σ-motor. This Σ-motor is built in such a way that the unique arrow from $\mathbf{M}_\Sigma$ gives the desired result. We start with a simple concept:

Definition 6.2 *We say that a tree t* absorbes *a tree $t' = \sigma_1 x_1 \oplus \sigma_2 x_2 \oplus \cdots \oplus \sigma_n x_n$, in symbols $t' \sqsubset t$, iff for each $1 \le i \le n$ there is in t a path $\tau^p \sigma_i \tau^q$ for some $p, q \in \mathbb{N}$ ending at x_i.*

Given a tree $t = \sigma_1 x_1 \oplus \sigma_2 x_2 \oplus \cdots \oplus \sigma_n x_n$, a path is a sequence $\sigma_k \alpha_1 \alpha_2 \cdots \alpha_p \in \Sigma^*$ such that $\alpha_1 \cdots \alpha_p$ is a path for x_n.

This definition is a rephrasing in our context of the case in which the *arc reduction* rule defined in [BK85] applies. This rule states that if two states s, s' are linked by an arc σ and by a sequence of arcs $\tau^p \sigma \tau^q$, then the former can be deleted without changing the congruence class. The following picture illustrates a typical case:

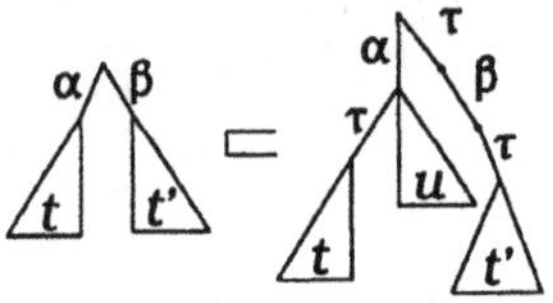

It is easy to check that

Proposition 6.1 *The absorption relation* $\sqsubset$ *is a preorder.*

Note that $\sqsubset$ is not generally an order (take σ and $\sigma \oplus \sigma$) but if we restrict ourselves to trees with a unique addendum in their decomposition, $\alpha x \sqsubset \beta y$ and $\beta y \sqsubset \alpha x$ implies $\alpha = \beta$, $x = y$.

The other property of absorption we will need is:

Proposition 6.2 *If* $t' \sqsubset t$ *then* $t' \oplus t =_{\mathcal{A}} t$.

Proof. The only case to be proved is $t' = \alpha x$. The basic idea behind the proof is that we can, *via* idempotence and two of the τ-laws, make a copy of x "running up" along the path $\tau^p \alpha \tau^q$. We assume $p, q \neq 0$. Thus, we get a situation like (1).

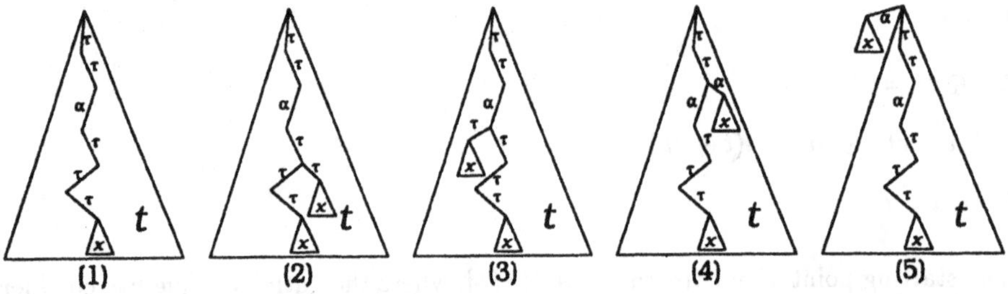

Since using the second τ-law and idempotence we can write

$$\tau(\tau x \oplus y) =_{\mathcal{A}} \tau(\tau x \oplus y) \oplus \tau x \oplus y =_{\mathcal{A}} \tau(\tau x \oplus y) \oplus \tau x \oplus \tau x \oplus y =_{\mathcal{A}} \tau(\tau x \oplus y) \oplus \tau x,$$

we can append a copy of τx aside the first τ-arc (2). This copy can be copied again (and deleted from the previous position), until we arrive just below the α-labelled arc (3). Now, the third τ-law allows us to skip α by applying α to x and copying it one level further (4). Finally, again using idempotence and the second τ-law, αx can run up until it pops out of t (5). Of course, when $p = 0$ or $q = 0$ some steps become unnecessary. Note also that when copying τx, idempotence and the second τ-law could be replaced by the special case of the third τ-law in which $\alpha = \tau$. $\square$

We could now take any tree $t = \sigma_1 x_1 \oplus \sigma_2 x_2 \oplus \cdots \oplus \sigma_n x_n$ and simply delete any addendum $\sigma_i x_i$ which is absorbed by another addendum $\sigma_j x_j$. Of course, this would not produce a canonical representative, because we would not take care of analyzing the subtrees. Actually, we need to rebuild entirely the tree paying attention to absorption and to the second rule we take from [BK85], namely *removal of a non-initial deterministic τ-step*. In our setting, this is equivalent to axiom 1 of $\mathcal{A}$, i.e., $\alpha\tau t = \alpha t$.

This considerations lead us to the idea of summing freely the trees, and then deleting anyone which can be absorbed elsewhere. This intuition is formalized the Σ-motor below:

Definition 6.3 *Let* $\mathbf{M}^c_\Sigma = \langle U(\mathbf{M}_\Sigma), 0, \oplus^c, f^c \rangle$, *where* f^c *is defined by* $t \mapsto \sigma t$ *if* $t \in \mathbf{M}_\Sigma \backslash \tau(\mathbf{M}_\Sigma)$, $\tau t \mapsto \sigma t$ *elsewhere, and if for* $t, t' \neq 0$ *we have* $t \oplus t' = \sum_j k_j \sigma_j x_j$, *then* $t \oplus^c t'$ *is defined to be* $\sum_m \sigma_{j_m} x_{j_m}$, *where* $\{\sigma_{j_m} x_{j_m}\}$ *is the subset of maximal elements (with respect to* $\sqsubset$*) of* $\{\sigma_j x_j\}$, *while* $t \oplus^c 0 = t$.

In other words, if $t = \alpha_1 x_1 \oplus \alpha_2 x_2 \oplus \cdots \oplus \alpha_j x_j$ and $t' = \beta_1 y_1 \oplus \beta_2 y_2 \oplus \cdots \oplus \beta_k y_k$ then $t \oplus^c t'$ can be defined by the following process: consider the sum $t \oplus t' = \sigma_1 z_1 \oplus \sigma_2 z_2 \oplus \cdots \oplus \sigma_n z_n$ (with the obvious assignments), and if for some $1 \leq p \neq q \leq n$ we have $\sigma_p z_p \sqsubset \sigma_q z_q$, then delete $\sigma_p z_p$ until $\sigma_q z_q$ does not absorbe $\sigma_p z_p$ for each $p \neq q$. Note that the order in which the terms are picked up for test is irrelevant, because $\sqsubset$ is a preorder. For the same reason, $\oplus^c$ is associative.

All said, still without taking any quotient, we have the following

Theorem 6.1 *The unique map*

$$\mathbf{M}_\Sigma \xrightarrow{!_c} \mathbf{M}^c_\Sigma$$

yields canonical representatives for the observational congruence as induced by $\mathcal{A}$.

Proof. Consider instead the quotient of $\mathbf{M}_\Sigma$ with respect to the congruence induced by $\mathcal{A}$: we have a unique map in Σ-**Mot**

$$p : \mathbf{M}_\Sigma \longrightarrow \mathbf{M}_\Sigma/\mathcal{A},$$

namely the projection on the quotient. Now, if we consider the epi-mono factorization of the map $\mathbf{M}_\Sigma \xrightarrow{!_c} \mathbf{M}^c_\Sigma$ as

$$\mathbf{M}_\Sigma \xrightarrow{e} !_c(\mathbf{M}_\Sigma) \xrightarrow{m} \mathbf{M}^c_\Sigma,$$

$!_c(\mathbf{M}_\Sigma)$ can be easily seen to satisfy the idempotence law and the τ-laws. Indeed, since an element $t \in !_c(\mathbf{M}_\Sigma)$ cannot be simplified *via* absorption by construction, $t \oplus^c t$ is trivially t. Also, if we write σ^c for $f^c(\sigma, -)$, $\alpha^c \tau^c t = \alpha^c t$ follows from the definition of the endofunctions on $\mathbf{M}^c_\Sigma$; the second law, $t \oplus^c \tau^c t = \tau^c t$, is true because $t \sqsubset \tau^c t$; the third one, $\alpha^c(t \oplus^c \tau^c t') \oplus^c \alpha^c t' = \alpha^c(t \oplus^c \tau^c t')$, holds because $\alpha^c t' \sqsubset \alpha^c(t \oplus^c \tau^c t')$. For the universal property of the quotient, this implies that the map $\mathbf{M}_\Sigma \xrightarrow{e} !_c(\mathbf{M}_\Sigma)$ factors through $\mathbf{M}_\Sigma/\mathcal{A}$, i.e., there exists a unique e' such that the following diagram commutes:

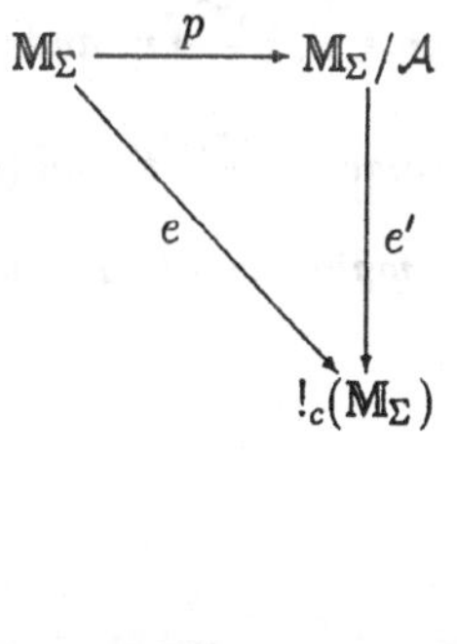

and consequently, by composition with m, this one commutes too:

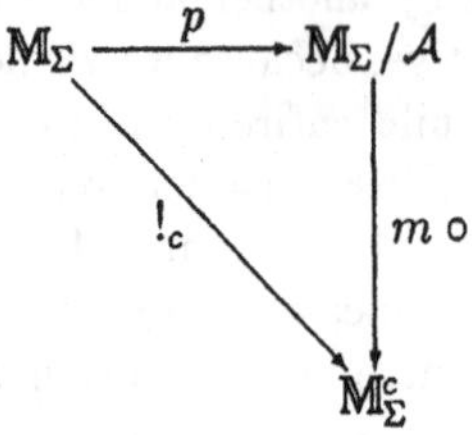

Therefore, if two trees are observationally congruent, they have the same image in $\mathbf{M}^c_\Sigma$, i.e., $p(t) = p(t') \Longrightarrow !_c(t) = !_c(t')$.

Consider now the map p as a (set) map $\mathbf{M}^c_\Sigma \longrightarrow \mathbf{M}_\Sigma/\mathcal{A}$. Trivially, $p(0) = 0$. Since by the first τ-law $\alpha\tau t = \alpha t$, we have $p(\sigma^c t) = p(\sigma t)$ (the only effect of σ^c can be the deletion of a τ-labelled arc outgoing from the root of t) and since $p(\sigma t) = \sigma p(t)$, we have $p(\sigma^c t) = \sigma p(t)$. Moreover, $p(t \oplus^c t') = p(t \oplus t')$, because if $u \sqsubset u'$ then $p(u \oplus u') = p(u')$. Finally, $p(t \oplus t') = p(t) \oplus p(t')$ because p is a Σ-motor morphism $\mathbf{M}_\Sigma \longrightarrow \mathbf{M}_\Sigma/\mathcal{A}$. But then p is a Σ-motor map from $\mathbf{M}^c_\Sigma$ to $\mathbf{M}_\Sigma/\mathcal{A}$, so that by initiality the following diagram commutes:

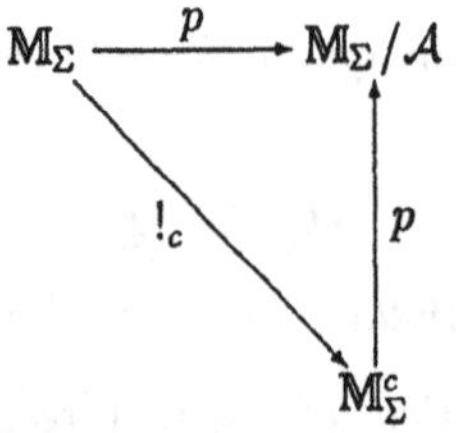

Thus, $!_c(t) = !_c(t')$ implies $p(!_c(t)) = p(!_c(t'))$ and $p(t) = p(t')$. □

Now, if $[\![-]\!]^c$ denotes the fully abstract *observational semantics* defined as the one of Section 4, where $\mathbf{M}^c_\Sigma$ is subsituted for $\mathbf{M}_\Sigma$ in all definitions (analogously for $[\![-]\!]^s$), for all processes P, $[\![P]\!]^c$ ($[\![P]\!]^s$, respectively) is the canonical representative of $[\![P]\!]$ up to observational (strong, respectively) congruence. Note that in order to define correctly restriction and relabelling, $\mathbf{M}^c_\Sigma$ has to be substituted for $\mathbf{M}_\Sigma$ as *codomain* of p and R_ϕ.

As we did for strong bisimulation, we will now state the "classical" definition of observational congruence, which passes through the definition of *weak bisimulation*, that we take again from [Mil89]:

Definition 6.4 *A binary relation* $R \subseteq U(\mathbf{M}_\Sigma) \times U(\mathbf{M}_\Sigma)$ *is a* weak bisimulation *if* $(t, u) \in R$ *implies, for all* $\sigma \in \Sigma$,

- *Whenever* $t \xrightarrow{\sigma} t'$ *then, for some* u', $u \xrightarrow{\tau^p \hat{\sigma} \tau^q} u'$ *and* $(t', u') \in R$
- *Whenever* $u \xrightarrow{\sigma} u'$ *then, for some* t', $t \xrightarrow{\tau^p \hat{\sigma} \tau^q} t'$ *and* $(t', u') \in R$

where $\hat{\sigma} = \sigma$ *if* $\sigma \neq \tau$, $\hat{\tau} = \varepsilon$ *(the empty string). The weak equivalence, denoted by* $\approx$, *is the greatest weak bisimulation.*

Now we can define the following equivalence, which can be easily proved to be a congruence:

Definition 6.5 $t =_c u$ *(t is observationally congruent to u) if*[6] *for all* $\sigma \in \Sigma$,

- *Whenever* $t \xrightarrow{\sigma} t'$ *then, for some* u', $u \xrightarrow{\tau^p \sigma \tau^q} u'$ *and* $t' \approx u'$
- *Whenever* $u \xrightarrow{\sigma} u'$ *then, for some* t', $t \xrightarrow{\tau^p \sigma \tau^q} t'$ *and* $t' \approx u'$

The only obvious property we will need is that

Proposition 6.3 *If* $t \approx u$, *then* $t =_c u$, *or* $\tau t =_c u$, *or* $t =_c \tau u$.

Our next theorem is that

Theorem 6.2 *The unique map*

$$\mathbf{M}_\Sigma \xrightarrow{!_c} \mathbf{M}^c_\Sigma$$

yields canonical representatives for the relation $=_c$.

Proof. Since $=_c$ is a congruence, $\mathbf{M}_\Sigma / =_c$ is endowed with naturally induced Σ-motor operations. Thus, we have a unique map in Σ-**Mot**

$$p : \mathbf{M}_\Sigma \longrightarrow \mathbf{M}_\Sigma / =_c .$$

With the same notations of the previous theorem, in order to prove that $!_c$ factors through p which have to prove that $p(t) = p(u) \Longrightarrow !_c(t) = !_c(u)$. But $p(t) = p(u)$ means that (writing t as $\sum_i \alpha_i x_i$ and u as $\sum_j \beta_j y_j$ with $1 \le i \le m$ and $1 \le j \le n$) there are two maps $a : \underline{m} \longrightarrow \underline{n}$ and $b : \underline{n} \longrightarrow \underline{m}$ such that $\beta_{a(i)} y_{a(i)} \xrightarrow{\tau^p \alpha_i \tau^q} x'_i$ and $\alpha_{b(j)} x_{b(j)} \xrightarrow{\tau^p \beta_j \tau^q} y'_j$, with $x'_i \approx x_i$ and $y'_i \approx y_i$. It is obvious that $\beta_{a(i)} y_{a(i)} \oplus \alpha_i x_i =_c \beta_{a(i)} y_{a(i)}$, so for any term $\alpha_{\bar{\imath}} x_{\bar{\imath}}$ such that $\beta_{a(\bar{\imath})} y_{a(\bar{\imath})} \xrightarrow{\tau^p \alpha_{\bar{\imath}} \tau^q} x'_{\bar{\imath}}$ with $p \vee q \neq 0$, we add $\alpha_{\bar{\imath}} x_{\bar{\imath}}$ to u (setting $\beta_{n+1} = \alpha_{\bar{\imath}}$). We modify h so that $\alpha_{\bar{\imath}} x_{\bar{\imath}}$ is mapped in its copy (i.e., $a(\bar{\imath}) = n + 1$), and we extend k so that the copy is mapped to the original term in t (i.e., $b(n+1) = \bar{\imath}$). After a finite number of steps, we get two trees t' and u' that are congruent to t and u, respectively. Their maps a and b are such that for each i, j $\beta_{a(i)} y_{a(i)} \xrightarrow{\alpha_i} x'_i \approx x_i$ and $\alpha_{b(j)} x_{b(j)} \xrightarrow{\beta_j} y'_j \approx y_j$ (note that there are no τ-derivations).

By Proposition 6.3, for each i $x'_i =_c x_i$, or $\tau x'_i =_c x_i$, or $x'_i =_c \tau x_i$ (and analogously for j). In any case, by induction on $h(t) + h(u)$, $!_c(x'_i) = !_c(x_i)$, or $\tau^c !_c(x'_i) = !_c(x_i)$, or $!_c(x'_i) = \tau^c !_c(x_i)$. After applying α^c_i, all these cases reduce to $\alpha^c_i !_c(x'_i) = \alpha^c_i !_c(x_i)$.

Summing in $\mathbf{M}^c_\Sigma$ on i (resp. j) we get

$$\begin{aligned} \sum_i{}^c \alpha^c_i !_c(x_i) &= \sum_i{}^c \beta^c_{a(i)} !_c(y_{a(i)}) \\ \sum_j{}^c \alpha^c_{b(j)} !_c(x_{b(j)}) &= \sum_j{}^c \beta^c_j !_c(y_j) \end{aligned}$$

If we now sum these two equations, the result is clearly

$$\sum_i{}^c \alpha^c_i !_c(x_i) = \sum_j{}^c \beta^c_j !_c(y_j)$$

[6]Notice that the only difference from the definition 6.4 is the absence of the ^ operator.

because any re-indexed term is already present in the original sum, or it is absorbed. Since $!_c$ is a Σ-motor morphism, we get

$$!_c(\sum_i \alpha_i x_i) = !_c(\sum_j \beta_j y_j).$$

Consider now the map p as a (set) map $\mathbf{M}_\Sigma^c \longrightarrow \mathbf{M}_\Sigma / =_c$. As in the proof above, p is also a Σ-motor map from $\mathbf{M}_\Sigma^c$ to $\mathbf{M}_\Sigma / =_c$, so that by initiality the following diagram commutes:

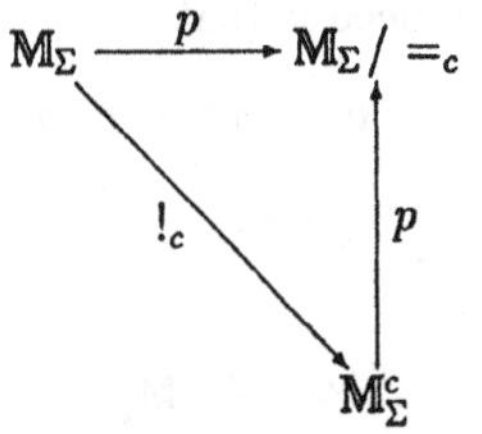

Thus, $!_c(t) = !_c(t')$ implies $p(!_c(t)) = p(!_c(t'))$ and $p(t) = p(t')$. $\square$

Corollary 6.1 *The equation set $\mathcal{A}$ is a sound and complete axiomatization for $=_c$.*

7 Weak equivalence

Finally, we want to discuss the less well-behaved case of the *weak equivalence*, and provide, for the sake of esthetics, a result analogous to Theorem 6.1. Although the weak equivalence is not a congruence, there is a "canonical" way to relate arbitrary maps with monoid maps on trees, called *parallel translation*, which was introduced by Bénabou and developed in [KV91]:

Theorem 7.1 *Given a motor $\langle M, \oplus, 0, f' \rangle$ and a map $u : \Sigma \times U(\mathbf{M}_\Sigma) \longrightarrow M$ there is a unique arrow $T(u)$ in $\mathbf{Mon}$ such that the following diagram commutes:*

$$
\begin{array}{ccc}
\Sigma \times U(\mathbf{M}_\Sigma) & \xrightarrow{\;f\;} & U(\mathbf{M}_\Sigma) \\[2mm]
{\scriptstyle (\pi_\Sigma, u)} \downarrow & & \downarrow {\scriptstyle U(T(u))} \\[2mm]
\Sigma \times M & \xrightarrow{\;f'\;} & M
\end{array}
$$

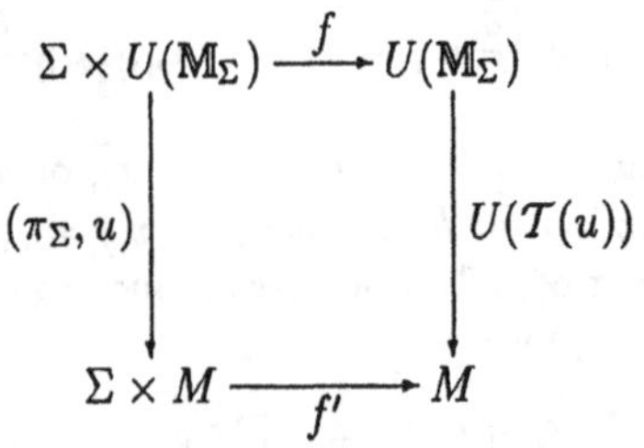

where $\pi_\Sigma : \Sigma \times U(\mathbf{M}_\Sigma) \longrightarrow \Sigma$ is the canonical projection. In symbols, $U(T(u)) \circ \sigma = \sigma \circ u(\sigma, -)$ for each $\sigma \in \Sigma$.

The basic idea of this property is that a given arbitrary map from $\mathbf{M}_\Sigma$ to a Σ-motor can be "made" into a monoid map by applying it not to the whole tree, but rather to the first-level subtrees. The original motivation for parallel translation was the invariant definition of maps such as $T_n = T^n(0)$, which cuts every tree at height n. We will drop the U's in what follows.

Theorem 7.1 can be applied to a map $u : \mathbf{M}_\Sigma \longrightarrow M$ by composing it with the projection $\pi_{\mathbf{M}_\Sigma} : \Sigma \times \mathbf{M}_\Sigma \longrightarrow \mathbf{M}_\Sigma$. What we are going to do is to "back-translate" the observational congruence canonical representative map $!_c$, and to show that the corresponding tree map $w : \mathbf{M}_\Sigma \longrightarrow \mathbf{M}_\Sigma$ yields the canonical representatives we look for.

Theorem 7.2 *The unique map* $w : \mathbf{M}_\Sigma \longrightarrow \mathbf{M}_\Sigma$ *such that* $T(w) = !_c$ *and such that the following diagram commutes*

$$\begin{array}{ccc} \Sigma \times \mathbf{M}_\Sigma & \xrightarrow{\ f\ } & \mathbf{M}_\Sigma \\ {\scriptstyle \mathbf{1}_\Sigma \times w}\downarrow & & \downarrow{\scriptstyle !_c} \\ \Sigma \times \mathbf{M}_\Sigma & \xrightarrow[\ f\ ]{} & \mathbf{M}_\Sigma \end{array}$$

yields canonical representatives for the weak equivalence.

Proof. It is immediate that two trees t, t' are weakly equivalent iff $!_c(\sigma t) = !_c(\sigma t')$. By commutativity of the diagram above, $\sigma w(t) = \sigma w(t')$, and by the unique decomposition theorem this happens iff $w(t) = w(t')$. □

A simple inspection of the commutativity condition shows that w can be obtained from $!_c$ simply by composition with the map which sends $\tau t \mapsto t$ and is the identity elsewhere.

8 Acknowledgments

The authors thank for substantial suggestions the referees and, of course, Jean Bénabou.

References

[Bén90] J. Bénabou. Lectures held at the Mathematics and Computer Science Departments of the University of Milano, and talk at the Category Theory '90 Conference in Como. 1990.

[BK85] J.A. Bergstra and J.W. Klop. Algebra of communicating processes with abstraction. *Theoretical Computer Science*, 37(1):77–121, 1985.

[Cas87] I. Castellani. Bisimulations and abstraction homomorphisms. *Journal of Computer and System Sciences*, 34:210–235, 1987.

[DD89] Ph. Darondeau and P. Degano. Causal trees. In *Proc. 11th Int. Coll. on Automata and Languages ICALP*, number 372 in Lecture Notes in Computer Science, pages 234–248, 1989.

[DDM90] P. Degano, R. De Nicola, and U. Montanari. A partial ordering semantics for CCS. *Theoretical Computer Science*, 75:223–262, 1990.

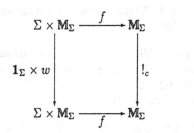

[HM85] M. Hennessy and R. Milner. Algebraic laws for nondeterminism and concurrency. *Journal of Assoc. Comput. Mach.*, 32:137–161, 1985.

[IN90] P. Inverardi and M. Nesi. A rewriting strategy to verify observational equivalence. *Info. Proc. Letters*, 35:191–199, 1990.

[Knu73] D.E. Knuth. *The Art of Computer Programming.* Addison-Wesley, 1973.

[KV] S. Kasangian and S. Vigna. Trees in a distributive category. To appear in Proceedings CT '90.

[KV91] S. Kasangian and S. Vigna. Introducing a calculus of trees. In *Proceedings of the International Joint Conference on Theory and Practice of Software Development (TAPSOFT/CAAP '91)*, number 493 in Lecture Notes in Computer Science, pages 215–240, 1991.

[Mil75] R. Milner. Processes, a mathematical model of computing agents. In *Logic Colloquium, Bristol 1973*, pages 157–174. North-Holland, 1975.

[Mil80] R. Milner. *A Calculus of Communicating Systems.* Number 92 in Lecture Notes in Computer Science. Springer-Verlag, 1980.

[Mil84] R. Milner. A complete inference system for a class of regular behaviours. *Journal of Computer and System Sciences*, 28:439–466, 1984.

[Mil89] R. Milner. *Communication and Concurrency.* International Series in Computer Science. Prentice Hall, 1989.

[MS89] U. Montanari and M. Sgamma. Canonical representatives for observational equivalence classes. In *Proceedings Colloquium on the resolution of equations in algebraic structures*, pages 293–319. Academic Press, Inc., 1989.

[Par81] D. Park. Concurrency and automata on infinite sequences. In *Proc. Theoretical Computer Science*, number 104 in Lecture Notes in Computer Science, pages 167–183. Springer-Verlag, 1981.

[Pet80] C.A. Petri. Concurrency. In *Net Theory and Applications*, number 84 in Lecture Notes in Computer Science, pages 1–19. Springer-Verlag, 1980.

[PQ68] C. Pair and A. Quere. Définition et étude des bilangages réguliers. *Information and Control*, 13:565–593, 1968.

[Pra86] V.R. Pratt. Modeling concurrency with partial orders. *International Journal of Parallel Programming*, 15(1):33–71, February 1986.

[Rut90] J.J.M.M. Rutten. Explicit canonical representatives for weak bisimulation equivalence and congruence. Technical Report CS–R9062, CWI, 1990.

[Wal89] R.F.C. Walters. Data types in a distributive category. *Bull. Austr. Math. Soc.*, 40:79–82, 1989.

[Win84] G. Winskel. Synchronization trees. *Theoretical Computer Science*, 34:33–82, 1984.

[Win87] G. Winskel. Petri nets, algebras, morphisms and compositionality. *Info. and Co.*, 72:197–238, 1987.

Lecture Notes in Computer Science

For information about Vols. 1–448
please contact your bookseller or Springer-Verlag

Vol. 449: M.E. Stickel (Ed.), 10th International Conference on Automated Deduction. Proceedings, 1990. XVI, 688 pages. 1990. (Subseries LNAI).

Vol. 450: T. Asano, T. Ibaraki, H. Imai, T. Nishizeki (Eds.), Algorithms. Proceedings, 1990. VIII, 479 pages. 1990.

Vol. 451: V. Marík, O. Stepánková, Z. Zdráhal (Eds.), Artificial Intelligence in Higher Education. Proceedings, 1989. IX, 247 pages. 1990. (Subseries LNAI).

Vol. 452: B. Rovan (Ed.), Mathematical Foundations of Computer Science 1990. Proceedings, 1990. VIII, 544 pages. 1990.

Vol. 453: J. Seberry, J. Pieprzyk (Eds.), Advances in Cryptology – AUSCRYPT '90 Proceedings, 1990. IX. 462 pages. 1990.

Vol. 454: V. Diekert, Combinatorics on Traces. XII, 165 pages. 1990.

Vol. 455: C.A. Floudas, P.M. Pardalos, A Collection of Test Problems for Constrained Global Optimization Algorithms. XIV, 180 pages. 1990.

Vol. 456: P. Deransart, J. Maluszyn´ski (Eds.), Programming Language Implementation and Logic Programming. Proceedings, 1990. VIII, 401 pages. 1990.

Vol. 457: H. Burkhart (Ed.), CONPAR '90 – VAPP IV. Proceedings, 1990. XIV, 900 pages. 1990.

Vol. 458: J.C.M. Baeten, J.W. Klop (Eds.), CONCUR '90. Proceedings, 1990. VII, 537 pages. 1990.

Vol. 459: R. Studer (Ed.), Natural Language and Logic. Proceedings, 1989. VII, 252 pages. 1990. (Subseries LNAI).

Vol. 460: J. Uhl, H.A. Schmid, A Systematic Catalogue of Reusable Abstract Data Types. XII, 344 pages. 1990.

Vol. 461: P. Deransart, M. Jourdan (Eds.), Attribute Grammars and their Applications. Proceedings, 1990. VIII, 358 pages. 1990.

Vol. 462: G. Gottlob, W. Nejdl (Eds.), Expert Systems in Engineering. Proceedings, 1990. IX, 260 pages. 1990. (Subseries LNAI).

Vol. 463: H. Kirchner, W. Wechler (Eds.), Algebraic and Logic Programming. Proceedings, 1990. VII, 386 pages. 1990.

Vol. 464: J. Dassow, J. Kelemen (Eds.), Aspects and Prospects of Theoretical Computer Science. Proceedings, 1990. VI, 298 pages. 1990.

Vol. 465: A. Fuhrmann, M. Morreau (Eds.), The Logic of Theory Change. Proceedings, 1989. X, 334 pages. 1991. (Subseries LNAI).

Vol. 466: A. Blaser (Ed.), Database Systems of the 90s. Proceedings, 1990. VIII, 334 pages. 1990.

Vol. 467: F. Long (Ed.), Software Engineering Environments. Proceedings, 1969. VI, 313 pages. 1990.

Vol. 468: S.G. Akl, F. Fiala, W.W. Koczkodaj (Eds.), Advances in Computing and Information – ICCI '90. Proceedings, 1990. VII, 529 pages. 1990.

Vol. 469: I. Guessarian (Ed.), Semantics of Systeme of Concurrent Processes. Proceedings, 1990. V, 456 pages. 1990.

Vol. 470: S. Abiteboul, P.C. Kanellakis (Eds.), ICDT '90. Proceedings, 1990. VII, 528 pages. 1990.

Vol. 471: B.C. Ooi, Efficient Query Processing in Geographic Information Systems. VIII, 208 pages. 1990.

Vol. 472: K.V. Nori, C.E. Veni Madhavan (Eds.), Foundations of Software Technology and Theoretical Computer Science. Proceedings, 1990. X, 420 pages. 1990.

Vol. 473: I.B. Damgård (Ed.), Advances in Cryptology – EUROCRYPT '90. Proceedings, 1990. VIII, 500 pages. 1991.

Vol. 474: D. Karagiannis (Ed.), Information Syetems and Artificial Intelligence: Integration Aspects. Proceedings, 1990. X, 293 pages. 1991. (Subseries LNAI).

Vol. 475: P. Schroeder-Heister (Ed.), Extensions of Logic Programming. Proceedings, 1989. VIII, 364 pages. 1991. (Subseries LNAI).

Vol. 476: M. Filgueiras, L. Damas, N. Moreira, A.P. Tomás (Eds.), Natural Language Processing. Proceedings, 1990. VII, 253 pages. 1991. (Subseries LNAI).

Vol. 477: D. Hammer (Ed.), Compiler Compilers. Proceedings, 1990. VI, 227 pages. 1991.

Vol. 478: J. van Eijck (Ed.), Logics in AI. Proceedings, 1990. IX, 562 pages. 1991. (Subseries in LNAI).

Vol. 480: C. Choffrut, M. Jantzen (Eds.), STACS 91. Proceedings, 1991. X, 549 pages. 1991.

Vol. 481: E. Lang, K.-U. Carstensen, G. Simmons, Modelling Spatial Knowledge on a Linguistic Basis. IX, 138 pages. 1991. (Subseries LNAI).

Vol. 482: Y. Kodratoff (Ed.), Machine Learning – EWSL-91. Proceedings, 1991. XI, 537 pages. 1991. (Subseries LNAI).

Vol. 483: G. Rozenberg (Ed.), Advances In Petri Nets 1990. VI, 515 pages. 1991.

Vol. 484: R. H. Möhring (Ed.), Graph-Theoretic Concepts In Computer Science. Proceedings, 1990. IX, 360 pages. 1991.

Vol. 485: K. Furukawa, H. Tanaka, T. Fullsaki (Eds.), Logic Programming '89. Proceedings, 1989. IX, 183 pages. 1991. (Subseries LNAI).

Vol. 486: J. van Leeuwen, N. Santoro (Eds.), Distributed Algorithms. Proceedings, 1990. VI, 433 pages. 1991.

Vol. 487: A. Bode (Ed.), Distributed Memory Computing. Proceedings, 1991. XI, 506 pages. 1991.

Vol. 488: R. V. Book (Ed.), Rewriting Techniques and Applications. Proceedings, 1991. VII, 458 pages. 1991.

Vol. 489: J. W. de Bakker, W. P. de Roever, G. Rozenberg (Eds.), Foundations of Object-Oriented Languages. Proceedings, 1990. VIII, 442 pages. 1991.

Vol. 490: J. A. Bergstra, L. M. G. Feljs (Eds.), Algebraic Methods II: Theory, Tools and Applicatlons. VI, 434 pages. 1991.

Vol. 491: A. Yonezawa, T. Ito (Eds.), Concurrency: Theory, Language, and Architecture. Proceedings, I989. VIII, 339 pages. 1991.

Vol. 492: D. Sriram, R. Logcher, S. Fukuda (Eds.), Computer-Aided Cooperative Product Development. Proceedings, 1989 VII, 630 pages. 1991.

Vol. 493: S. Abramsky, T. S. E. Maibaum (Eds.), TAPSOFT '91. Volume 1. Proceedings, 1991. VIII, 455 pages. 1991.

Vol. 494: S. Abramsky, T. S. E. Maibaum (Eds.), TAPSOFT '91. Volume 2. Proceedings, 1991. VIII, 482 pages. 1991.

Vol. 495: 9. Thalheim, J. Demetrovics, H.-D. Gerhardt (Eds.), MFDBS '91. Proceedings, 1991. VI, 395 pages. 1991.

Vol. 496: H.-P. Schwefel, R. Männer (Eds.), Parallel Problem Solving from Nature. Proceedings, 1991. XI, 485 pages. 1991.

Vol. 497: F. Dehne, F. Fiala. W.W. Koczkodaj (Eds.), Advances in Computing and Intormation - ICCI '91 Proceedings, 1991. VIII, 745 pages. 1991.

Vol. 498: R. Andersen, J. A. Bubenko jr., A. Sølvberg (Eds.), Advanced Information Systems Engineering. Proceedings, 1991. VI, 579 pages. 1991.

Vol. 499: D. Christodoulakis (Ed.), Ada: The Choice for '92. Proceedings, 1991. VI, 411 pages. 1991.

Vol. 500: M. Held, On the Computational Geometry of Pocket Machining. XII, 179 pages. 1991.

Vol. 501: M. Bidoit, H.-J. Kreowski, P. Lescanne, F. Orejas, D. Sannella (Eds.), Algebraic System Specification and Development. VIII, 98 pages. 1991.

Vol. 502: J. Bārzdiņš , D. Bjørner (Eds.), Baltic Computer Science. X, 619 pages. 1991.

Vol. 503: P. America (Ed.), Parallel Database Systems. Proceedings, 1990. VIII, 433 pages. 1991.

Vol. 504: J. W. Schmidt, A. A. Stogny (Eds.), Next Generation Information System Technology. Proceedings, 1990. IX, 450 pages. 1991.

Vol. 505: E. H. L. Aarts, J. van Leeuwen, M. Rem (Eds.), PARLE '91. Parallel Architectures and Languages Europe, Volume I. Proceedings, 1991. XV, 423 pages. 1991.

Vol. 506: E. H. L. Aarts, J. van Leeuwen, M. Rem (Eds.), PARLE '91. Parallel Architectures and Languages Europe, Volume II. Proceedings, 1991. XV, 489 pages. 1991.

Vol. 507: N. A. Sherwani, E. de Doncker, J. A. Kapenga (Eds.), Computing in the 90's. Proceedings, 1989. XIII, 441 pages. 1991.

Vol. 508: S. Sakata (Ed.), Applied Algebra, Algebraic Algorithms and Error-Correcting Codes. Proceedings, 1990. IX, 390 pages. 1991.

Vol. 509: A. Endres, H. Weber (Eds.), Software Development Environments and CASE Technology. Proceedings, 1991. VIII, 286 pages. 1991.

Vol. 510: J. Leach Albert, B. Monien, M. Rodríguez (Eds.), Automata, Languages and Programming. Proceedings, 1991. XII, 763 pages. 1991.

Vol. 511: A. C. F. Colchester, D.J. Hawkes (Eds.), Information Processing in Medical Imaging. Proceedings, 1991. XI, 512 pages. 1991.

Vol. 512: P. America (Ed.), ECOOP '91. European Conference on Object-Oriented Programming. Proceedings, 1991. X, 396 pages. 1991.

Vol. 513: N. M. Mattos, An Approach to Knowledge Base Management. IX, 247 pages. 1991. (Subseries LNAI).

Vol. 514: G. Cohen, P. Charpin (Eds.), EUROCODE '90. Proceedings, 1990. XI, 392 pages. 1991.

Vol. 515: J. P. Martins, M. Reinfrank (Eds.), Truth Maintenance Systems. Proceedings, 1990. VII, 177 pages. 1991. (Subseries LNAI).

Vol. 516: S. Kaplan, M. Okada (Eds.), Conditional and Typed Rewriting Systems. Proceedings, 1990. IX, 461 pages. 1991.

Vol. 517: K. Nökel, Temporally Distributed Symptoms in Technical Diagnosis. IX, 164 pages. 1991. (Subseries LNAI).

Vol. 518: J. G. Williams, Instantiation Theory. VIII, 133 pages. 1991. (Subseries LNAI).

Vol. 519: F. Dehne, J.-R. Sack, N. Santoro (Eds.), Algorithms and Data Structures. Proceedings, 1991. X, 496 pages. 1991.

Vol. 520: A. Tarlecki (Ed.), Mathematical Foundations of Computer Science 1991. Proceedings, 1991. XI, 435 pages. 1991.

Vol. 521: B. Bouchon-Meunier, R. R. Yager, L. A. Zadek (Eds.), Uncertainty in Knowledge-Bases. Proceedings, 1990. X, 609 pages. 1991.

Vol. 522: J. Hertzberg (Ed.), European Workshop on Planning. Proceedings, 1991. VII, 121 pages. 1991. (Subseries LNAI).

Vol. 523: J. Hughes (Ed.), Functional Programming Languages and Computer Architecture. Proceedings, 1991. VIII, 666 pages. 1991.

Vol. 524: G. Rozenberg (Ed.), Advances in Petri Nets 1991. VIII, 572 pages. 1991.

Vol. 525: O. Günther, H.-J. Schek (Eds.), Large Spatial Databases. Proceedings, 1991. XI, 471 pages. 1991.

Vol. 526: T. Ito, A. R. Meyer (Eds.), Theoretical Aspects of Computer Software. Proceedings, 1991. X, 772 pages. 1991.

Vol. 527: J.C.M. Baeten, J. F. Groote (Eds.), CONCUR '91. Proceedings, 1991. VIII, 541 pages. 1991.

Vol. 528: J. Maluszynski, M. Wirsing (Eds.) Programming Language Implementation and Logic Programming. Proceedings, 1991. XI, 433 pages. 1991.

Vol. 529: L. Budach (Ed.) Fundamentals of Computation Theory. Proceedings, 1991. XII, 426 pages. 1991.

Vol. 530: D. H. Pitt, P.-L. Curien, S. Abramsky, A. M. Pitts, A. Poigné, D. E. Rydeheard (Eds.), Category Theory and Computer Science. Proceedings, 1991. VII, 301 pages. 1991.

Vol. 531: E. M. Clarke, R. P. Kurshan (Eds.), Computer-Aided Verification. Proceedings, 1990. XIII, 372 pages. 1991.

Vol. 532: H. Ehrig, H.-J. Kreowski, G. Rozenberg (Eds.), Graph Grammars and Their Application to Computer Science. Proceedings, 1990. X, 703 pages. 1991.

Vol. 533: E. Börger, H. Kleine Büning, M. M. Richter, W. Schönfeld (Eds.), Computer Science Logic. Proceedings, 1990. VIII, 399 pages. 1991.

Vol. 534: H. Ehrig, K. P. Jantke, F. Orejas, H. Reichel (Eds.), Recent Trends in Data Type Specification. Proceedings, 1990. VIII, 379 pages. 1991.

Vol. 535: P. Jorrand, J. Kelemen (Eds.), Fundamentals of Artificial Intelligence Research. Proceedings, 1991. VIII, 255 pages. 1991. (Subseries LNAI).

Vol. 536: J. E. Tomayko, Software Engineering Education. Proceedings, 1991. VIII, 296 pages. 1991.

Vol. 539: H. F. Mattson, T. Mora, T. R. N. Rao (Eds.), Applied Algebra, Algebraic Algorithms and Error-Correcting Codes. Proceedings, 1991. XI, 489 pages. 1991.

Vol. 540: A. Prieto (Ed.), Artificial Neural Networks. Proceedings, 1991. XIII, 476 pages. 1991.

Vol. 541: P. Barahona, L. Moniz Pereira, A. Porto (Eds.), EPIA '91. Proceedings, 1991. VIII, 292 pages. 1991. (Subseries LNAI).

Vol. 543: J. Dix, K. P. Jantke, P. H. Schmitt (Eds.), Nonmonotonic and Inductive Logic. Proceedings, 1990. X, 243 pages. 1991. (Subseries LNAI).